FIRST STEPS IN THE ORIGIN OF LIFE IN THE UNIVERSE

Sponsors

The Abdus Salam International Centre for Theoretical Physics, Trieste, Italy
European Commission
The SETI Institute
Consiglio Nazionale delle Ricerche, Rome, Italy
The International Centre for Genetic Engineering and Biotechnology, Trieste, Italy
European Space Agency (ESA)
National Aeronautics and Space Administration (NASA)
Université Paris 12
The ICTP Public Information Office
Fondazione Internazionale Trieste per il Progresso e la Libertà delle Scienze
Laboratorio dell'Immaginario Scientifico

International Advisory Committee

Previous proceedings of conferences and colleges in the Trieste Series on Chemical Evolution and the Origin of Life in the Universe

Ponnamperuma, C. and Chela-Flores, J. (eds.), (1993) *Chemical Evolution: Origin of Life,* A. Deepak Publishing, Hampton, Virginia, USA.

Chela-Flores, J., Chadha, M., Negron-Mendoza, A. and Oshima, T. (eds.), (1995) *Chemical Evolution: Self-Organization of the Macromolecules of Life,* A. Deepak Publishing, Hampton, Virginia, USA.

Ponnamperuma, C. and Chela-Flores, J. (eds.), (1995) *Chemical Evolution: The Structure and Model of the First Cell,* Kluwer Academic Publishers, Dordrecht.

Chela-Flores, J. and Raulin, F. (eds.), (1996) *Chemical Evolution: Physics of the Origin and Evolution of Life,* Kluwer Academic Publishers, Dordrecht.

Chela-Flores, J. and Raulin, F. (eds.), (1998) *Chemical Evolution: Exobiology. Matter, Energy, and Information in the Origin and Evolution of Life in the Universe,* Kluwer Academic Publishers, Dordrecht.

Chela-Flores, J., Lemarchand, G. A. and Oro, J. (eds.), (2000) *Astrobiology From the Big Bang to Civilization,* Kluwer Academic Publishers, Dordrecht.

FIRST STEPS IN THE ORIGIN OF LIFE IN THE UNIVERSE

Proceedings of the Sixth Trieste Conference on Chemical Evolution
Trieste, Italy
18–22 September, 2000

Edited by

JULIÁN CHELA-FLORES
The Abdus Salam International Centre for Theoretical Physics, Trieste, Italy and Instituto de Estudios Avanzados (IDEA), Caracas, Venezuela

TOBIAS OWEN
Institute for Astronomy, Honolulu (Hawaii), United States of America

and

FRANÇOIS RAULIN
LISA, Universités Paris 12 & Paris 7, Faculté des Sciences et Technologie, Creteil Cedex, France

KLUWER ACADEMIC PUBLISHERS
DORDRECHT / BOSTON / LONDON

Library of Congress Cataloging-in-Publication Data

ISBN 1-4020-0077-4

Published by Kluwer Academic Publishers,
P.O. Box 17, 3300 AA Dordrecht, The Netherlands.

Sold and distributed in North, Central and South America
by Kluwer Academic Publishers,
101 Philip Drive, Norwell, MA 02061, U.S.A.

In all other countries, sold and distributed
by Kluwer Academic Publishers,
P.O. Box 322, 3300 AH Dordrecht, The Netherlands.

Cover picture – An artist's view of Mars Express,
a European orbiter and lander mission "to the Red Planet" or "to Mars".
(Image courtesy of the European Space Agency (ESA), © 2001.)

Printed on acid-free paper

Printed in the Netherlands.

DEDICATED TO

Giordano Bruno, for his contribution to the contemporary debate on astrobiology. His intuition anticipated concepts which are still relevant to the first steps of life in the universe: an infinite cosmos and the plurality of inhabited worlds. Bruno's cosmological vision matured long before present day astronomical observations led to the discovery of the new extrasolar planets.

DR

A seventeenth century representation of the Solar System, in which both planets and some of its satellites are present; but also neighboring stars are drawn with other worlds circling around them. This image was to a certain extent due to the intuition of Giordano Bruno. (Courtesy of Professor Francesco Bertola, form an original image reproduced form his book "Imago Mundi", Biblos, Cittadella PD, Italy, 1995, p. 174.)

PREFACE

The Sixth Trieste Conference on Chemical Evolution for the first time has also been a Euroconference. This year we focused on "First Steps in the Origin of Life in the Universe". (Both this Centre and the ICGEB have been sponsors since we first started planning the series with Professors Abdus Salam and Cyril Ponnamperuma.)

The conference had the following 11 sponsors:

The Abdus Salam International Centre for Theoretical Physics,
The European Commission,
The SETI Institute,
Consiglio Nazionale delle Ricerche, Rome, Italy,
The International Centre for Genetic Engineering and Biotechnology, Trieste,
The European Space Agency (ESA),
National Aeronautics and Space Administration (NASA)
Université Paris 12,
The ICTP Public Information Office,
La Fondazione Internazionale Trieste per il Progresso
e la Liberta della Scienza, and
Il Laboratorio dell'Immaginario Scientifico.

Over 90 participants allowed us to make a truly comprehensive review of our subject. In 82 presentations we had contributions from experts in questions related to the origin, evolution, distribution and destiny of life in the universe, topics that are known together as the science of astrobiology. The publication of these proceedings was largely due to a generous grant form the Consiglio Nazionale delle Ricerche, Rome, Italy.

The general topic of the conference concerned the origin, evolution, distribution and destiny of life in the Universe, a subject referred to as astrobiology and also as exobiology. In this report we prefer the synonym Astrobiology. It aimed to discuss the transition from inert matter to cellular life, as we know it on Earth. It also proposed to discuss the possibility of life occurring elsewhere as well, particularly in other environments in our own and other solar systems. The theoretical framework may be probed with forthcoming series of space missions, which at the time of writing are being planned for the next ten to 15 years.

Advanced extraterrestrial life can also be probed by means of radio astronomy in the well-established project of search for extraterrestrial intelligence. We aimed to bring together researchers in all of these fields, both from the European Community, particularly form the host country, from the United States and Japan; finally an important component were those scientists whose expertise is in this interdisciplinary area, who are currently working in the Third World.

The progress of the main space agencies the European Space Agency (ESA) and the National Aeronautics and Space Administration (NASA) was highlighted. The main sessions of the Conference were the following:

1. Life without starlight:
   - Can life originate in the absence of starlight?
   - Can life evolve to the point where it is truly independent of starlight?
   - How important is the "invention" of photosynthesis for the evolution of life?

2. Is there indigenous organic material on Mars?
3. The search for prebiotic and biological indicators in the satellites of the outer solar system.
4. New Approaches to the Detection of Extraterrestrial Intelligent Radio Signals and Some of its Implications.

The year 2000 presented us with the unique opportunity for highlighting the current interest in this most interdisciplinary subject: Four years ago David McKay (who was present at the conference) and co-workers at the Lyndon Johnson Space Center in Houston, and in several other research centers in the USA, announced the possible discovery of ancient life in a Martian meteorite. Independently of its eventual confirmation, his contribution stimulated a renewed robust effort in solar system exploration, including the leading European Space Agency (ESA), with the Beagle 2 project, as well as NASA, the American Space Agency, in their series of Mars missions, which need to review the Viking results. Sadly, there has been no consensus on the significance of the 1976 results.

One of the highlights of the conference was the Evening Lecture of Stanley Miller. He discussed the first steps in chemical evolution, namely what came before the much discussed RNA world, the leading candidate for the pre-cellular first stages of organized replication. Miller argued that peptide nucleic acids, rather than RNA, may have been the first genetic molecule.

From the origin of genetic information, the first indicators of cellular life on Earth was the subject discussed by J. William Schopf in the Second Cyril Ponnamperuma Lecture: Over the past few decades, after a century of trial and error, search and eventual discovery - life's earliest history has finally been uncovered as the documented fossil record has been extended to some 3,500 million years ago, an age more than three-quarters that of the planet itself.

Not only the first indicators of life on Earth were highlighted during the conference: in the Opening lecture Cynthia Phillips discussed the prospects of life in the proposed ocean of Europa, of the second Galilean moon of Jupiter.

The discussion was taken further into the second most attractive possibility of extraterrestrial life in the solar system by Schopf in a well-attended Public Lecture on Life on early Mars.

Some of the deeper questions raised by the search for the first steps of life in the universe were approached in the Abdus Salam Lecture, in which Paul Davies was the speaker. He considered the relevance of Physics in the life sciences in a richly illustrated lecture. Davies discussed the burgeoning field of quantum computation, which is providing new insights into the relationship between information, computation and physics, with hints that life may already be exploiting these deep linkages.

Frank Drake, in his Pre-Dinner Speech brought the conference up to date with the latest technology that is a available to radio astronomers and optical astronomers for answering the question whether the first steps of life have opened the way to other intelligence.

Finally, the Closing Lecture of Francesco Bertola set the present progress in the discovery of extrasolar planets in a historical context, in his excellent lecture on "The Plurality of Worlds".

Aspects of space technology were a prominent part throughout our discussions. These included the Galileo, Cassini-Huygens, Mars Express and Rosetta Missions, as well as proposed missions that are only at the preliminary level of planning for Mars and Europa. Marcello Coradini, Coordinator, Solar System Missions, Directorate of Scientific Program ESA gave a comprehensive review of this important aspect of astrobiology.

Altogether twenty nine keynote addresses by leading researchers went deeper into the broad framework with emphasis on the origin, evolution, distribution and destiny of life in the universe ('astrobiology', or 'exobiology'). Many young researchers from highly industrialized nations completed the review of the latest progress in research. This was possible due to the representation of 12 contributions and 15 posters form the European Union and the United States, as well as that form from the less industrialized nations. The broad review was possible thanks to a well-supported and high-level poster session with a total of 33 entries.

Participants came from: Africa, Asia, Australia, Eastern Europe, the European Union, Latin America, and also from Israel, Russia, Switzerland, Japan, the United States of America. With the current participants, the Trieste conferences have brought together almost 500 scientists, philosophers and theologians, since they began in October 1992.

Julian CHELA-FLORES
Tobias OWEN
François RAULIN,
May, 2001.

Conference participants

1. Bahn
2. Pappelis
3. Coliolo
4. Ascheri
5. Pagan
6. Mrs. Schopf
7. Schopf
8. Miller.
9. Chela-Flores
10. Raulin
11. Owen
12. Navarro-Gonzalez
13. Raulin-Cerceau
14. Miss Raulin
15. Master Raulin
16. Colli
17. Mrs. Chadha
18. Wang
19. Oro
20. Mrs. Oro
21. Seckbach
22. Shah
23. Gallori
24. Tewari
25. Bonnacorsi
26. Horneck
27. Musso
28. Lazcano
29. B. Prieur
30. Cosmovici
31. Asuzu
32. Mendez
33. Trigo
34. Guimaraes
35. Chadha
36. Cairns-Smith
37. Negron-Mendoza
38. Drake
39. Islam
40. Mrs. Islam
41. Genta
42. Caprara
43. Kamaddulin
44. Chandra
45. Kletcheska
46. Wanda Davis
47. Turian
48. Mrs. Turian
49. Moorbath
50. Brack
51. Ramos
52. Corso
53. Miniussi
54. Ollivier
55. Simakov
56. Participant
57. M. Baltscheffsky
58. 59. H. Baltscheffsky
60. Kritskiy
61. Vicuna
62. Ruiterkamp
63. D. Prieur
64. Matsuno
65. Patel
66. Corliss
67. Participant
68. D. McKay
69. Mukerjee
70. Participant
71. Coyne
72. Greenberg
73. C. McKay
74. Russell
75. Whitehead
76. Gatta
77. Vuitton
78. Collis
79. Laogun

CONTENTS

Contributions

Contributions

Contributions

Contributions

Contributions

Special Evening Lecture

Speaker:

Stanley Miller

PEPTIDE NUCLEIC ACIDS AS A POSSIBLE PRIMORDIAL GENETIC POLYMER

KEVIN E. NELSON [1], ANTONIO LAZCANO [2], AND STANLEY L. MILLER[1]

[1]*Department of Chemistry and Biochemistry, University of California, San Diego, La Jolla, CA 92093-0506, USA*
[2]*Facultad de Ciencias, UNAM, Apdo. Postal 70-407, Cd. Universitaria, México 04510 D.F., MEXICO*

1. Introduction

Essential to all contemporary life forms is an intracellular genetic apparatus capable of expressing and, upon reproduction, transmitting to the progeny information capable of undergoing evolutionary change. How this ubiquitous genetic system originated is one of the major problems in contemporary biology. The discovery of ribozymes has given considerable acceptance to the concept of an RNA world, an era in which RNA molecules are proposed to have acted as both the genetic material and the catalysts, i.e., combined genotype and phenotype in the same molecule (Gilbert, 1986).

However, the assumption that life began with RNA molecules brewed in the prebiotic environment from random abiotic reactions faces major problems, including (a) the instability of ribose (with a half-life of 73 minutes at 100 °C, and of 44 years at 0 °C) and other sugars (Larralde et al., 1995); (b) the difficulties of robust prebiotic synthesis of the glycosidic bonds of activated nucleotides (Fuller et al., 1972a,b); and (c) the lack of efficient two-way non-enzymatic template-directed polymerizations (Orgel 1987; Joyce, 1987). Although there are alternative views (Ertem and Ferris, 1998; Reimann and Zubay, 1999; Zubay and Mui, 2001), the difficulties associated with the prebiotic origin of RNA have prompted the search for alternative genetic macromolecules, i.e., for a pre-RNA world where the informational polymers had a backbone different from ribose-phosphate and perhaps even with other nucleobases. The pre-RNA world is assumed to have the same essential characteristic of the RNA world, with phenotype and genotype both residing in the same polymer, so neither proteins or related catalysts are required (Lazcano and Miller, 1996). As elegantly demonstrated by Eschenmoser and his

J. Chela-Flores et al. (eds.), First Steps in the Origin of Life in the Universe, 3–10.

coworkers, a wide variety of informational macromolecules are possible, even when restricted to sugar-phosphate backbones (Schöning et al., 2000). Here we review recent experimental evidence supporting the possibility that the components of peptide nucleic acid, i.e., PNA (which has a backbone related to those of proteins and has been consider as an alternative nucleic acid analog), were present in the primitive environment as the result of prebiotic synthetic reactions.

2. Peptide nucleic acids as genetic polymers

The search for antisense DNA for gene therapeutics has prompted the syntheses of many modified nucleic acid backbones. However, most of these have been modifications of the phosphodiester backbone by replacing it, for instance, with an amide or ester linkage, or by modifying the furan ring of the ribose (cf. Nielsen et al.,1991). By contrast, PNA has a considerably greater change in its backbone, which consists of poly-2-aminoethyl glycine (AEG), which is sometimes referred to as ethylenediamine monoacetic acid (Figure 1). The nucleobases are attached to the backbone by the base-substituted acetyl units shown in Figure 2.

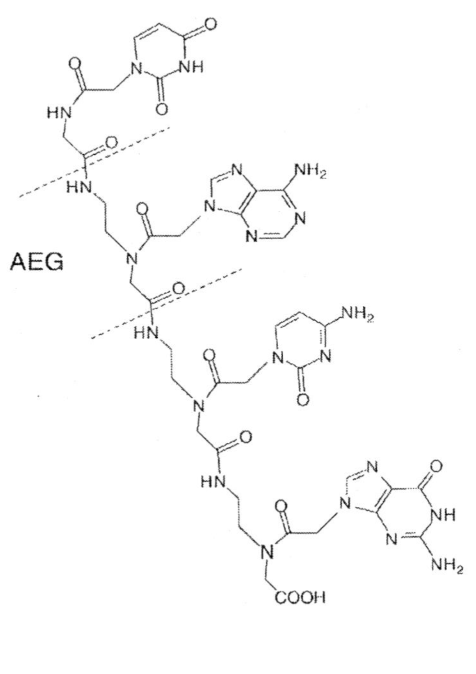

Figure 1. Peptide Nucleic Acid (PNA)

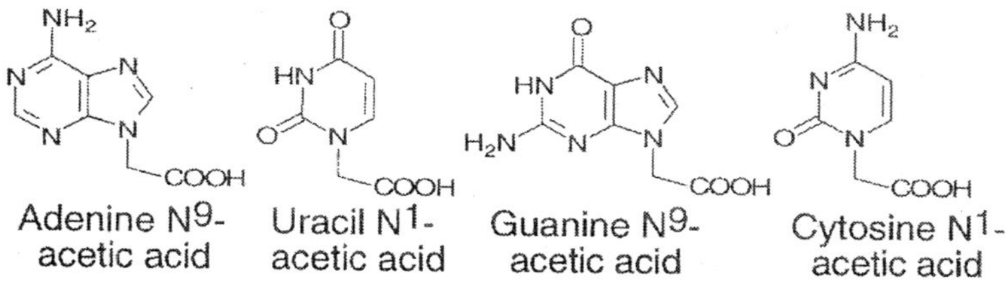

Figure 2. Acetic acid derivatives of adenine, guanine, uracil, and cytosine

Although the possibility that AEG could replace the ribose-phosphate backbone had been envisoned by Westheimer (1987), the report that PNA could bind DNA in spite of their chemical differences came as a complete surprise (Nielsen et al., 1991). Oligo-PNAs act as templates for the polymerization of the complementary nucleotide phosphorimidazolides, and the oligo-DNA template catalyzes the polymerization of PNA dimers (Bohle et al., 1995). Due to the absence of charges on the PNA backbone, the PNA-DNA duplex, which is similar to the A-form DNA, is more stable that the DNA-DNA or DNA-RNA helices, which tend to separate the two strands (Ericksson and Nielsen, 1996). A PNA-PNA double helix has also been developed (Rasmussen et al. 1997), and although it differs considerably from the B-forms of DNA, the Gibbs free-energy differences among the various structures appear to be small, suggesting that the changes required to transform the PNA-PNA duplex into the B-configuration are small (Miller, 1997).

Although PNA was originally deviced as a therapeutic antisense DNA analogue, it was inmediatly apparent that it might also be a candidate as a prebiotic macromolecule and potentially the first genetic material that preceded the RNA world (Nielsen, 1993). The simplicity of the components of PNA suggested that their prebiotic syntheses were feasible. We have therefore examined a number of abiotic syntheses including electric discharges and ammonium cyanide polymerizations for ethyleneamide (ED) and AEG, as well the nucleobase acetic derivatives , adenine- and guanine-N^9-acetic acids and cytosine- and uracil-N^1-acetic acids. As reviewed here, the available experimental evidence shows that the components of PNA are prebiotic compounds.

3. The prebiotic synthesis of AEG and ethylenediamine

N-(2-aminoethyl) glycine and ethylenediamine have been obtained in low yields in spark discharge experiments at high- (100 °C) and low temperature (25 °C) conditions (Nelson et al., 2000). The polymerization of NH_4CN, which produces adenine and amino acids, was also studied. The results of these experiments are summarized in Table 1. As shown in Table 1, the polymerization of NH_4CN gave 0.03% adenine, 2.1% glycine, and 4.4×10^{-5}% ED, but no AEG ($<1 \times 10^{-5}$%). These yields of ED and AEG are low, but they may be sufficient given the high solubility of AEG, which would allow it to be concentrated on drying lagoons.

Table 1. Percent yields of glycine, AEG, and ED from electric discharge experiments and NH_4CN polymerization

	Percent yield		
	Glycine	AEG	ED
Spark discharge (25 °C)	0.23	1.5×10^{-5}	7.5×10^{-5}
Spark discharge (100 °C)	0.58	6.2×10^{-6}	5.5×10^{-5}
NH_4CN polymerization	2.1	$<1.0 \times 10^{-5}$	4.4×10^{-5}
NH_4CN + 10% H_2CO	9.1	$<1.0 \times 10^{-5}$	0.05

The spark discharge yields are based on the CH_4, the NH_4CN polymerization yields are based on the carbon in CN^-, and the $NH_4CN + H_2CO$ on the concentration of formaldehyde.

Since formaldehyde has been shown to catalyze the production of adenine from NH_4CN (Schwartz and Goverde, 1982), it seemed likely that a more efficient synthesis of AEG would be possible under these conditions. Polymerization of 8.3M NH_4CN in the presence of 0.8M H_2CO gave 0.05% yields of ED, together with 9.5% glycine, 0.1% diaminopropionic acid, and 0.041% adenine. Both NH_4CN and H_2CO are prebiotic reagents which could have achieved high concentrations on the primitive Earth by freezing dilute solutions of ammonia, formaldehyde, and hydrogen cyanide.

Although no AEG was detected in the NH_4CN-H_2CO reaction, it can be produced via the Strecker synthesis:

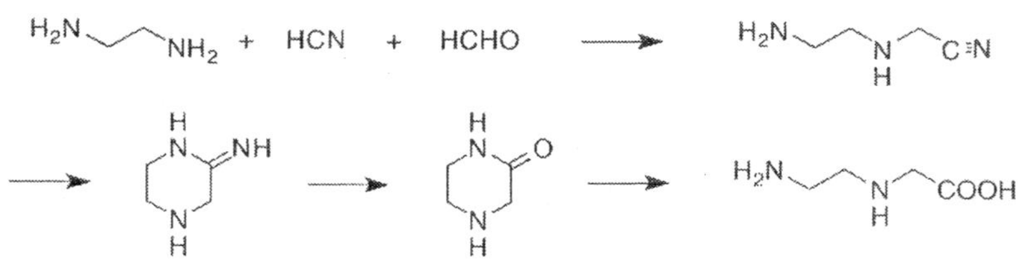

The above reaction is quite robust and efficient even at high dilutions. At 25 °C and pH7, a 21% yield of AEG can be obtained from a 10^{-1}M concentration of ED, HCN, and formaldehyde. Moreover, Ogura et al. (1989) have shown that photolysis of mixtures of methane, ammonia, and water gives 33%yields of ED and 66% yield of methylamine. These experiments suggest that AEG may have been more abundant on the primitive Earth than is indicated by the spark discharge experiments (Table 1).

4. The abiotic syntheses of acetic acid derivatives of nucleobases

The synthesis of adenine and guanine-N^9-acetic acids via an internal Cannizaro rearrengement using glyoxal gave very low yields. Equally low outpùts were observed with a Strecker synthesis at room temperature with 1mM concentrations of each base, with 30-70 mM HCN and H_2CO at pH 9.6, gives low yields ranging from 0.04% to 0.15% (Nelson et al., 2000):

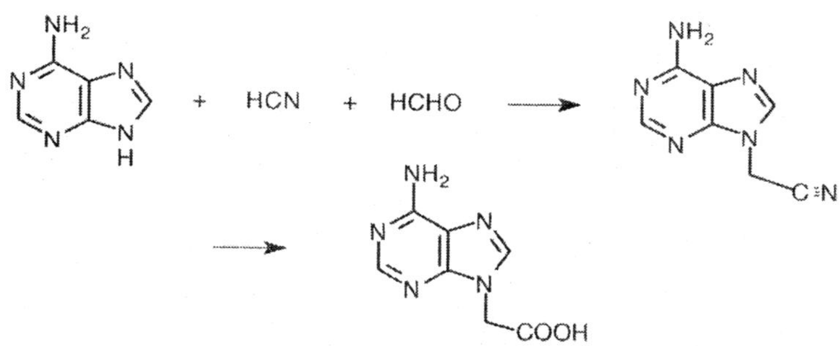

The production of adenine and guanine N^9-acetic acid derivatives could be expected from the polymerization of cyanide in the presence of glycine. Heating mixtures of 2.5M glycine, 7.5M NH_4OH and 10M HCN at 80 C, all of which can be considered prebiotic reagents. This reaction produced 0.0062% adenine N^9-acetic acid, 0.0013% adenine N^6-acetic acid, and 0.0051% adenine, and 0.011% guanine N^9-acetic acid and 5 x 10^{-4}% guanine. The purine N^9-acetic acids from HCN and glycine may have been produced via the mechanism proposed in Figure 3.

On the other hand, the synthesis of pyrimidine N^1-acetic acids is a very efficient process that resulted from a modification of pyrimidine base at syntheses under dry beach conditions from the reaction of cyanoacetaldehyde and urea (Robertson and Miller, 1995). The reaction, shown in Figure 4, proceeds from enolic cyanoacetaldehyde with hydantoic acid, which is produced in the polymerization of HCN (Ferris et al., 1974) and has been identified in the Murchison meteorite (Cooper and Cronin, 1995).

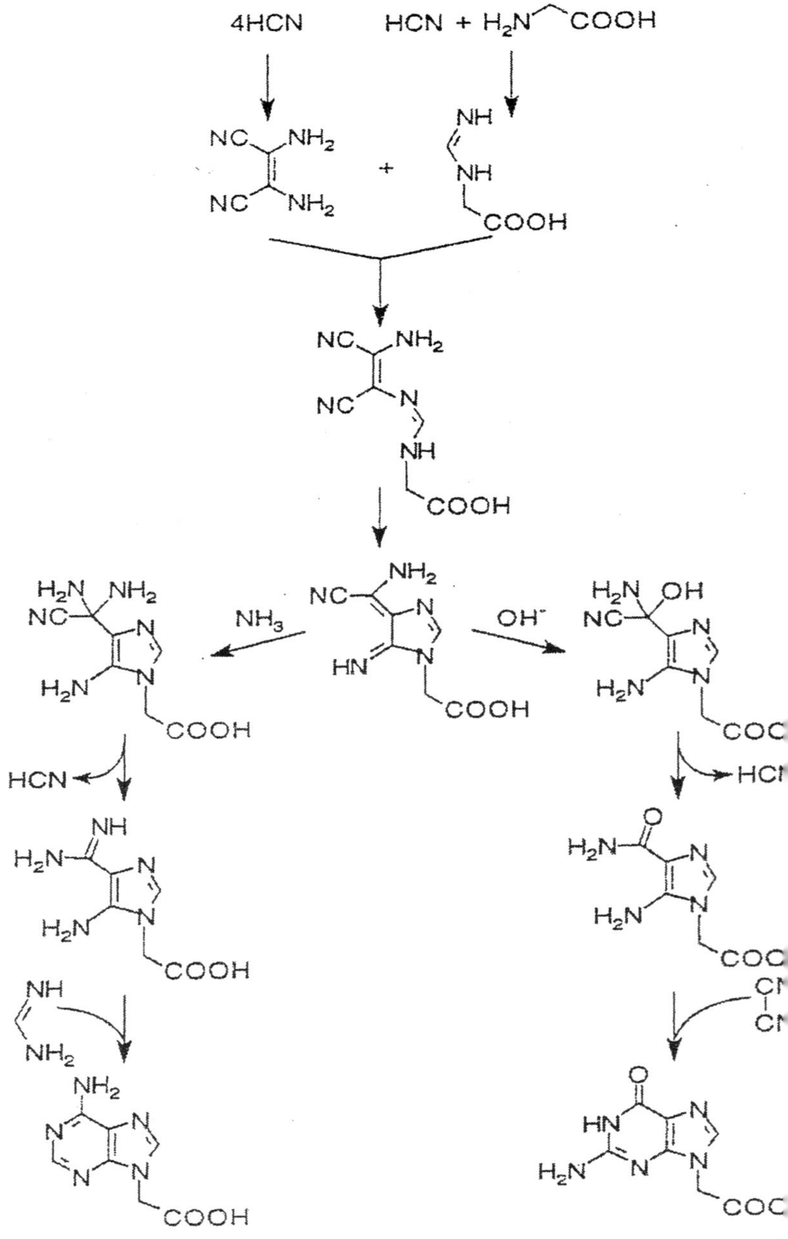

Figure 3. A possible mechanism for the production of acetic acid purine derivatives

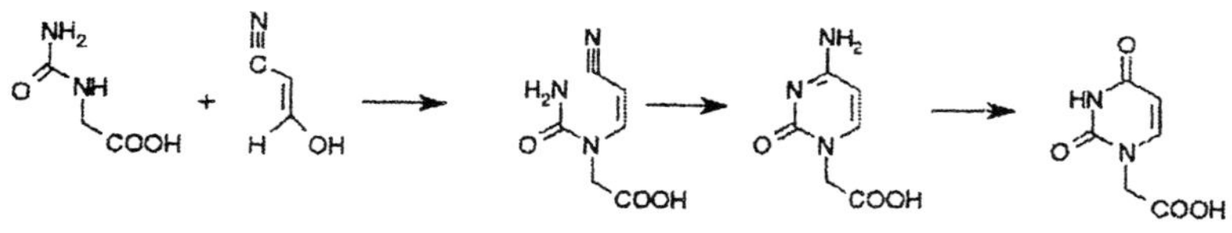

Figure 4. Synthesis of pyrimidine acetic acid derivatives from cyanoacetalhdeyde and hidantoic acid.

The reaction of 1 mM cyanoacetaldehyde with 2M hydantoic acid at 100 °C produced high yields of 18% for cytosine-N1-acetic acid, and 1.8% for uracil-N1-acetic acid based on cyanoacetaldehyde. Using 2M urea and 2M hydantoic acid gives a ratio of 8:1 cytosine to cytosine-N^1-acetic acid.

5. Conclusions

Although AEG cyclizes rapidly at 100 °C and pH7 to form the lactam, monoketopiperazine, the results reviewed here show that the components of PNA are likely prebiotic compounds that, under favorable conditions, may have reached high concentrations. While it is true that PNA is the first example of a possible RNA precursor for which the components are prebiotic, it also must be kept in mind that there are no obvious prebiotic polymerization processes for PNA or other macromolecules. Moreover, peptide nucleic acids also hydrolyze rather rapidly unless the free amino group at its end (Figure 1) is blocked (Eriksson et al., 1998), and its overall stability needs to be assesed. Thus, although the results reported here support the possibility that PNA or other similar molecule may have been the first genetic polymer, the alternative that other backbones and nucleobases may have been more abundant and efficient deserve further attention.

Acknowledgments

We thank the NASA Specialized Center of Research and Training for fellowships (K.E.N. and A.L.) and grant support (S.L.M.)

6.References

Bohle, C., Nielsen, P. E., and Orgel, L. E. (1995) Template switching between PNA and RNA oligonucleotides, *Nature* **376**, 578-581

Cooper, G. W. and Cronin, J. R. (1995) Linear and cyclic aliphatic carboxamides of the Murchison meteorite: hydrolyzable derivatives of amino acids and other carboxylic acids, *Geochim. Cosmochim. Acta* **59**, 1003-1005

Eriksson, M. and Nielsen, P. E. (1996) Solution structure of a peptide nucleic acid-DNA duplex, *Nature Struct.. Biol.* **3**, 410-413

Eriksson, M., Christensen, L., Schmidt, J., Haaima, G., Orgel, L., and Nielsen, P. E. (1998) Sequence dependent N-terminal rearrangement and degradation of peptide nucleic acid (PNA) in aqueous solution, *New J. Chem.* **22**, 1055-1059

Ertem, G. and Ferris, J. (1998) Synthesis of RNA oligomers and their template properties, *Viva Origino* **26**, 203-218

Ferris, J. P., Wos, J. D., and Lobo, A. P. (1974) Chemical evolution XXII. The hydantoins released on hydrolysis of HCN oligomers, *J. Mol. Evol.* **3**, 311-316

Fuller, W.D., Sanchez, R.A., and Orgel, L. E. (1972) Studies in prebiotic synthesis VII. Solid-state synthesis of purine nucleosides, *J. Mol. Evol.* **1**, 249-257

Fuller, W.D., Sanchez, R.A., and Orgel, L. E. (1972) Studies in prebiotic synthesis VI. Solid-state synthesis of purine nucleotides, *J. Mol. Biol.* **67**, 25-33

Gilbert, W. (1986) The RNA world, *Nature* **319**, 618

Joyce, G. F. (1987) Nonenzymatic template-directed synthesis of informational macromolecules, *Cold Spring Harbor Symp. Quant. Biol.* **52**, 41-51

Larralde, R., Robertson, M.P., and Miller, S. L. (1995) Rates of decomposition of ribose and other sugars: implications for chemical evolution, *Proc. Natl. Acad. Sci. USA* **92**, 8158-8160

Lazcano, A. and Miller, S. L. (1996) The origin and early evolution of life: prebiotic chemistry, the pre-RNA world, and time, *Cell* **85**, 793-798

Miller, S. L. (1997) Peptide nucleic acids and prebiotic chemistry, *Nature Struct. Biol.* **4**, 167-169

Nelson, K. E., Levy, M., and Miller, S. L. (2000) Peptide nucleic acids rather than RNA may have been the first genetic molecule, *Proc. Natl. Acad. Sci. USA* **97**, 3868-3871

Nielsen, J. E. (1997) Peptide nucleic acid (PNA): a model structure for the primordial genetic material? *Origins of Life Evol. Biosph.* **23**, 323-327

Nielsen, P. E., Egholm, M., Berg, R. H. and Buchardt, O. (1991) Sequence-selective recognition of DNA by strand-displacement with a thymine-substituted polyamide, *Science* **254**, 1497-1500

Orgel, L. E. (1987) Evolution of the genetic apparatus: a review, *Cold Spring Harbor Symp. Quant. Biol.* **52**, 9-16

Ogura, K., Migita, C. T., and Yamada, T. (1989) Photolysis of CH_4-NH_3-H_2O mixture: formation of methylamine and ethylenediamine , *J. Photochem. Photobiol.* A**49**, 53-61

Rasmussen, H., Kastrup, J. S., Nielsen, J. N., Nielsen, J. M., and Nielsen, P. E. (1997) Crystal structure of a peptide nucleic acid (PNA) duplex at 1.7 A resolution, *Nature Struct. Biol.* **4**, 98-100

Reimann, R. and Zubay, G. (1999) Nucleotide phosphorylation: a feasible step in the prebiotic pathway to RNA, *Orig. Life Evol. Biosph.* **29**, 229-247

Robertson, M. P. and Miller, S. L. (1995) An efficient prebiotic synthesis of cytosine and uracil, *Nature* **375**, 772-774

Schöning, K.-U., Scholz, P., Guntha, S., Wu, X., Krishnamurthy, R., and Eschenmoser, A. (2000) Chemical etiology of nucleic acid structure: the □ -threofuranosyl-(3'->2') oligonucleotide system, *Science* **290**, 1347-1351

Schwartz, A. W. and Goverde, M. (1982) Acceleration of HCN oligomerization by formaldehyde and related compounds: implications for prebiotic syntheses, *BioSystems* **15**, 191-193

Westheimer, F. H. (1987) Why nature chose phosphates, *Science* **235**, 1173-1178

Zubay, G. and Mui, T. (2001) Prebiotic synthesis of nucleotides, *Orig. Life Evol. Biosph.* 31, 87-102

The Abdus Salam Lecture

Speaker:

Paul Davies

PHYSICS AND LIFE

Lecture in honour of Abdus Salam

PAUL DAVIES

Physics Department
Imperial College
London SW7 2AY

Fifty years ago, physicists seemed on the verge of solving the problem of life. Inspired by Erwin Schrödinger's book *What is Life?* (1944), researchers began unraveling the molecular basis of the living cell, in the belief that a solution to both the nature and the origin of life would soon be found. Today, these hopes seem very naïve. Indeed, physics is regarded by most investigators as not especially relevant to the problem of life.

To be sure, physics plays an *indirect* role in life, in two ways. First, life operates in accordance with the laws of physics, but so does everything else, including, for example, the Italian Constitution. Physics is universal, its laws simple, mathematical and general. But life, like the Italian Constitution, is complex, non-mathematical, and very special. Precisely because physics is universal, it is unusual that its principles are crucial for explaining any *particular* physical system.

The second way in which physics has an indirect bearing on life is by providing the basic 'tool kit' which biology requires. It is well known that the existence of life (at least as we know it) depends rather sensitively on the precise nature of the laws of physics, and particularly on the numerical values that nature assigns to various coupling constants and particle masses. This topic often goes under the name 'the anthropic principle' and has been much reviewed (Barrow & Tipler, 1986). Here I shall restrict myself to two simple examples, the first due to Fred Hoyle (Hoyle, 1954). Life is based on carbon. This element did not exist at the birth of the universe, but was made inside large stars. A carbon nucleus forms from the fusion of three helium nuclei. This reaction has a tiny cross-section, and the carbon yield would be paltry were it not for the fortuitous existence of a resonance at just the right energy to correspond to the temperatures of stellar cores. If the strong interaction did not have the strength it does, the universe could have been largely devoid of this life-giving substance.

The second example has been discussed by Freeman Dyson (Dyson, 1971), and also concerns the strength of the strong nuclear force. If it were just two per cent stronger, then it would be possible for two protons to overcome their electric repulsion and bind to form a di-proton. This would soon decay via the weak interaction to a deuteron. Dyson considers the consequences of this for the hot big bang theory of the universe. One second after the origin of the universe, the cosmic material was a soup of free protons, neutrons, electrons, etc. During the subsequent few minutes, about 25 per cent of the nuclear matter was converted into helium, the rest being left as hydrogen. If di-protons were possible, the leftover protons would rapidly convert into deuterons, which would then combine to form more helium. The result would be that all the hydrogen would be consumed, leaving a

J. Chela-Flores et al. (eds.), First Steps in the Origin of Life in the Universe, 13–20.

universe composed entirely of helium. There would then be no stable stars like the sun and no water. Life would almost certainly be impossible in such a universe. There is a long list of similar happy accidents in particle physics that make for a remarkably bio-friendly universe – analogous to what science administrators call 'a well-found laboratory'.

All this is well and good, but is physics relevant to life's *specific* and peculiar properties? One person who was convinced that physics plays a direct role in life, at least in its genesis, was Abdus Salam, in whose honour this paper is delivered. His work on molecular chirality (Salam, 1991, 1992) was a bold attempt to trace the well-known handedness of biological molecules to parity violation in electroweak interactions, a subject that he himself founded. If this link is correct, it provides a key biological role for one the most fundamental aspects of particle physics. Whilst not wishing to revisit the topic of chirality here, I shall argue, in the spirit of Salam's work, that physics may enter into biology at a basic level in a number of important ways.

In approaching this problem, it is helpful to be reminded of Jacques Monod's celebrated distinction between chance and law, or necessity as he termed it (Monod, 1972). All physical systems come about from a combination of these factors. Some, like crystals, are almost completely determined in their structure and properties by the laws of physics alone, which embody the relevant crystal symmetries. Others, like clouds in the sky, are shaped mostly by chance. In between are systems such as snowflakes, for which the overall hexagonal structure is determined by physics, but the specific filigree details are a matter of happenstance. Where, on this spectrum from pure chance to pure law, does life lie? Opinions vary widely. Monod himself argued that life was pretty much the product of chance, a stupendous chemical fluke unique in the observable universe. By contrast, Christian de Duve (de Duve, 1995) thinks life is a 'cosmic imperative,' more or less bound to occur wherever earthlike conditions prevail.

The belief that life is 'written into' the laws of nature is sometimes called biological determinism (Shapiro, 1986). In its most extreme form, as advocated for example by Sidney Fox (Fox, 1988), it asserts that the laws of the universe are cunningly rigged to coax life into being from lifeless chemicals, by favouring the production of just those molecules that life needs. On this manifestly teleological view, life's information content derives from the physical laws that generate the informational molecules. This view is hard to sustain, since the information content, as measured algorithmically (Chaitin, 1990), of the known laws of physics at least, is demonstrably low (Yockey, 1992). That is why crystals, which are determined by those laws, have low information content, being just regular arrays of atoms. By contrast, DNA is a random string of atoms - the 'aperiodic crystal' famously predicted by Schrödinger (1944) - and so has high algorithmic information content; it is then hard to see how such an entity could be a product of law alone. In this respect it is worth noting that although the backbone structure of DNA is determined by the laws of physics and chemistry, the precise sequence of nucleotides – the 'letters of the genetic alphabet' – are not. There are no chemical bonds between successive nucleotides; chemistry is indifferent to the sequence chosen.

Those like de Duve who espouse a less teleological and conspiratorial form of biological determinism argue that because the stuff of life is common in the universe, then so must life be. Oft-cited is the commonness of the life-giving elements C,H,N,O,P,S, and the ubiquity of simple organic building blocks like methane, formaldehyde, alcohol and even amino acids. These molecules are easy to make, and are found across the universe, in meteorites, comet tails and even interstellar clouds. Therefore, the argument goes, life should be common too.

This argument is, however, flawed. The building blocks of life are easy to make because their synthesis is thermodynamically favoured. But stringing them together in an aqueous environment into complex molecular chains like proteins and RNA is thermodynamically 'uphill.' Just as a pile of bricks alone don't make a house, so organic building blocks alone don't make life. Put a stick of dynamite under a pile of bricks, and you don't make a house, you just make a mess. In the same way, merely throwing energy willy-nilly at a collection of amino acids, for example, to drive it against the thermodynamic gradient, won't produce a protein. Just as a house requires the delicate assembly of bricks into an elaborate and specific arrangement, so amino acids need to be carefully linked in a precise way to make a functional protein, rather than gunk. The same goes for nucleic acids.

A hundred years ago, it was commonly supposed that life is some sort of magic matter, and that life's origin would be analogous to baking a cake. All it needs is the right ingredients mixed in the right order under the right conditions. Today we know that the living cell is less magic matter, more a supercomputer; i.e. it is an information processing and replicating system. The key property that distinguishes life from other forms of complexity is the informational aspect, the message in the genes. Chemistry cannot explain information. Chemistry is the medium of life, but one must not confuse the medium with the message.

In the living cell, nucleic acids and proteins, which are scarcely on nodding terms chemically, deal with each other via an information channel, i.e. using software rather than hardware, written in a triplet mathematical code. The advantage of life 'going digital' in this way is much greater flexibility and fidelity (as is also the case with digitization in electronic devices). The situation can be likened to flying a kite versus a radio-controlled plane. A kite is hard-wired to the controller, and is clumsy to control by pulling on the strings. By contrast, a radio-controlled plane is easier to fly because the controller's instructions are digitized and transmitted to the plane, where they are decoded and used to harness local energy sources. The radio waves themselves do not push and pull the plane around; they merely convey the information. Analogously, nucleic acids do not themselves assemble proteins, they relay the instructions for ribosomes to do it. This frees protein assembly from the strictures of chemistry, and permits life to choose whatever amino acid sequences it needs. So, far from deriving from physics and chemistry, biological information is quasi-independent of it. To explain the origin of this information-based control, we need to understand how mere hardware (atoms) wrote its own software.

Note that we must do more than simply explain where information per se came from. A gene is a set of coded instructions (e.g. for the manufacture of a protein). To be effective, there must exist a molecular milieu that can decode and interpret the instructions, and carry them out, otherwise the sequence information in the DNA is just so much gobbledygook. The information is therefore *semantic* in content, i.e. it must *mean* something (Küppers, 1985). So we are faced with the task of understanding the nature and origin of semantic, or meaningful, information. Since the very concept of information emerged from communication theory in the realm of human discourse, this is no trivial matter. Information is not like mass or energy: you can't tell by looking whether a molecule has it or not. As yet, there is no 'info-dynamics' comparable to the dynamics of matter, let alone an understanding of how 'meaning' emerges in nature.

The central puzzle, as it seems to me (Davies, 1998), is that life possesses two apparently contradictory properties. The first is that the key informational content demands *randomness* (in the algorithmic sense) since order is by definition low in information content. The second is specificity; arbitrary randomness is no good. A protein, for example, is a *specific* random sequence of amino acids; any old higgledy-piggledy sequence will almost certainly be biologically useless. Individually, randomness and specificity are not hard to create. Chance (in the form of thermal chaos say) generates randomness, and law generates specificity. But what combination of chance and law generates *specific randomness*?

One solution is Darwinism. Chance mutation and lawlike natural selection has what it takes to produce the right mix of randomness and specificity. But Darwinism kicks in only after life gets going; it cannot be used to explain the origin of life. Some researchers suggest defining life as any system that undergoes replication, variation and selection, and argue for 'Darwinism all the way down'. The system may be simply a collection of replicating molecules that could plausibly form by chance in a prebiotic soup.

Can molecular Darwinism explain biogenesis? Maybe, but we have scant idea what those first replicating molecules might be. Examination of real organic replicator systems like RNA/proteins indicates that even the simplest replicators are extremely large and complex molecules, unlikely to form by chance. Moreover, the smaller the molecules the sloppier they copy, suggesting that molecules small enough to form by chance would be very bad at replicating information, and thus subject to Eigen's error catastrophe (Eigen & Schuster, 1979), whereby information is eroded by the inaccurate copying process faster than natural selection can inject it.

I concede that if something like the RNA world (Cech, 1986) were given to us ready-made, it has the capacity to evolve into life as we know it. But it strains credulity to suppose that the RNA world sprang into being in one huge chemical transformation. Likely it would be the product of a long series of steps. We can liken the situation to a vast decision tree of chemical reactions, with the RNA world as one tiny twig on the tree. (There is the question of whether there are other twigs that could lead to life, but I shall assume here that the RNA route is the only one.) So we need to understand how a hypothetical class of simple, small replicators navigated through that decision tree and 'found' the RNA twig. Was this just a lucky fluke, or is there something other than a random walk involved?

Now searching databases and navigating decision trees is an interesting branch of science that we might term informational physics. I wish to conjecture that some new discoveries in this field just might help explain how life's decision tree was navigated.

The first point I want to make is that informational physics encompasses mechanisms capable of converting random motion into directed motion. An example of current interest is the so-called Brownian ratchet, based on a device first studied in detail by Smoluchowski (1912). It consists of a ratchet and pawl connected via a rod to a set of vanes, and immersed in a gas in thermodynamic equilibrium. The ratchet allows the vanes to rotate in one direction, but not the reverse. The random motion of the molecules bombarding the vanes will cause the system to rotate, thereby apparently converting undirected chaotic molecular motion into directed macroscopic motion. This seems to violate the second law of thermodynamics, because the rotation could be used to perform work, e.g. by lifting a weight. The resolution of the paradox was essentially spotted by Smoluchowski (1912) and refined by Feynman (Feynman, Leighton & Sands, 1963) and Abbott (Abbott, Davis & Parrondo, 1999), in which it was pointed out that in thermodynamic equilibrium the position of the pawl will fluctuate due to thermal noise, and allow the ratchet to slip backwards as often as it is driven forwards. There is then no net rotation on average.

Moreover, a type of Brownian ratchet that serves to convert random into directed motion has been devised by Magnasco (1993), and studied by Doering (1995) and Harmer & Abbott (1999). In this system, an ensemble of randomly bouncing balls can be made to diffuse *uphill* if driven by a tilted sawtooth forcing potential that flashes on and off - the so-called flashing ratchet (Ajari & Prost 1993). The ratchet thus drives the system 'the wrong way' from a thermodynamic viewpoint (though there is no violation of the second law because the system is not closed on account of the external potential). The relevance of this discussion to life is that Darwinian evolution is an example of a ratchet, because advantageous random changes are locked in, thus also giving a superficial appearance of going against the second law of thermodynamics.

Derived from the physical example of the Brownian ratchet is the curious paradox of Parrondo (Harmer & Abbott, 1999; Parrondo et. al., 2000), involving games of chance. Parrondo has proved that two fair games that individually have an expectation of loss to the player can be played in combination with an expectation of gain! Again, the relevance of this to biological evolution is clear: Darwinism is a type of game of chance in which the winners, driving against the thermodynamic gradient ('climbing Mount Improbable', to use Richard Dawkins' evocative description (Dawkins, 1996)) are the survivors. If chance variations could lead to ordered evolution as opposed to random diffusion, then canalization within the chemical decision tree may result. (Of course, if nature obligingly directs the activity preferentially *towards* the RNA world we are back to teleology again.)

My remaining examples concern the possibility that quantum mechanics may have a more direct role to play in life than merely providing the mechanism of chemical bonding. The founders of quantum mechanics generally believed that life required some extraordinary physics to explain it. Thus Schrödinger wrote (Schrödinger, 1944, p. 81), 'We must be prepared to find a new kind of physical law prevailing.' Several researchers have suggested that quantum mechanics might be biologically relevant. An early conjecture along these lines is Fröhlich's theory (Fröhlich, 1983) that collective vibrational modes (coherent phonons) in biological membranes can create conditions similar to a Bose-Einstein condensate, leading to ordered, cooperative behaviour in which the vibrational energy is concentrated into the lowest mode.

A more recent example has been given by McFadden (2000), who points out that certain mutations occur as a result of quantum tunneling events in the pair bonds within DNA. He conjectures that the biological environment might 'select' certain mutations by affecting the tunneling probabilities. Is this credible? Certainly the theory of quantum transitions involving strong coupling to the environment involves some unusual features. For example, in the watchdog or quantum Zeno effect (Itano et. al., 1990), continuous measurement-like interaction with by the environment can serve to paralyze a quantum system in its initial state. The inverse watchdog or Zeno effect (Altenmuller & Schenzle, 1993; Kofman & Kurizki, 2000) can amplify certain transitions and 'steer' a quantum system through a sequence of states by environmental interactions. McFadden conjectures that competing quantum transitions with biochemically very distinct consequences might have very different transition rates, so that adaptive mutations might be quantum mechanically favoured. Applying this to biogenesis, it is possible to imagine that states that are in some sense 'more lifelike' (e.g. more complex, more organized, more information rich) might also be favoured. The trouble is, it's very hard to pin down a precise attribute for 'lifelike' that can exercise a well-defined physical effect. The most obvious candidate is replication, which has a clear physical basis. In a quantum system with feedback, it may be that the production of a replicator in a complicated network of chemical reactions acts like an attractor, with the feedback amplifying, via something like the inverse watchdog effect, the transition probabilities leading to replicating molecules.

These ideas hint that maybe quantum mechanics can 'fast-track' a chemical soup to complex biologically-relevant states. Since the object of the exercise is to explain the origin of biological information, the appropriate theoretical framework would seem to be quantum information theory. This subject is currently of intense interest because of the possibility of constructing a quantum computer (Milburn, 1998; Bennett & DiVincenzo, 2000). The key property of quantum information processing is that it is far more powerful than classical information processing. That is because the wavefunction of a collection of entangled particles can store information in the phases. So long as quantum coherence is maintained, transformations of the wavefunction can simultaneously process exponentially more information than the corresponding classical system. Farhi and Gutmann (1998) have applied quantum information theory to decision trees, and found an exponential improvement in the search time. Treating the biogenesis problem as the need to navigate the molecular decision tree to 'find' the RNA world, or something similar, then a quantum search would obviously be vastly quicker.

Building upon these ideas, a fruitful line of investigation would be to apply quantum information theory to ratchets. Quantum ratchets might combine quantum search efficiency with the directionality property of ratchets. Another related field under active investigation is quantum game theory (Meyer, 1999; Eisert et. al., 1999). This is closely related to the molecular evolution decision tree problem: if competing chemical reactions are regarded as participants in a game, with the 'winner' being life (or simply a replicator), then quantum strategies are expected to be much more efficient than their classical counterparts.

Hameroff (1998), and more recently Nanopoulos (Mershin et. al., 2000), have suggested that quantum information processing may play an important role in protein folding – another famous decision tree problem, where this time the branches of the tree are alternative conformational states. These researchers point out that the protein tubulin can undergo quantum flips between two specific conformational states, and thus form a binary quantum switch – the basic component of a quantum computer. In a microtubule of the sort found within living cells, ordered arrays of tubulin molecules constitute a sort of quantum cellular automaton, potentially capable of prodigious information processing. Penrose and Hameroff (Penrose, 1994) have also suggested that quantum information processing takes place in microtubules, and, more controversially, that this process may be involved in the phenomenon of consciousness.

There is some circumstantial evidence in favour of the theory that quantum computation plays a crucial role in life. Grover's algorithm was devised to apply quantum information processing to search an unsorted database of N objects by posing Q yes-no questions. Grover (1999) proved that this would produce a $N1/2$ improvement in the search time. The relationship between N and Q in Grover's algorithm is

$$(2Q + 1)\ \text{sin-1}(N\text{-}1/2) = \pi/2$$

which has the intriguing solutions $Q = 1$, $N = 4$, and $Q = 3$, $N = 20.2$. Patel (2000) has suggested that these numbers could explain the genetic code. N = 4 corresponds to the four nucleotide bases, $Q = 3$ to the triplet code and $N \cong 20$ to the twenty amino acids life uses. He has developed a scenario of molecular assembly using quantum interrogation in which these numbers may crop up naturally, as a consequence of quantum mechanics.

Another hint of quantum physics at work in the genetic code is the discovery that the coding assignments possess a compact description in terms of supersymmetry (Bashford et. al., 1999). Supersymmetry arises in particle physics as a unified description of fermions and bosons, and is a subject to which Salam made important contributions. To find supersymmtery appearing in a biological context is remarkable, and still somewhat mysterious. Unless it is a weird coincidence, it points to a deep link between the quantum realm of particle physics and the quasi-classical realm of protein assembly.

Exciting though these various quantum conjectures may be, they all come up against a major obstacle - decoherence. Explicitly quantum effects may be manifested only so long as the phase relationships between various branches of the wave function are maintained. But these phase relationships are exceedingly delicate, and will be disrupted by even slight interactions with a noisy environment (Zurek, 1991). A simple-minded calculation (Tegmark, 1999) for the conditions inside a living cell, for example, indicates decoherence timescales of 10-13 s or less - too fast to be biochemically relevant, and far too fast to navigate a decision tree of significant complexity.

If this obstacle is to be circumvented, there have to be special reasons why certain organic systems are screened from decohering influences. There is a claim that water and proteins could have a shielding effect on electromagnetic disturbances (Mershin et. al., 2000). Moreover, most decoherence calculations assume linear, near-equilibrium systems. In biology one is dealing with highly nonlinear systems involving strong feedback loops, often driven far from equilibrium by an energy throughput. The quantum theory of such systems is more or less nonexistent; greatly extended decoherence times may not be impossible under such conditions. Also, decoherence calculations are applied primarily to electromagnetic disturbances on charged particles. In the case of coherent vibrational modes of bio-polymers and membranes, *phonons* are the relevant quantum particles, and these are likely to have much longer decoherence times than electrons and ions.

Although the case for quantum information processing in living systems is far from proved, I think Salam would have approved of the following philosophical observation. Given that quantum mechanics provides the possibility of stupendous information processing power, why does nature have need of it? To what use is it put? Does this extraordinary power just go to waste, or is it harnessed somewhere? I believe it is indeed harnessed, in bringing life into existence, and maybe mind too. That is not a scientific conclusion, of course, but the history of science does show that what *can* happen in physics usually *does* happen somewhere in nature. If quantum computation turns out to technologically feasible, I would find it hard to believe that nature didn't get there first.

Acknowledgements

I should like to thank Derek Abbott, Carlton Caves, Johnjoe McFadden, Peter Jarvis, Gerard Milburn, Lee Smolin and Duncan Steel for their help and encouragement in preparing this paper.

References

Abbott, D., Davis, B. and Parrondo, J.M.R. (1999) The problem of detailed balance for the Feynman-Smoluchowski engine (FSE) and the multiple pawl paradox, in Proceedings of the Unsolved Problems of Noise (UpoN99), *American Inst. Phys.* **511**, 213.

Ajari, A. & Prost, J. (1993) Mouvement induit par un potentiel periodique de basse symmetrie: dielectrophorese pulsee, *C.R. Acad. Sci. Paris II* **315**, 1635.

Altenmuller, T.P. & Schenzle, A. (1993) Dynamics of measurement: Aharonov's inverse quantum Zeno effect, *Phys. Rev.* **A48**, 70.

Barrow, J.D. & Tipler, F.J.(1986) *The Anthropic Cosmological Principle,* Clarendon Press, Oxford.

Bashford, J.D., Jarvis, P.D. & Tsohantjis, I. (1998) Supersymmetry in the genetic code, in *Physical Applications and Mathematical Aspects of Geometry,* eds. H.-D. Doebner, P. Nattermann, W. Scherer and C. Schulte, World Scientific Press, Singapore.

Bennett, C.H. & DiVincenzo, D.P. (2000) Quantum information and computation, *Nature* **404** (2000).

Cech. T. (1986) RNA as an enzyme, *Scientific American* **255**, No. 5, 64.

Chaitin, G. (1990) *Information, Randomness & Incompleteness: Papers on Algorithmic Information Theory,* second edition, World Scientific Press, Singapore.

Davies, P. (1998) *The Fifth Miracle: The Search for the Origin of Life*, Penguin, London.

Dawkins, R. (1996) *Climbing Mount Improbable*, Viking, London.

De Duve, C. (1995) *Vital Dust*, Basic Books, New York.

Doering, C.R. (1995) Randomly rattled ratchets, *Nuovo Cimento,* **17D**, 685.

Dyson, F. (1971) *Scientific American* **225** (September issue), 25.

Eigen, M. & Schuster, P. (1979) *The Hypercycle: The Principle of Natural Self-Organization,* Springer-Verlag, Berlin..

Eisert, J., Wilkens, M. & Lewenstein, M. (1999) Quantum games and quantum strategies, LANL preprint quant-ph/9806088.

Farhi E. & Gutmann, S. (1998) Quantum computation and decision trees, *Phys. Rev.* **A58**, 915.

Feynman, R.P., Leighton, R.B. and Sands, M. (1963) *The Feynman Lectures on Physics,* Addison-Wesley, Reading, Mass., vol. 1, sec. 46.1.

Fox, S. (1988) Prebiotic roots of informed protein synthesis, in *The Roots of Modern Biology*, ed. H. Kleinkauf et.al., de Gruyter, Berlin, p. 897.

Fröhlich, H. (1983) *Coherent Excitations in Biological Systems*, Springer-Verlag, Berlin.
Grover, L. (1999) Quantum computing, *The Sciences,* July/August edition, 24.
Hameroff, S.R. (1998) Quantum computation in brain microtubules? The Penrose-Hameroff "Orch OR" model of consciousness, *Phil. Trans. Royal Soc. (London)* **A356**, 1869.
Harmer, G.P. & Abbott, D. (1999) Parrondo's paradox, *Statistical Science*, **14**, 206.
Hoyle, F. (1954) *Astrophys. J. Supplement* **1**, 121.
Itano, W.M., Heinzen, D.J., Bollinger, J.J. & Weinland, D.J. (1990) Quantum Zeno effect, *Phys. Rev.* **A41**, 2295.
Kofman, A.G. & Kurizki, G. (2000) Acceleration of quantum decay processes by frequent observations,' *Nature* **405**, 546.
Küppers, B.-O. (1985) *Molecular Theory of Evolution*, Springer-Verlag, Berlin.
Magnasco, M.O. (1993) Forced thermal ratchets, *Phys. Rev. Lett.* **71**, 1477.
Mershin, A., Nanopoulos, D.V. & Skoulakis, E.M.C. (2000) Quantum brain?, LANL preprint quant-ph/0007088n.
McFadden, J. (2000) *Quantum Evolution*, HarperCollins, London.
Meyer, D.A. (1999) Quantum strategies, *Phys. Rev. Lett.* **82**, 1052.
Milburn, G. (1998) *The Feynman Processor*, Perseus Books, Reading, Mass.
Monod, J. (1972) *Chance and Necessity*, trans. A. Wainhouse, Collins, London.
Parrondo, J.M.R., Harmer, G.P. & Abbott, D. (2000) New paradoxical games based on Brownian ratchets, *Phys. Rev. Lett.* **85**, 3386.
Patel, A. (2000) Quantum algorithms and the genetic code, LANL preprint quant-ph/0002037.
Penrose, R. (1994) *Shadows of the Mind*, Oxford University Press, Oxford.
Salam, A. (1991) The role of chirality in the origin of life, *J. Mol. Evol.* **33**, 105.
Salam, A. (1992) Chirality, phase transitions and their induction in amino acids, *Phys. Lett.* **B288**, 153.
Schrödinger, E. (1944) *What is Life?*, Cambridge University Press, Cambridge.
Shapiro, R. (1986) *Origins: A Skeptic's Guide to the Creation of Life on Earth*, Summit Books, New York.
Smoluchowski, M. (1912) *Experimentall nachweisbare, der üblichen Thermodynamic widersprechende Molekularphänomene, Phys. Z.* **13**, 1069.
Tegmark, M. (1999) The quantum brain, LANL preprint quant-ph/9907009.
Yockey, H. (1992) *Information Theory and Molecular Biology*, Cambridge University Press, Cambridge.
Zurek, W.H. (1991) Decoherence and the transition from quantum to classical, *Physics Today,* **44**, No. 10, 36.

The Cyril Ponnamperuma Lecture

Speaker:

J. William Schopf

THE SECOND CYRIL PONNAMPERUMA LECTURE

SOLUTION TO DARWIN'S DILEMMA: DISCOVERY OF THE MISSING PRECAMBRIAN RECORD OF LIFE

J. WILLIAM SCHOPF

Department of Earth and Space Sciences,
Institute of Geophysics and Planetary Physics
(Center for the Study of Evolution and the Origin of Life),
and Molecular Biology Institute,
University of California, Los Angeles, CA 90095-1567
PHONE: (310) 825-1170
FAX: (310) 825-0097
E-MAIL: Schopf@ess.ucla.edu

ABSTRACT: In 1859, in On the Origin of Species, Darwin broached what he regarded to be the most vexing problem facing his theory of evolution -- the lack of a rich fossil record predating the rise of the first primitive animals that marks the beginning of the Cambrian Period of geologic time (~550 million years ago), an "inexplicable" absence that could be "truly urged as a valid argument" against his all-embracing synthesis. For more than 100 years, the "missing Precambrian history of life" stood out as one of the greatest unsolved mysteries in natural science. Over the past few decades, however -- after a century of trial and error, search and eventual discovery -- life's earliest history has finally been uncovered as the documented fossil record has been extended to some 3,500 million years ago, an age more than three-quarters that of the planet itself. An immense early fossil record, unknown and assumed unknowable, has been unearthed to reveal an evolutionary progression dominated by microbes that stretches seven times farther into the geologic past than had previously been thought. Life began far earlier, and evidently evolved initially much farther and faster than anyone had imagined. Perhaps the origin of life is "easier" -- and perhaps life itself is cosmically more widespread -- than conventional views have led us to suspect.

J. Chela-Flores et al. (eds.), First Steps in the Origin of Life in the Universe, 23.

Opening Lecture

Speaker:

Cynthia Phillips

EUROPA: PROSPECTS FOR AN OCEAN AND LIFE

CYNTHIA B. PHILLIPS AND CHRISTOPHER F. CHYBA
Center for the Study of Life in the Universe
SETI Institute, 2035 Landings Drive
Mountain View, California, 94043 USA

1. Introduction

Life as we know it depends on the presence of three things: liquid water, certain "biogenic" elements (including carbon as well as oxygen, nitrogen, hydrogen, phosphorous, sulfur, and others), and a useful source of free energy. Jupiter's satellite Europa may meet these criteria, and thus be a possible location for a non-terrestrial biosphere. Recent theoretical and observational evidence has suggested that Europa may possess a subsurface "ocean" of liquid water beneath its icy surface. Theoretical and spectral studies of Europa provide evidence of biogenic compounds, and theoretical models, coupled with measurements of the radiation environment of Europa, are beginning to examine possible sources of free energy. We will consider these requirements for life on Europa, how Europa may or may not meet them, and what the implications are for a possible Europan biosphere. We also discuss the future exploration of Europa.

2. Europa Overview

Europa, the second of the four Galilean satellites in its distance from Jupiter, is an ice-covered body with a radius of 1560 km, about the size of Earth's moon. Europa is locked in a tidal resonance with its neighboring satellites, volcanic Io on the inside and large Ganymede (which is larger than Mercury) on the outside. This resonance keeps the satellites from perfect circular orbits, inducing a forced eccentricity that varies their distances from Jupiter and results in tidal flexing and internal dissipation of heat and energy. This heating is most intense at Io, due to its proximity to the giant planet Jupiter and comparatively high eccentricity, and results in constant volcanic activity on that small moon. A substantial amount of heat is dissipated at Europa's distance as well (Cassen *et al.* 1979; 1980). Gravity measurements taken by the Galileo spacecraft (Anderson *et al.* 1998) show that Europa is an internally differentiated rocky body with about 100 km of water and / or ice at its surface. The center of the current debate about Europa is the past and current state of this water, which if predominantly liquid would exceed the volume of all of Earth's oceans combined. The first study of the habitibility of Europa was by Reynolds *et al.* (1983).

J. Chela-Flores et al. (eds.), First Steps in the Origin of Life in the Universe, 27–34.

3. Evidence for Liquid Water

A number of lines of evidence have been suggested for the presence of liquid water beneath the icy surface of Europa (Pappalardo *et al.* 1999; Stevenson 2000). These include theoretical studies, spectroscopic measurements, observations of Europa's geology, gravity measurements, and magnetic field measurements.

3.1. THEORETICAL EVIDENCE

Europa's surface was known to contain water ice from ground-based spectroscopic measurements (e.g. Pilcher *et al.* 1972) even before the Voyager spacecraft arrived at Jupiter in the late 1970's. Models of satellite accretion (Stevenson *et al.* 1986) suggest that Europa probably consists of silicates, metals, and water, and most models begin with chondritic abundances. Early thermal models (Lewis 1971; Consolmagno and Lewis 1976) using conductive cooling and radiogenic heating predicted the possibility of a liquid water layer beneath a frozen ice crust on Europa. Later models (Reynolds and Cassen 1979; Cassen *et al.* 1982) showed that such a configuration was unstable due to convection as the ice thickened, and the ice layer would freeze solid only about 100 Myr after formation. However, when tidal dissipation was added as a heat source (Cassen *et al.* 1979, 1980), the possibility remained that there could be a balance between tidal heating, radiogenic heating, and conductive and convective cooling that could allow a portion of the subsurface water layer to remain liquid over geologic time.

Current models have failed to resolve this issue completely (Ojakangas and Stevenson 1989; McKinnon 1999), mostly because the rheology of ice is poorly known at the temperature of Europa and at the long frequencies associated with tidal flexing (Durham *et al.* 1997). Also unknown is the exact composition of Europa's ice: the addition of small amounts of other volatiles such as ammonia or salts to the water could dramatically change the viscosity of the ice (Deschamps and Sotin 1998; McCord *et al.* 1999). Another factor which could substantially affect the strength of the ice, and thus its response to tidal flexing, is the physical state of the ice, including its grain size and degree of fracturing. Thus, theoretical models of tidal heating and deformation suggest that liquid water could have been or currently be present beneath the icy surface of Europa, but cannot confirm or deny its actual presence.

Gravity measurements of Europa's moment of inertia are based on radio Doppler data from the Galileo spacecraft's close flybys of Europa (Anderson *et al.* 1998). They suggest that Europa is completely differentiated, with an Fe or Fe-S central core, an anhydrous rocky mantle, and a surface layer of material with a density of around 1000 kg/m^3 that is between 80 and 170 km thick. The most likely thickness of the outer layer is about 100 km, and the only cosmochemically plausible material with this density is water, either solid or liquid. However, the gravity models are sensitive only to the densities of the internal layers, and the difference in density between solid and liquid water is too small to be detected from the geometries of Galileo's Europa flybys. Thus, gravity models can only state that Europa has a surface layer of about 100 km of some combination of solid water ice and/or liquid water. Higher-order gravity terms may be measured by an orbiting spacecraft such as the planned Europa Orbiter, and may be able to resolve Europa's internal structure with greater detail.

3.2. VOYAGER AND GALILEO IMAGES OF EUROPA'S SURFACE

Before the Voyager flybys of Europa in the late 1970's, little was known about Europa's surface except for its anomalously high albedo and the spectroscopic suggestion of the presence of water there. Images of Europa taken by Voyager revealed a surface covered with crack-like features, and very few impact craters (Malin and Pieri 1986). The lack of craters was surprising, since all bodies in the solar system are continually hit by debris. This results in a pock-marked surface like the Moon's unless geologic activity takes place to remove craters from the surface. The relative lack of craters on Europa means that the surface is young, perhaps as young as a few million years (Zahnle *et al.* 1998). A young surface requires geologic activity, which is more likely in the presence of liquid water. However, a search for signs of geologic activity on Europa's surface between the Voyager and Galileo observations, over a 20-year period, has shown no detectable changes (Phillips *et al.* 2000).

Images of Europa's surface taken by the Galileo spacecraft have shown surface features which could be consistent with the presence of liquid water beneath Europa's surface, but again do not prove it (Figure 1). Europa's surface is primarily covered by a vast set of interconnecting cracks and ridges. Also present are areas of disrupted "chaotic terrain", where the surface appears to have been broken up into coherent iceberg-like blocks that seem to have "rafted" into new positions (Carr *et al.* 1998). Such areas can be reconstructed by fitting the preexisting features on the blocks back together (Spaun *et al.* 1998). Other features of interest on Europa's surface include regions that could possibly be low viscosity surface flows, and impact craters that are anomalously shallow.

Many models have been proposed for the formation of the variety of features visible in Galileo images of Europa's surface. There is currently no consensus among these often contradictory views of Europa's geophysics. For example, the morphology of Europa's impact craters suggests that they formed within a solid target, but their shallow depths are suggestive of post-formation viscoelastic relaxation (Moore *et al*, 1998). Such models suggest that most craters on Europa formed in a 5-15 km thick brittle surface layer, overlying a lower-viscosity subsurface layer. This subsurface material, however, could either be liquid water or warm, low-viscosity ice. Cryovolcanic surface flows would be intriguing, but there are very few regions where the morphology is suggestive of flows (Fagents *et al.* 2000; see Figure 1 for one example). There is also a substantial buoyancy problem in their formation, as it is difficult to get liquid water to the surface of Europa since it is denser than ice.

Models of ridge formation range from cryovolcanism to tidal squeezing to linear diapirs to compression and plastic deformation (Pappalardo *et al.* 1999). These models range in requirements from a very thin crust overlying liquid water (the tidal squeezing model) to completely solid-state models with a thin brittle crust on top of a lower-viscosity, warm ice layer (diapirism or compression). For instance, cycloidal ridges have been shown (Hoppa *et al.* 1999) to correspond in orientation and location to cracking of the surface in response to the changing diurnal tidal stresses. Hoppa *et al.* argue that this model requires the existence of a global ocean near the surface to obtain sufficient tidal stresses to crack the ice. Clearly, current models in the literature are contradictory and have very different implications for Europa's subsurface structure.

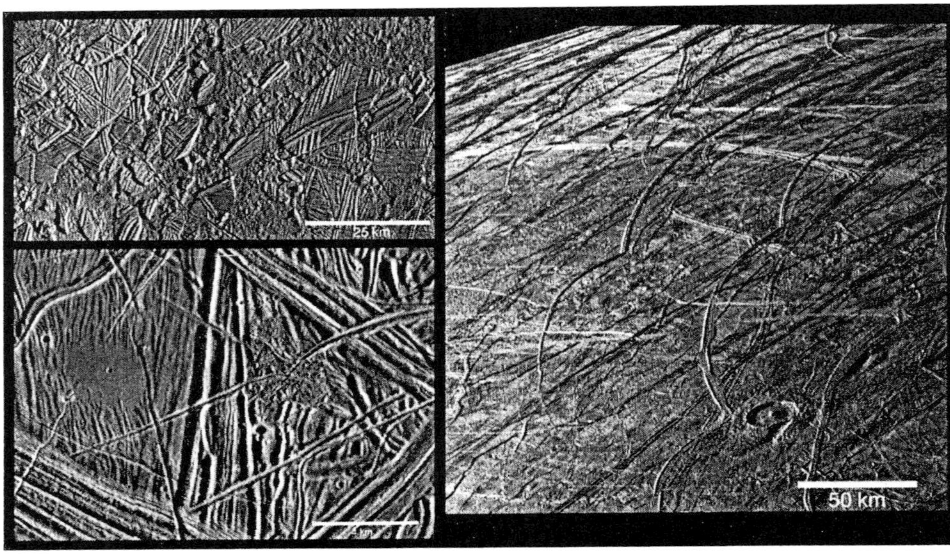

Figure 1: Galileo images of geological features on Europa. Chaotic terrain (upper left), ridges and a possible cryovolcanic flow (lower left), arc-shaped cycloidal ridges (right) and a shallow impact crater (right).

Chaotic terrain was first seen as a "smoking gun" for the presence of liquid water beneath Europa's surface, but solid-state models are also possible (Carr *et al.* 1998; Greenberg *et al.* 1998; Pappalardo *et al.* 1998). At the liquid end of the spectrum, regions of chaotic terrain are seen as areas of localized heat flow where the ice layer melted all the way to the surface. In this model, the blocks are buoyant remnants of the preexisting icy crust that move about in a slushy matrix, both translating and tilting. Eventually the matrix freezes solid, ending the blocks' motion and preserving their final positions. This model requires localized heating of the crust, but it may be difficult to concentrate the heating in both space and time. The solid-state formation model suggests that ice rises to the surface in a diapir, eventually disrupting the brittle surface. These convective upwellings make it difficult to tilt the blocks as observed, however. A third, intermediate model (Head and Pappalardo 1999) suggests that runaway melting within rising diapers produces chaos. Thus, models of chaos formation do seem to favor the existence of either water or an ice-water slurry at shallow depths near the surface, but such patches could be localized and not require the existence of a global liquid ocean layer.

3.3. EUROPA'S INDUCED MAGNETIC FIELD

Perhaps the most convincing current evidence for an ocean of liquid water beneath Europa's surface comes from magnetic field results. Recent evidence from Galileo's magnetometer (Kivelson *et al.* 2000) shows that Europa has an induced magnetic field, which varies in direction and strength in response to Europa's position within Jupiter's strong magnetic field. The strength and response of the induced field at Europa require a near-surface, global conducting layer. The most likely layer that meets these requirements is a global layer of salty water, with a salinity similar to that of Earth's oceans. The data cannot be explained by localized pockets of salty water, and require a nearly complete spherical shell. The thickness of the water layer would need to be greater than about 10 km. The induced field cannot be the result of a frozen ice layer, even if it has pockets of briny water, since ions in solid ice would not be mobile enough (Stevenson 2000). It is possible that a type of conducting layer other than a global salty ocean could account for the induced magnetic field, but the salty ocean explanation seems the most plausible.

3.4. DIRECT DETECTION OF AN OCEAN

The previous arguments for the presents of an ocean of liquid water beneath Europa's surface are all indirect in nature. Direct evidence of an ocean on Europa requires a spacecraft orbiting Europa with a dedicated instrument suite. The proposed Europa Orbiter spacecraft, currently planned by NASA to launch later this decade, would meet these requirements. The most definitive answer will likely come from measurements of the tidal deformation of Europa's surface. The tidal bulge of Europa will flex by about 1 meter for a solid ice shell, vs. as much as 30 meters for 10's of kilometers of ice over a liquid water layer. An orbiting laser altimeter, coupled with sensitive gravimetric measurements, will be able to track Europa's time-varying tidal bulge in real time and should determine whether or not an ocean is present. Another instrument proposed for the Europa Orbiter spacecraft is a radar sounder, which could provide information about Europa's subsurface structure and may be able to detect the ice-water interface, depending on the state of the ice. Such measurements will have to await the arrival of a dedicated spacecraft, however, as they cannot be performed from Earth or with the instrument suite on the Galileo spacecraft.

4. Evidence for Biogenic Compounds

After liquid water, the second requirement for life as we know it is a suite of appropriate biogenic elements. We have no direct samples of Europa's surface or in-situ measurements, and spectral measurements taken from orbit can only sample the top few microns. Globally, the surface is predominantly water ice, but the Galileo Near-Infrared Mapping Spectrometer (NIMS) instrument has also directly detected absorption features on Europa due to SO_2 and H_2O_2. There has also been the indirect detection of hydrated compounds from the presence of asymmetric water bands, which are suggestive of the presence of materials such as sulfuric acid, magnesium and sodium sulfates, and/or sodium carbonates (McCord *et al.* 1999; Carlson *et al.* 1999b). Such hydrated salts are consistent with theoretical studies that assume evolution from an initial carbonaceous chondrite composition (Kargel *et al.* 2000). The locations of these potential salts are mostly along ridges, in craters, or in the matrix of chaotic terrain in between the blocks. The composition of the salts is remarkably consistent from place to place on Europa, suggesting that they come from a globally mixed reservoir. One possible explanation is a global salty water ocean, which reaches the surface in places of recent geologic activity and leaves salts behind. However, the presence of these salts does not require a global ocean. In addition to salts, a variety of simple organic compounds have been detected on Ganymede and Callisto, and there are spectral hints that these are also present on Europa (McCord *et al.* 1998).

Another possible delivery mechanism is comets, which are rich in biogenic elements and could deliver large volumes of carbon and nitrogen, among other things (Pierazzo and Chyba 2000). Significant survival of organics is possible in large impacts, and given cometary densities of 1g cm^{-3}, substantial amounts of organics could be accreted over geologic time. However, cometary densities are poorly constrained, and if they are highly porous much more material could be lost on impact.

5. Sources of free energy

The most severe constraint on a putative Europan biosphere may be finding a sufficient source of free energy (Gaidos *et al.* 1999). Photosynthesis would not be possible beneath an ice layer more than 1 km thick (Reynolds *et al.* 1983). Other possible energy sources include tidal energy, hydrothermal vents, and chemical energy due to radiation processing at the surface.

Tidal-tectonic processes may concentrate tidal heating in cracks in the ice (Greenberg *et al.* 2000). This model assumes that as a result of tides, liquid water regularly bathes crustal cracks and surfaces with heat and nutrients from a subsurface ocean. This may create a variety of habitable environments, but requires a thin crust, a near-surface liquid ocean, and a frequent interchange of materials between the two.

Volcanism at the base of the ocean is possible if tidal heat is dissipated in Europa's rocky mantle, and such volcanism could spawn hydrothermal vents. Terrestrial hydrothermal ecology depends mostly on transported surface oxidants, however, though a simple ecosystem based on methanogenesis, for example, could survive in the absence of surface-produced oxidants (McCollom 1999). One speculative alternative energy source is photosynthesis performed by organisms near the vent using near-IR light (Van Dover *et al.* 1996).

Another possibility is a radiation-driven ecosystem (Chyba 2000a,b; Chyba and Phillips 2001). In this scenario, disequilibrium chemistry in the ice at Europa's surface, driven by charged particles accelerated by Jupiter's magnetosphere, could produce enough organic and oxidant molecules to drive a Europan biosphere. This requires a substantial exchange of material between the surface and the ocean, however. In this model, radiation chemistry produces oxidants and organics in H_2O and CO_2 ices in the upper 1 mm of Europa's surface; these can be mixed through the upper ~1 m by impact gardening (Chyba and Phillips, 2001). Charged particles interacting with water ice can produce molecular oxygen, hydrogen peroxide, and other oxidants in comparable abundance, and H_2O_2 has been detected on Europa (Carlson *et al.* 1999a). CO_2 has been found on Callisto, and appears to be present on Europa as well. Radiation can drive cycling among CO_2, CO, and organics in the ice, and produce simple organic species like formaldehyde (HCHO). It is possible that this potential source of free energy could sustain a microbial ecosystem.

Radiation products produced at the surface must reach a subsurface ocean to be biologically relevant. Crustal recycling mechanisms are currently unclear. Melt-throughs are possible in chaotic terrain, and microbial blooms might occur wherever the surface communicates with the ocean, such as in chaos regions or along active cracks. Sputtering and impact gardening are two competing processes that occur at the surface (Cooper *et al.* 2000), and serve to create, remove, and bury organics. Depending on the relative rates of sputtering and gardening, different results are possible. Impact gardening can mix these products below the radiation processing depth, preserving them for future interaction with a subsurface ocean (Chyba and Phillips 2001).

6. Prospects for life and future exploration

In summary, Europa may meet the three criteria for life as we know it, but this is far from certain at the present time. An ocean of liquid water beneath Europa's surface is possible, but not proven. The composition of Europa is unknown, but biogenic elements are likely. Sufficient free energy sources may be the largest current unknown.

It is unclear if life would require a global ocean to originate and survive, or if it could have done so in isolated melt pockets (Gaidos and Nimmo 2000). If tidal tectonic motion produces partial melting in favorable locations within Europa's near-surface crust (Greenberg *et al.* 2000), it is possible that these melt pockets could have 30-year lifetimes. If transient water or brine pockets are present near Europa's surface, it is possible that organisms located in these pockets could perform photosynthesis. The organisms would have to be able to survive dormant in the ice for millennia between brief blooms when favorable conditions were present, but recent discoveries of terrestrial bacterial spores that may have been reanimated after up to 250 million years suggest that this could be possible (Vreeland *et al.* 2000). Even if life could survive in isolated melt pockets, with brief periods of activity interspersed with long dormant periods, it is unclear whether life could originate under such conditions.

The Europa Orbiter mission is currently under review by NASA. The launch date is uncertain due to funding and technology issues, but is likely sometime this decade. The primary goals of the Europa Orbiter mission are to determine definitively whether or not Europa currently possesses an ocean of liquid water beneath its surface, to characterize Europa's subsurface structure, and to pave the way for subsequent exploration. If an ocean is detected, followup missions are likely, depending on how thick the ice layer is. The next mission would likely be a lander, which could look for evidence of life while sampling the surface composition and possibly performing seismic studies of the subsurface structure. Subsequent missions may try to melt down to the ocean or even perform a sample return.

References

Anderson, J.D., *et al.* (1998) Europa's differentiated internal structure, *Science* **281**, 2019-2022.
Carlson, R. W., *et al.* (1999a) Hydrogen Peroxide on the Surface of Europa, *Science* **283**, 2062-2064.
Carlson, R. W., *et al.* (1999b) Sulfuric acid on Europa and the radiolytic sulfur cycle. *Science* **286**, 97-99.
Carr, M. H., *et al.* (1998) Evidence for a subsurface ocean on Europa, *Nature* **391**, 363-365.
Cassen, P., *et al.* (1979) Is there liquid water on Europa?, *Geophys. Res. Lett.* **6**, 731-734.
Cassen, P., *et al.* (1980) Tidal dissipation in Europa: A correction, *Geophys. Res. Lett.* **7**, 987-988.
Cassen, P., S. J. Peale, and R. T. Reynolds (1982) Structure and thermal evolution of the Galilean satellites, in

D. Morrison (ed.), *Satellites of Jupiter*, Univ. of Ariz. Press, Tucson, pp. 93-128.
Chyba, C.F. (2000a) Energy for microbial life on Europa, *Nature* **403**, 381-382.
Chyba, C.F. (2000b) Energy for microbial life on Europa: A Correction, *Nature* **406**, 368.
Chyba, C.F. and C. B. Phillips (2001) Possible ecosystems and the search for life on Europa, *Proc. Natl. Acad. Sci. USA*, submitted.
Consolmagno, G. J. and J. S. Lewis (1976) Structural and thermal models of icy Galilean satellites, in T. Gehrels (ed.), Jupiter, Univ. of Ariz. Press, Tucson, pp. 1035-1051.
Cooper, J.F., *et al.* (2000) Energetic Ion and Electron Irradiation of the Icy Galilean Satellites, *Icarus*, in press.
Deschamps, F. and C. Sotin (1998) Thermal convection in the outer ice I shell of icy satellites, in *The Jovian System After Galileo; The Saturnian System Before Cassini-Huygens*, Lab. De Géophys. Et Planétol., Nantes, France, p. 51.
Durham, W.B., *et al.* (1997) Creep of water ices at planetary conditions, *J. Geophys. Res.* **102**, 16,293-16,302.
Fagents, S. A., *et al.* (2000) Cryomagmatic mechanisms for the formation of Rhadamanthys Linea, triple band margins, and other low albedo features on Europa, *Icarus* **144**, 54-88.
Gaidos, E. J., K.H. Nealson, and J.L. Kirschvink (1999) Life in Ice-Covered Oceans, *Science* **284**, 1631-1633.
Gaidos, E.J. and F. Nimmo (2000) Tectonics and water on Europa, *Nature* **405**, 637.
Greenberg, R., *et al.* (1998) Chaos regions: Widespread melt-through to the surface of Europa? *Bull. Am. Astron. Soc.* **30**, 1086.
Greenberg, R., Geissler, P., Tufts, B.R. and Hoppa, G.V. (2000) Habitability of Europa's crust: The role of tidal-tectonic processes, *J. Geophys. Res.* **105**, 17551-17562.
Head, J. W., and R. T. Pappalardo (1999) Brine mobilization during lithospheric heating on Europa, *J. Geophys. Res.* **104** , 27,143 – 27,156.
Hoppa, G.V., *et al.* (1999) Formation of Cycloidal Features on Europa, *Science* **285**, 1899-1902.
Kargel, J.S., *et al.* (2000) Europa's Crust and Ocean, *Icarus* **148**, 226-265.
Kivelson. M.G., *et al.* (2000) Galileo Magnetometer Measurements: A Stronger Case for a Subsurface Ocean at Europa, *Science* **289**, 1340-1343.
Lewis, J. S. (1971) Satellites of the outer planets: Their physical and chemical nature, *Icarus* **15**, 174-185.
Malin, M. C., and D. C. Pieri (1986) Europa, in J. A. Burns and M. S. Matthews (eds.), *Satellites*, Univ. of Ariz. Press, Tucson, pp. 689-717.
McCollom, T. M. (1999) Methanogenesis as a potential source of chemical energy for primary biomass production by autotrophic organisms in hydrothermal systems on Europa, *J. Geophys. Res.* **104** , 30,729.
McCord, T. B., *et al.* (1999) Hydrated salt minerals on Europa's surface from the Galileo near-infrared mapping spectrometer (NIMS) investigation, *J. Geophys. Res.* **104**, 11,827-11,851.
McCord, T. B., *et al.* (1998) Non-water-ice constituents in the surface material of the icy Galilean satellites from the Galileo NIMS investigation , *J. Geophys. Res.* **103**, 8603.
McKinnon, W.B. (1999) Convective instability in Europa's floating ice shell, *Geophys. Res. Lett.* **26**, 951-954.
Moore, J. M., *et al.* (1998) Large impact features on Europa, *Icarus* **135**, 127-145.
Ojakangas, G. W., and D. J. Stevenson (1989) Thermal state of an ice shell on Europa, *Icarus* **81**, 220-241.
Pappalardo, R. T., *et al.* (1998) Geological evidence for solid-state convection in Europa's ice shell, *Nature* **391**, 365-368.
Pappalardo, R.T., *et al.* (1999) Does Europa have a subsurface ocean? Evaluation of the geological evidence, *J. Geophys. Res.* **104**, 24,015-24,055.
Phillips, C. B., *et al.* (2000) The Search for Current Geologic Activity on Europa, *J. Geophys. Res.* **105**, 22,579-22,598.
Pierazzo, E., and C. F. Chyba (2000) Impact delivery of organics to Europa, LPSC XXXI, abs. 1656.
Pilcher, C. B., *et al.* (1972) Galilean Satellites: Identification of Water Frost, *Science* **178**, 1087-1089.
Reynolds, R T., *et al.* (1983) On the habitability of Europa, *Icarus* **56**, 246-254.
Reynolds, R. T., and P. M. Cassen (1979) On the internal structure of the major satellites of the outer planets, *Geophys. Res. Lett.* **6**, 121-124.
Spaun, N. A., J. W. Head, G. C. Collins, L. M. Prockter, and R. T. Pappalardo (1998) Conamara Chaos Region, Europa: Reconstruction of mobile polygonal ice blocks, *Geophys. Res. Lett.* **25**, 4277-4280.
Stevenson, D. J., A. W. Harris, and J. I. Lunine (1986) Origins of satellites, in J. A. Burns and M. S. Matthews (eds.), *Satellites*, Univ. of Ariz. Press, Tucson, pp. 39-88.
Stevenson, D. (2000) Europa's Ocean - the Case Strengthens, *Science* **289**, 1305-1307.
Van Dover, C. L., *et al.* (1996) Light at deep-sea hydrothermal vents, *Geophys. Res. Lett.* **23**, 2049-2052.
Vreeland, R. H., W. D. Rosenzweig, and D. W. Powers (2000) Isolation of a 250 million-year-old halotolerant bacterium from a primary salt crystal, *Nature* **407**, 897-900.
Zahnle, K, L. Dones, and H. F. Levison (1998) Cratering rates on the Galilean Satellites, *Icarus* **136**, 202-222.

Section 1:

Historical aspects

Sidney Fox during his first visit to the Fourth Trieste conference in 1995.

(Courtesy of ICTP Photo Archives.)

SIDNEY WALTER FOX :

The Father of Thermal Proteins and Protocells Therefrom

ARISTOTEL PAPPELIS[1], PETER BAHN[2], AND RANDALL GRUBBS[3]

1. *Department of Plant Biology, Southern Illinois University, Carbondale, Illinois 62901 USA;*
2. *Bahn Biotechnology Co., RR2 Box 239A, Mount Vernon, Illinois 62864 USA*
3. *Department of History, Southern Illinois University, Carbondale, Illinois 62901 USA*

Sidney Walter Fox was born in 1912 and died in 1998. He was a prolific writer of matters scientific. His envious record and sharp mind made him a respected researcher. He spoke of himself as a chemist: "Those of us who experiment with a retracement of aspects of the emergence of life find that the experiments themselves often contribute richly to identifying new processes, new phenomena, and new philosophy (Fox, 1988)."

Chemistry, biochemistry, physiology, molecular biology, genetics, and evolution led to the characterization of life, the identification of DNA genes, and the identification of their proteins products (central dogma). An early attempt to synthesize a naturally occurring protein (pre-1950s) was, in essence, an attempt to synthesize a gene since the true nature of the mechanism of inheritance was not yet acknowledged (until 1953, although the bacterial evidence that DNA was part of cellular inheritance had been reported). Some scientists proposed that the acellular (prebiotic) synthesis of DNA led to the origin of the first living cell. The period between 1940 and 1965 was critical to those in all fields focusing on the understanding of the origin and evolution of life. Sidney Fox believed it was possible to polymerize amino acids in the laboratory as a simulation of a naturally occurring acellular synthesis of protein. Only he expected to find the simultaneous answer to the origin of proteins and cellular life through an organic synthesis mechanism. By 1980, he believed that he had succeeded (synthesis of thermal proteins) (Fox 1980, 1988).

That Oparin tried to synthesize a cell is to his credit but the coacervate droplet was hopelessly short of the goal. Colloidal life was a topic for speculation from the 1860s. It may have been the work of Buchner with his cell-free metabolic activities (enzymes) in a test tube that induced Oparin to put together a colloidal system. After all, no one at that time knew about the cellular membrane with any certainty. Cellular metabolism was dependent upon proteins. They were essential for life (agents for molecular transformation). Fifty years of study on metabolic pathways laid the basis for how to study proteins. Even those proposing liposomes (or lipid vesicles) or RNA fell short in describing the pathway to cellular life. The golden age of electron microscopy lead to the viewing of the cell boundary and, soon thereafter, the fluid mosaic model of membranes. Was life cellular? Was Oparin's work the search for chemical, acellular life?

J. Chela-Flores et al. (eds.), First Steps in the Origin of Life in the Universe, 37–38.

The work of Morgan, one of Sidney's Ph.D. mentors, pushed for a protein explanation for the inheritance mechanism. "Fox, all the important problems of biology are problems of protein (Fox 1988)." Proteins "were informed." In 1953, DNA was found to be the inheritance molecule. But, how did proteins originate? Sidney then answered the question: "Proteins originated through the synthetic activities of thermal protein protocells." Sidney, with his associates, published about 386 manuscripts and nine books. His efforts changed the question "How did cellular life originate?" into an answer: "How cellular life emerged (Fox 1980, 1988)." The amino acids in a "melt" self-ordered (stereo-electronic reactions) and, when moistened, they self-assembled to form protocells. These properties of thermal protein synthesis were the "birth of the genetic code of life." Protocells, prokaryotic cells, and eukaryotic cells were the simplest units of life. In this volume we (Pappelis, *et al.* 2000) extend the work of Fox (1995) and Fox, *et al.* (1995, 1996, 1998). Sidney liked the NASA definition of life. We modified it: TPs (chemical life) → protocells (biological life).

Sidney served as our teacher (AP, in Intermediary Metabolism at Iowa State, 1954; PB, as a Post Doctoral Student for three years, 1978-1981; RG, as a research assistant, University of South Alabama, 1994-1998). His impact on us was beyond our ability to describe. Before the Moscow meeting in 1994, AP introduced Sidney as "my teacher." After his presentation at Moscow, Sidney turned the trick around and began introducing him as "his teacher." AP (in 1988) and Sidney (in 1990) had by-pass heart surgeries. The delay in their deaths was always a source of jokes. Sidney was sincerity, warmth, honor, decency, humor, and true class. He and Raia loved jazz concerts, dancing, and traveling. We miss them.

References

Fox, S. W. 1980. New missing links. How did life begin? The Sciences 20 (1): 18-2

Fox, S. W. 1988. *Emergence of Life, Darwinian Evolution From the Inside.* Basic Books, New York.

Fox, S. W. 1995. To cellular life and neurocellular assemblies. In *Evolutionary Biochemistry and Related Areas of Physicochemical Biology*, B. F. Poglazov, B. I. Kurganov, M. S. Kritsky, and K. L. Gladilin, Eds., Bach Institute of Biochemistry and ANKO, Moscow, pp. 161-175.

Fox, S. W., with 20 coworkers. 1995. Experimental retracement of the origins of a protocell: It was also a protoneuron. In *Chemical Evolution: Structure and Model of the First Cell*, C. Ponnamperuma and J. Chela-Flores, Eds., Kluwer Academic Publishers, Dordrecht, pp. 17-36.

Fox, S. W., P. R. Bahn, A. Pappelis, and B. Yu. 1996. Experimental retracement of the terrestrial origin of an excitable cell: Was it predictable? In *Chemical Evolution: Physics and the Origin and Evolution of Life*, J. Chela-Flores and F. Raulin, Eds., Kluwer Academic Publishers, Dordrecht, pp. 21-32.

Fox, S. W., A.. Pappelis, and R. Grubbs. 1998. The origin of mind and life. In *Exobiology: Matter, Energy, and Information in the Origin and Evolution of Life in the Universe.* J. Chela-Flores an F. Raulin, Eds., Kluwer Academic Publishers, Dordrecht, pp. 103-106.

Pappelis, A., P. Bahn, R. Grubbs, J. Bozzola, and P. Cohen. 2001. From inanimate macromolecules to the animate protocell: In search of thermal protein phase-shifting. This volume.

THEORIES ON ORIGINS OF LIFE BETWEEN 1860 AND 1900

The spontaneous generation controversy years post Darwin and Pasteur's works

F. RAULIN-CERCEAU

Centre Alexandre Koyré & Muséum national d'Histoire naturelle
Grande Galerie de l'Evolution, 36 rue Geoffroy st Hilaire
75005- Paris-France- e-mail : raulin@mnhn.fr

During the second part of the XIXth century, the spontaneous generation controversy was at its height. Louis Pasteur (1822-1895) concluded from his experiments that the appearance of microorganisms was produced not by spontaneous generation but rather by germs present in the atmosphere (Pasteur, 1860). However the French *coup de grâce* of Pasteur did not completely kill the doctrine of spontaneous generation, which continued to be debated as far as 1900. In particular, this new theory didn't solve the problem of the origin of the first living organisms, within the context of the theory of transformism and biological evolution. If life has always derived from living organisms, where did primitive life come from ?

In the middle of such a complex situation, a few theories on origin of life began to emerge, showing that life could have arisen from inorganic matter, a long time ago. This paper presents the main ideas that have been proposed during the second half of the nineteenth century, a period specially interesting for studying the first steps of the science of origins of life and the last ones of the dogma of spontaneous generation.

Between 1860 and 1900, the question of origins of life was a very complex one. The very strong dogma of spontaneous generation remained on certain conditions a valid theory to explain primitive life. Pasteur himself wondered : « *Can matter organize itself ? Can organisms come into the world without parents, without ancestors ? There's the question* » (Farley, 1977).

Pasteur did not attempt to give any answer to the problem of origins of life by means of his experiments. However he was aware of the importance of the concept of evolution and he followed somewhere the idea of archeobiosis (even if the terms of « creation » and « spontaneoulsy » are used in his writings): « *Take a drop of sea water containing some nitrogenous material, some mucus, some fertile jelly as it is called, and in the midst of it the first beings of creation take birth spontaneoulsy. Little by little they transform themselves and climb from rank to rank, for example, to insects after 10 000 years and no doubt to monkeys and man after 100 000 years* » (Farley, 1977). Otherwise, Charles Darwin (1809-1882) proposed in the same time, and explained, the theory of biological evolution by means of natural selection. And the theory of biological evolution, regarded as a phenomena based on natural causes, demanded an abiogenic origin of life.

The question of Origins (of life, of matter) appeared to Darwin as a very important problem to be solved, now that the question of the origin of species had become clearer

J. Chela-Flores et al. (eds.), First Steps in the Origin of Life in the Universe, 39–42.

(Raulin-Cerceau *et al,* 1998). Darwin expected that the formation of complex organic matter from inorganic starting compounds was a prerequisite for the primordial generation of living organisms, a phenomena no longer observed in nature : any organic matter newly formed in nature now would be rapidly devoured by preexisting organisms. Nevertheless, the theory of Darwin added to the results of Pasteur converge to a dilemma: a single theory of origins of life couldn't emerge from these two concepts, because on one hand we could accede to a belief in abiogenesis (Darwin's theory), and on the other hand experimental findings seemed to prove heterogenesis (Pasteur's theory). « Life always comes from life »(suggesting homogenesis and heterogenesis) and in the same time « where and when do come from primitive ancestors ? » (suggesting abiogenesis or archeobiosis).

One of the main difficulties for scientists at that time was to go from contemporary spontaneous generation to primeval spontaneous generation. Of course, the materialists saw very quickly the implications of Darwin's theory, especially the unification of organic and inorganic world.

The first main problem however for the one who believed in the theory of evolution was the long time-span for the transformation of inorganic into organic matter and the organisation of the latter into living organisms. Archeobiosis was necessarily included in the darwinian theory of evolution. Acceptance of life emergence merely in the past, a long time ago on the primitive earth following an evolution from mineral matter to life by purely physical causes, required the integration of a very large time-scale. Moreover, very little was known concerning the physical and chemical conditions of the Earth during the period when life made its first appearance. Consequently, the problem of the beginning of life couldn't be solve with the only arguments of that time.

If life emergence was considered as an every-day process, most of the scientists accepted the idea of two distinct processes for simple organisms: *via* contemporary spontaneous generation from another living matter(heterogenesis), or *via* transition from mineral matter to life (abiogenesis). But both are relevant to spontaneous generation. For instance, the biologist Henry C. Bastian (1837-1915) considered four different ways in which bacteria could come into being: direct homogenesis, indirect homogenesis, heterogenesis and archeobiosis (Bastian, 1871). Direct homogenesis was the reproductive method whereby bacteria are derived from pre-existing bacteria of the same kind by fission. Indirect homogenesis referred to a process whereby bacteria were derived from more complex organisms. Heterogenesis involved the reorganisation of particules of living matter derived from other organisms. Archebiosis represented a process which gave rise to bacteria independently of pre-existing living things (Kamminga,1980). It meant that chemical transformation of mineral matter into living systems could give answer both to the origin of life and of primitive organisms such as bacteria. Ernst Haeckel (1834-1919) proposed the process of plasmagonia, a sort of heterogenesis : matter deriving from pre-existing, decayed organisms, and plasma becoming more and more individual, finally alive. He suggested also autogonia, a sort of abiogenesis : direct transition from inorganic substance into individualised organic substance, and plasma evolving from very simple carbonaceous compounds (theory called « carbon theory »).

This necessity of a primeval abiogenesis led some scientists to reconstruct the first steps of chemical evolution on the early earth. The first attempts to formulate some steps of chemical evolution arising before biological evolution came from a few chemists such as Pflüger. Pflüger realized that the key to understand the origin of life in the past would necessarily be in connection with organics chemical syntheses. He proposed that this key could be the synthesis of cyanogen radical, added to the synthesis of other organics when very warm conditions of temperature ruled on early earth.

Proteins could be formed later from these chemical reactions when the earth was cooler (Tirard, 1996).

A very different way to answer to the question of origins was the theory of panspermia proposed by the physicist Hermann Richter, who supported the idea of eternal life in an eternal universe. In that case, living matter was always arising from former life. It could accommodate Darwin's theory of evolution without having to explain the origin of a primordial cell from inorganic matter on primitive earth. However, it's a statement of the obvious to say that Richter's theory left the problem of origins incomplete.

The years post Darwin and Pasteur's works showed that a very few scientists and philosophers pushed definitively back the theory of spontaneous generation. The physicist John Tyndall (1820-1893) suggested that all natural phenomena could be explained in terms of properties of matter and that life could arise from inorganic matter by purely physical causes. Tyndall concluded that the theory of evolution proposed by Darwin was incomplete. The philosopher Charles Spencer (1820-1903) envisaged a complex and prolonged series of events preceding the generation of the first organisms on Earth. Evolution of living forms was only possible after a long evolution of organic matter. However his position became quite ambiguous when he started on the problem of origins and he did not discuss the question of the origin of life in detail. In his opinion, there was no absolute beginning of life: « *The affirmation of universal evolution is in itself the negation of an absolute commencement of anything* » (Farley, 1977). He believed in long and gradual processes following global advances from the simple to the complex (changes of the Universe, geological and climatic evolution of the earth, evolution of organisms, and even of humanity). But Spencer's views were not widely followed in the nineteenth century.

Clearly it appeared that the second very hard problem to be solved was the one of « origin ». Scientists could easily speak about creation, beginning, or emergence of living organisms. But the concept of « origin » as a « starting point » of a very specific organisation of matter such as life, with no return to primeval conditions, was more difficult to apprehend. Haeckel tried to built the history of life and he placed the ancestors of the *Monera* before the point of differentiation into plants and animals in phylogenetic history of life (Haeckel, 1874). As he suggested, the primordial *Monera* arose by autogony as a result of forces inherent in matter present in primeval seas. He felt that the primeval conditions must have been very different from the present ones.

Haeckel believed that his hypothesis on the origin of life bridged the gap between the Kant-Laplace hypothesis on the formation on the earth and Darwin's theory of evolution, and hence provided further theoretical support for the unity of nature. Bastian proposed that a gradual formation of organic matter on the primeval earth had preceded the appearance of living things (Bastian, 1871).

Bastian differed from the evolutionists, however, in that he believed that the generation of living forms from non-living matter was an every-day process that had occurred ever since the appearance of the first living forms. This theory could explain the persistence of lower organisms throughout the course of evolution. Most of scientists followed this opinion.

In the 1860's, that the simplest organisms were generally regarded merely as naked lumps of protoplasm added credence to the belief that they, too, could be produced spontaneously. The discovery of *Bathybius haeckelii* by Thomas H. Huxley (1825-1895) in 1857 seemed to consolidate the possibility of present-day life emergence, a sort of spontaneous generation starting from chemical elements in the deep oceans. Huxley examined some mud dredged from the sea bottom during the voyage of H.M.S. Cyclops.

After some examinations under microscope, he reported that the mud contained « innumerable lumps of a transparent, gelatinous substance » in which were embedded « granules, coccoliths and foreign bodies » (Girard, 1874). Huxley saw in the *Bathybius haeckelii* as one of the Monera described by Haeckel. Huxley believed that he had discovered an exceedingly simple organism but was careful not to speculate about its origin. This enigmatic matter might represent primitive life from which evolution started. But this caution proved justified seven years later, when the chemist Buchanan found in 1876 that *Bathybius* was an inorganic precipitate of calcium sulphate, precipitated by the alcohol that had been used as a preservative. Huxley immediately admitted his error (Kamminga, 1980).

Darwin's theory of evolution had left the question of the origin of the first organisms unanswered. Pasteur's experiments were concerned with matter derived from pre-existing organisms. However, indirectly both of these new paradigms built the foundations of the concept of chemical evolution which will be clearly formulated in the early XXth century. For the first time, it became possible to investigate a primeval kind of spontaneous generation, namely the generation of complex organic molecules by chemical synthesis.

References

Bastian H.C.(1871) *The modes of Origin of Lowest Organisms,* London.

Farley J. (1977) *The spontaneous generation controversy from Descartes to Oparin*, The John Hopkins University Press, Baltimore and London, pp. 108-109.

Girard J. (1874) *Les Explorations sous-marines*, Paris, p.89.

Haeckel E. (1874) *Histoire de la création des êtres organisés d'après les lois naturelles*, Reinswald, Paris, p. 303.

Kamminga H. (1980) *Studies in the history of ideas on the origin of life from 1860*, thesis of the University of London, Chelsea College (History and Philosophy of Science), p. 55.

Pasteur L. (1860) Expériences relatives aux générations dites spontanées, *C.R. Acad. Sci.*, 50, pp.303-307.

Raulin-Cerceau F., Maurel M.C. and Schneider J. (1998) Impacts of Darwin's Theory on « Origins of Life » and « Extraterrestrial Life » debates, in Chela-Flores J. and Raulin F. (eds.), Exobiology : Matter, Energy, and Information in the Origin and Evolution of Life in the Universe, Kluwer Academic Publishers, Dordrecht, pp.175-180.

Tirard S. (1996) *Les travaux sur l'origine de la vie de la fin du XIXème siècle jusqu'aux années 1970*, thèse de l'Université Paris VII (Epistémologie et histoire des sciences), Paris, p. 82.

REMINISCENCES-PONT-A-MOUSSON-1970 TO TRIESTE-2000 (*)

MOHINDRA S. CHADHA

C-6, Beach House , Juhu
Mumbai- 400 049,INDIA
(e-mail: chadhams@mail.com)

The First International Conference on the origin of Life on Earth was held in Moscow under the auspices of the International Union of Biochemistry in 1957. This was followed by the Second Conference held in Wakulla Springs, Florida in 1963 under the auspices of the Institute of Space Biosciences, and the U.S. National Aeronautics and Space Administration.

As the interest in the all important subject – Chemical Evolution and Origin of Life was growing, the Third International Conference on the origin of Life was convened by Rene Buvet of France and Cyril Ponnamperuma of U.S.A. in 1970 at Pont-a-Mousson, France. Sessions were held on the synthesis of small molecules, oligomers and polymers, photochemical processes, origin of biological molecules, biological structures, primitive biochemistry and biology and exobiology. Also discussed were some general and theoretical problems related to the question of Origin of Life. Besides the father figure in this area of scientific activity, namely Alexander Ivanovich Oparin, several eminent scientists like I. Prigogine, M. Florkin, A. Rich, F. Lipmann, S.W. Fox, Cyril Ponnamperuma, S.L. Miller, J. Oro, H. Baltscheffsky, I.S. Kulaev, M.O. Dayhoff, A.A. Krasnovsky, L. Margulis and others contributed significantly to the conference. This became the fore-runner of the International Conferences on the Origin of Life that followed and the seed was sown for the formation of the International Society for the Study of the Origin of Life (ISSOL).

I was one of the fortunate ones to be a participant in the all important Pont-a-Mousson Conference and am pleased to see some of the contributors there viz., S.L. Miller, J. Oro, M.S. Kritsky and H. Baltscheffsky to be present at the Trieste-2000 conference also. In my personal scientific life, my participation in all the International Conferences on the Origin of Life and ISSOL Meetings (except for two), starting with the first one in Barcelona in 1973 (following the Pont-a-Mousson Conference) and all the six Trieste Conferences starting with the first one in 1992 has given me a great sense of scientific satisfaction.

The 4th ICOL and the 1st ISSOL Meeting were held in Barcelona, Spain in 1973.This meticulously organized meeting , with outstanding scientific content and generous hospitality was organized by Prof. John Oro, who also organized the 10th ICOL and 7th ISSOL Meeting in Barcelona in 1993. At the 1st ISSOL, the Hon. Chairman of the Conference Prof S.Ochoa had the following to say: **"Let us enjoy the pleasure of**

(*) The presentation was made using a large number of pictures taken at the ISSOL Meetings and Trieste Conferences. This may be viewed at the website: http://www.geocities.com/issol_trieste_in_pictures

J. Chela-Flores et al. (eds.), First Steps in the Origin of Life in the Universe, 43–46.

meeting with our friends and colleagues from all over the world in the spirit of understanding and co-operation. We all pursue the same goal to pursue life at the molecular level. Let us hope that we may pursue this goal in peace and harmony for the benefit of mankind"

The first ISSOL Meeting was held in honor of **Profs. A. I. Oparin, Melvin Calvin and H. C. Urey.** Medals and titles of the Hony. Counselors of the Higher Council of Scientific Research in Spain were conferred on: **Oparin, Calvin, Urey and Miller.** Proceedings were dedicated to the memory of several pioneering scientists-who had passed away and had contributed to the Origin of Life problem. They were **J.B.S. Haldane (1964), J.D.Bernal (1971), A.Katzir Katchalsky(1972) and W.Vishnaic(1973).** Enthusiastic participation by **M.Calvin,W.F.Libby, S.Ochoa, Sir Robert Robinson, H.C.Urey and G.Wald** was noteworthy.

The ISSOL Meetings that followed were characterized by their own flavor and brought almost invariably leaders and enthusiasts in the area of Origin of Life together. The participation has been global and these Conferences have catalyzed activities in many countries besides U.S.A. and European Countries. Today there are active programs in China, India, Israel, Mexico and Brazil.

For me personally, the Pont-a-Mousson conference and some of the International Conferences on the Origin of Life and the ISSOL Meetings are of great deal of significance as my close friend, the late Cyril Ponnamperuma, with whom I collaborated at Ames, NASA was the moving spirit for many of these. The Trieste Conferences are of equal significance to me as along with Professor Cyril Ponnamperuma, the outstanding theoretical physicist Professor Abdus Salam, who also had his roots in the Third World, was the co-founder of this unique series of Conferences in Trieste. Unfortunately, they have both physically left this world and we miss them greatly, but their spirit of inquiry and their quest to gain insights into questions related to the Origin of Life are all pervasive. Fortunately, Professor Chela-Flores is continuing the good work with distinction.

It is my privilege to share with you some precious moments, scientific and social which I and on occasions my wife, have experienced with many of the participants of the ISSOL Meetings and Trieste Conferences. It has been a pleasant journey during the last 30 years and I am grateful to all those who have contributed to some of the cherished memories.

In this paper I will focus on the Trieste Conferences. The first Trieste Conference in October 1992 was held under the joint umbrella of the International Atomic Energy Agency , and the United Nations Educational, Scientific and Cultural Organization and was attended by both Cyril Ponnamperuma and Abdus Salam, cofounders of the Trieste Conferences.

The second Trieste Conference (Oct, 1993) was entitled Chemical Evolution: Self-organization of the Macromolecules of Life and was dedicated to Cyril Ponnamperuma on his 70th birthday. The focal theme of self-organization was sub-divided into chemical aspects, geophysical aspects, biochemical aspects, biophysical aspects and chirality. At this conference many of the erstwhile collaborators of Cyril Ponnamperuma participated and the proceedings of the conference were edited by Julian Chela-Flores and Cyril

Ponnamperuma's former collaborators namely: Mohindra Chadha, Alicia Negron-Mendoza and Tairo Oshima.

The Third Trieste Conference (Sept, 1994) was entitled Chemical Evolution: structure and Model of the First Cell and was designated as Alexander Ivanovich Oparin 100th Anniversary Conference. The topics covered were: Beginings of Cellular Organization (the early paleontological record), physical, chemical and biological aspects of the origin and structure of membrane and origin and structure of the cell. This was the best attended of the three Trieste Conferences till then and a lot of new information was shared. The presentations on planetary extraterrestrial and interstellar conditions broke a lot of new ground.

The Fourth Trieste Conference (Sept, 1995) was the first such conference held after the untimely and sudden demise of Cyril Ponnamperuma, the co-founder of this series of conferences. He was involved in the early part of the preparation of this conference. However, he was not to be present to participate in the deliberations! Such is life. The conference was entitled Chemical Evolution: Physics of the Origin and Evolution. The conference was aptly dedicated to the Memory of Cyril Ponnamperuma. There were general overview presentations by John Oro on Cosmic Evolution and Sidney Fox entitled Experimental Retracement of Terrestrial Origin of an Excitable Cell: Was it Predictable? The sessions covered a broad range of topics – e.g. Origins; From Geophysics to Prebiotic Chemistry; Physiochemical Aspects; Biophysical Aspects – General Problems and Biomolecular Chirality; Evolutionary Aspects; Information Theory; Communication and Instrumentation in Exobiology and Mars Exploration. In a special session, homage was paid to Cyril Ponnamperuma and excerpts of letters received from various institutions, academics and admirers of Cyril Ponnamperuma were read. A touching note received from wife (Valli) and daughter (Roshini) was also shared.

The Fifth Trieste Conference (Sept, 1997) was entitled Exobiology: Matter, Energy and Information in the Origin and Evolution of Life in the Universe and was dedicated to the memory of Abdus Salam, the co-founder of the Trieste Conferences, who had unfortunately passed away in 1996 after prolonged illness. This conference had as many as 12 sponsors and was largely attended. The highlights were talks by Chela-Flores: Abdus Salam From Fundamental Interactions To the Origin of Life. The Abdus Salam lecture by John Oro; Cosmocological Evolution: A Unifying and Creative Process in the Universe. The Cyril Ponnamperuma lecture by Frank Drake; The Search for Intelligent Life in the Universe. The opening lecture: The Theory of Common Descent by Richard D. Keynes and a Public Lecture: Are We Alone In the Universe by Paul Davies.

The sessions at this conference dealt with – Matter in the Origin; Energy from Inert to Living Matter; Information; Early Evolution; Exobiology – The General Perspectives; Exobiology on Mars and Europa; The Interstellar Medium, Comets and Chemical Evolution; Exobiology on Titan; Extrasolar Planets and Search for Extraterrestrial Intelligence.

This conference was unique in focussing on Exobiology in general and on Comets, Planets and the Interstellar Medium in particular.

I compliment the organizers of the present conference, Profs. Julian Chela-Flores, Tobias Owen, and Francois Raulin for their untiring efforts to put the Sixth Trieste Conference together and entitle it as FIRST STEPS IN THE ORIGIN OF LIFE IN THE UNIVERSE. While major strides have been taken in understanding many aspects of this all important topic, but the answers are far from complete and may continue to remain so for long. However, our quest to find logical answers to this question will unfold many fascinating chapters, and some of these will be presented here itself in the course of the present conference!!

Section 2:

The origin of life in the universe

MATERIALS FOR "FIRST STEPS" IN EVOLUTION

A. G. CAIRNS-SMITH
Department of Chemistry
University of Glasgow
Glasgow G12 8QQ,
Scotland, U.K.

In the game of "Twenty Questions" one of the players thinks of some object, and the others try to discover what this object is by asking questions and being given only yes/no/don't-know answers. Beginners soon learn that it is a mistake to start with specific guesses. Rather one should start with general questions such as: "Is it in this room?", "Do you think I have one?", and so on, gradually narrowing the field. It is amazing how often seemingly obscure objects can be uncovered in this way, within the allowed "twenty questions".

Of course science is a more difficult game, but here too it is a mistake to suppose that the only way to arrive at specific conclusions is with specific information. This is just as well because with many of the most interesting problems, such as the problem of the origin of life, specific information is lacking.

We have no specific information on the very first evolving forms on Earth, but we know some general things. For example, like any system subject to natural selection they must have been able to pass on characteristics to offspring. And yet there are other general things that we are inclined to believe for no good enough reason. Some say that we know about the basic materials: the first organisms would have been made from "molecules of life" (that is to say the molecules that are "building blocks" of all life on Earth today). But we do not know this. What we *can* be pretty sure of is that the last common ancestor of all forms of life so far discovered on Earth was made of much the same materials as we are all made of now. But that is not to say that the very first evolving forms contained *any* of these materials, since it is becoming increasingly evident just what a sophisticated and complex piece of engineering this last common ancestor was. There was much evolving between the ultimate ancestor and the last common ancestor, and who is to say that materials did not change in the process? Who is to say that the very first steps were not made with altogether different kinds of materials, much less efficient, no doubt, but easier to come by and much more readily assembled into evolving entities? It seems like common sense to suppose that in the beginning there were such "low tech" life systems, easy to start; and then an evolution to the high-tech winner, the system which is now found everywhere; and that as part of this early evolution, as the now universal central biochemical machinery was being "invented", new materials were being brought in.

The problem of the origin of life may then seem more difficult than ever. Yet the first evolving forms had to conform at least to one current dogma about the nature of Darwinian systems in general: their evolution would have depended on the transmission of information between generations through replication of genetic information (of some

J. Chela-Flores et al. (eds.), First Steps in the Origin of Life in the Universe, 49–53.

sort — call it G) that could affect the chances of its own survival through creating phenotypes (P). That is to say the genetic information should have direct or indirect effects on its surroundings. Furthermore the replication process would have to be pretty accurate, although with occasional "mistakes" which were then also replicable:

$$\begin{array}{ccccccccc} & & P & & P & & P' & & P' \\ & \nearrow & & \nearrow & & \nearrow & & \nearrow & \\ \rightarrow G & \rightarrow & G & \rightarrow & G' & \rightarrow & G' & \rightarrow & \end{array}$$

Now think about it this way: once a sustained evolution of *any* sort had got going the world would change: there would be machines in operation making consistent supplies of new kinds of materials. This would create new possibilities for holding and transmitting information, possibilities for new kinds of genetic materials.

Thus we can imagine in abstract terms how genetic materials could change radically during early evolution, through *genetic takeover* [1]. Here is a formal representation of this idea, with just one takeover:

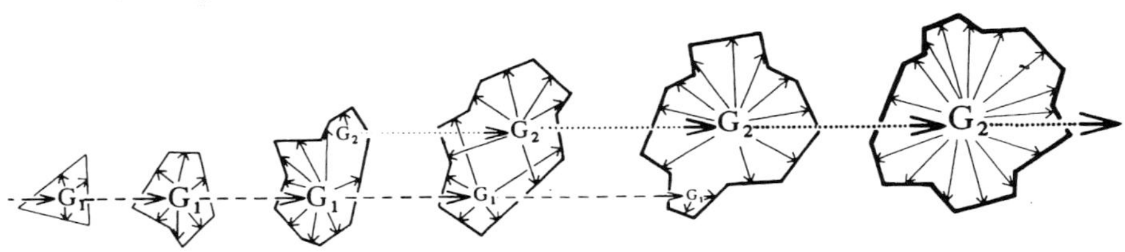

G_1 is a primitive genetic material able to replicate its information reasonably accurately, and exert some sort of control on its surroundings, so that in its evolving phenotypes (surrounding polygons) there was increasing competence in acquiring and producing a variety of materials. One of these turned out to be capable of replication as an alternative means of synthesis. Now there would be a second genetic material (G_2) present, because this replicative synthesis would be subject to evolution through natural selection. G_2 need have had no chemical resemblance to G_1: Mutual usefulness would be enough to explain their co-evolution. And then finally G_2 proves altogether more versatile and efficient, and G_1 becomes redundant.

The overall reason to expect that there was such a takeover (if not several) is that, as one finds in many branches of engineering, materials suitable for making primitive mechanisms, for "first steps", are not likely to be the materials of choice for advanced systems. (Think of quill pens and word processors.) So if there exists some evolutionary means through which a first genetic material (low tech, but able to operate in the wild) could be updated with something completely different (much more efficient but needing "high tech" support), then this is likely to have been what actually happened.

The bad news is that if takeover(s) took place during early evolution this will obscure the nature of the very the first genetic materials. The good news is that with genetic takeover as a possibility we have a much wider chemical landscape to explore.

Schrödinger applied general ideas to suggest that the genes in organisms today are like "aperiodic crystals", and held together by strong covalent bonds [2]. In the event DNA turned out to be indeed aperiodic and covalent, although not a crystal as such.

In looking for possible primitive genetic materials — the key to the origin of evolution — we may take a similar approach. The most critical competence required is accurate replication of information. And here aperiodic crystals are still on the agenda.

At Glasgow and Paisley we have been studying crystal growth of barium ferrites [3]. These materials grow a high temperatures, but provide a model for the kind of thing we are looking for. They are strong-bonded complex oxides containing oxygen, iron, barium and smaller amounts of other metals. Their structure can be likened to a stack of cards firmly glued together, each card about a nanometre thick. Two kinds of layers are common. One, a little thicker than the other, has twice as much barium in it. Crystals may have both kinds of layer in them, in a somewhat random sequence. Such crystals could in principle hold information in the form of a particular sequence (rather as DNA holds information as a sequence of base pairs). Amazingly, long random-like sequences, of thirty layers and more, have often been found repeated on top of each other again and again [4]. Is this due to a kind of *replication* during crystal growth?

It seems so. Our electron microscope studies have suggested that in the early stages these crystals grow in a manner illustrated in the following cartoon:

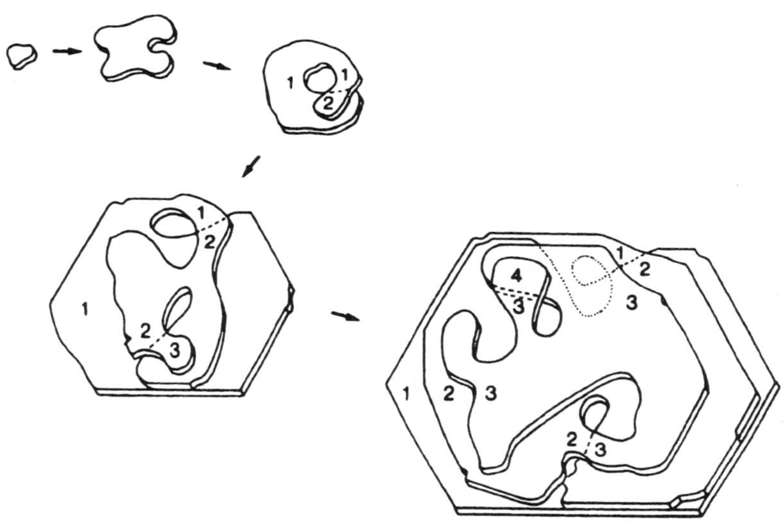

Here we imagine an initial small crystal consisting of a stack of the two kinds of layer arranged in some arbitrary sequence. This stack grows exclusively sideways, atoms being added only to its edges, extending the stack and maintaining the sequence. But the growth is uneven, spasmodically faster in different directions at different moments, in a manner that is typical of growing crystals. This leads to irregular, sometimes branched structures — but still flat, still the same thickness as the original "seed". We see crystals of this sort as growing like fronds of seaweed. Because they are tough and flexible different parts of the same "frond" can come to overlap each other, and then stack together to make thicker crystals. The random-like arrangement of thick and thin layers in the initial "seed" thus comes to be repeated within these thicker crystals (four times in the bottom right drawing of the cartoon). The sequence is repeated because it is the *same* connected multi-layer that is re-appearing at different levels [3], [5].

This is still some way from being a working genetic material, even a primitive one. In the first place pieces of crystal would have to break off from time to time to complete a replication cycle. This is not difficult to imagine, but it has yet to be demonstrated. What is most significant about our ferrite model is in the implied fidelity with which long random-like stacking sequences can be extended through sideways growth, and hence copied. Fidelity of copying some pre-existing random-like structure is a *sine qua non* for any kind of genetic system. It is a function that organic molecules (so far DNA and RNA) can only competently achieve with elaborate enzymatic assistance. Organic molecules have many virtues, but they are not "naturally" replicators.

The inorganic world of strong bonded crystal structures is more promising. Barium ferrite is not an isolated case. Mixed-layer structures are common in the inorganic world, among clay minerals for example, and repetition of stacking sequences is found here too [6].

Polytypism is a related phenomenon, and very common indeed. Here variety in stacking patterns may arise even although all the unit layers — the "cards" — are identical. A rich potential source of permutation arises because the unit layers can be stacked on top of each other with different orientations. Mica minerals provide a well studied example. Here the "cards", about a nanometre thick, have a directionality. It is as if any one "card" has arrows printed on it, all pointing the same way. In building a stack of mica unit layers there are three almost equally stable ways in which the next layer can be added: its "arrows" can be in the same direction as the layer underneath or orientated $\pm$ 120°. So there is a huge opportunity for permutations, and some startlingly long-range repeats have been found [7]. Possibly "seaweed growth" is part of this story too. Polytypism is also found among clay minerals, whose unit layers are similar to those in mica.

Returning now to the idea of genetic takeover, and the simplest case of a single takeover illustrated in the previous sketch, I have suggested that G_1 was a clay, or some similar mineral, and that G_2 was RNA [1]. But the large number of potential inorganic replicators [8], together with the likelihood that there were immediate organic precursors of RNA [9], raises the possibility of multiple genetic materials in early organisms, operating together in a kind of symbiosis. We might imagine early organisms which used several different kinds of mineral replicators, for controlling different phenotypic functions; and that later, before and during the invention of our current biochemical control systems, organic replicators were joining in.

This is all very general and speculative. We would like know how organic molecules became involved, how the set of molecules that we now have was chosen, and so on [10]. But we are still at the beginning of our game of "twenty questions". Let us be content for the moment with three general points:

1. Both clay minerals and organic molecules were likely to have been produced on the primitive Earth, for example in hydrothermal environments [11].

2. There is ample evidence that clay minerals can affect organic molecules and their reaction, in soils and elsewhere, by adsorption and catalysis.

3. Organic molecules can help to catalyse the synthesis of clay minerals [12].

A final comment. There are those who would put "metabolism" first in thinking about the origin of life. But it is unclear what this means. Uncontrolled organic transformations, tend to create increasingly complex products, culminating in tar: to produce organic molecules consistently enough and pure enough to build biochemical machinery — replicators, enzymes, and so on — would seemingly need an organic chemist, or an evolved cell. "Metabolism" implies control and should be defined, I think, as "chemical reactions taking place under genetic control". On such a view "metabolism first", before a genetic system of *some* sort, is a contradiction in terms. With the genetic takeover idea we can see, if still only in general terms, how it could have been that the whole of metabolism, including the primary metabolism of life now, could have been a *product* of evolution through natural selection, not a pre-condition for it. A crystal gene does not need any "metabolism" to begin with. That is its great virtue.

References

1. Cairns-Smith, A. G.: *Genetic Takeover,* Cambridge University Press, Cambridge, 1982;
——— *Seven Clues to the Origin of Life.* Cambridge University Press, Cambridge, 1985;
——— & Hartman, H. (eds.). *Clay Minerals and the Origin of Life*, Cambridge University Press, Cambridge, 1986;
—— Argille e origine della vita, in *Frontiere della Vita, Vol. I,* Istituto della Enciclopedia Italiana, (1998), 171-191.
2. Shrödinger, E.: *What is Life?,* Cambridge University Press, Cambridge, 1944.
3. Turner, G., Stewart, B., Baird, T., Peacock, R. D. & Cairns-Smith, A. G.. Layer morphology and growth mechanisms in barium ferrites. *J. Crystal Growth,* **158** (1996), 276-283.
4. Kohn, J. A., Eckart, D. W. &, Cook, Charles F., Jr.: Crystallography of the Hexagonal Ferrites. *Science,* **172** (1971), 519–525.
5. This is a similar situation to the standard explanation for repeat sequences in layered crystals which is through a screw dislocation, but such dislocations are apparently not present in barium ferrite crystals showing really long repeats: the more chaotic "seaweed" mechanism appears to operate here and, we think, provides a better mechanism for the replication of really long repeat sequences.
6. Reynolds, R. C. (1980) Interstratified clay minerals, in G. W. Brindley & G. Brown (eds.), *Crystal Structures of Clay Minerals and their X-ray Identification,* Mineralogical Society, London, pp 249-303,.
7. Baronnet, A.: *Current Topics in Materials Science,* **5** (1980), 447.
8. Cairns-Smith, A. G. The chemistry of materials for artificial Darwinian systems. *Internat. Revs. in Phys. Chem.* **7** (1988), 209-250.
9. Joyce, G. F., Schwartz, A. W., Miller, S. L. & Orgel, L. E.. *Proc. Natl. Acad. Sci. U.S.A.,* **84** (1987), 4398-4402
10. See reference [1], especially *Genetic Takeover,* chapters 8 & 9, for more detailed speculations.
11. Holm, N. G. (ed): Marine Hydrothermal Systems and the Origin of Life. Special issue of *Origins of Life and Evolution of the Biosphere,* **22** (1992) 191–242
12. Siffert, B. (1986) Clay synthesis: the role of organic complexing agents, in A. G. Cairns-Smith & H. Hartman (eds.), *Clay Minerals and the Origin of Life,* Cambridge University Press, Cambridge, pp. 75–78

ORIGIN AND PERSISTENCE OF GENETIC MATERIAL IN PREBIOTIC HABITATS. THE POSSIBLE ROLE OF NUCLEIC ACID-CLAY COMPLEXES

ENZO GALLORI and MARCO FRANCHI
Department of Animal Biology and Genetics, University of Florence. Via Romana, 17. 50125 Florence, Italy. E-mail: gallori@dbag.unifi.it

1. Introduction

Everyone agrees that certain specific conditions were required for the evolution of life from non-life, e.g. energy/catalysis sources, a concentration of small precursors of biomolecules, the protection of complex organic molecules once formed, etc. These and other conditions represent a sort of constraint for the possible environment in which life arose. Indeed, it is difficult to conceive that life could have originated in the presence of a high concentration of water (55 M), like that of a primordial ocean, because in these conditions hydrolysis and not polymerization is favoured (Pace, 1991).

We believe that it is crucial to consider surface organic chemistry. As early as 1951, J.D. Bernal suggested, on the basis of thermodynamic and kinetic considerations, that mineral surfaces (specifically clay minerals) could have played an important role in the prebiotic formation of the biomolecules basic to life.

In recent years, numerous observations have reinforced this hypothesis. Polymerization of different biochemical monomers, nucleotides and amino acids, has been obtained in the presence of clay minerals (Ertem and Ferris, 1996; Ferris *et al.*, 1996), which may also allow the assembly of small organic molecules into self-replicating biomolecules (Smith, 1998; Parsons *et al.*, 1998; Luther *et al.*, 1998). Moreover, experimental data on gene transfer in the natural environment agree nicely with this hypothesis. Studies of the fate of genetic material in a natural habitat, namely soil, indicated that DNA, originating from lysis of dead cells or excretion by living cells, can persist for a long time in the environment. Its interaction with soil particles protects the DNA from numerous degrading factors and allows it to maintain its biological activity (Stotzky *et al.*, 1996). Together, these observations suggest the possible involvement of mineral surfaces in the chemical evolution of life on earth.

2. Physical chemical characteristics of nucleic acid-clay complexes

Clay-nucleic acid complexes were prepared by reacting different nucleic acid molecules (DNA, RNA) with the clay minerals, montmorillonite (M) and kaolinite (K) (Fig. 1.a). The complexes were then studied by different analytical techniques, including X-ray diffractometry (X-RD), electron microscopy (TEM and SEM), and Fourier-transform infrared (FT-IR) absorption spectroscopy (Franchi *et al.*, 1999).

J. Chela-Flores et al. (eds.), First Steps in the Origin of Life in the Universe, 55–58.

TEM and SEM micrographs of DNA-clay complexes (Fig. 1.b) revealed the presence of long-sticked DNA strands that appeared to be associated with several points on the edges of clay particles, from which they spread in a large loop. In the case of small DNA molecules, such as plasmid DNA, the molecules appeared to lie on the planar surface of the clay particle.

The lack of penetration of nucleic acid molecules into M and K was also suggested by X-RD analysis. X-ray diffractograms of the complexes showed no expansion of the basal spacing of Al-Si layers when compared with the clays alone.

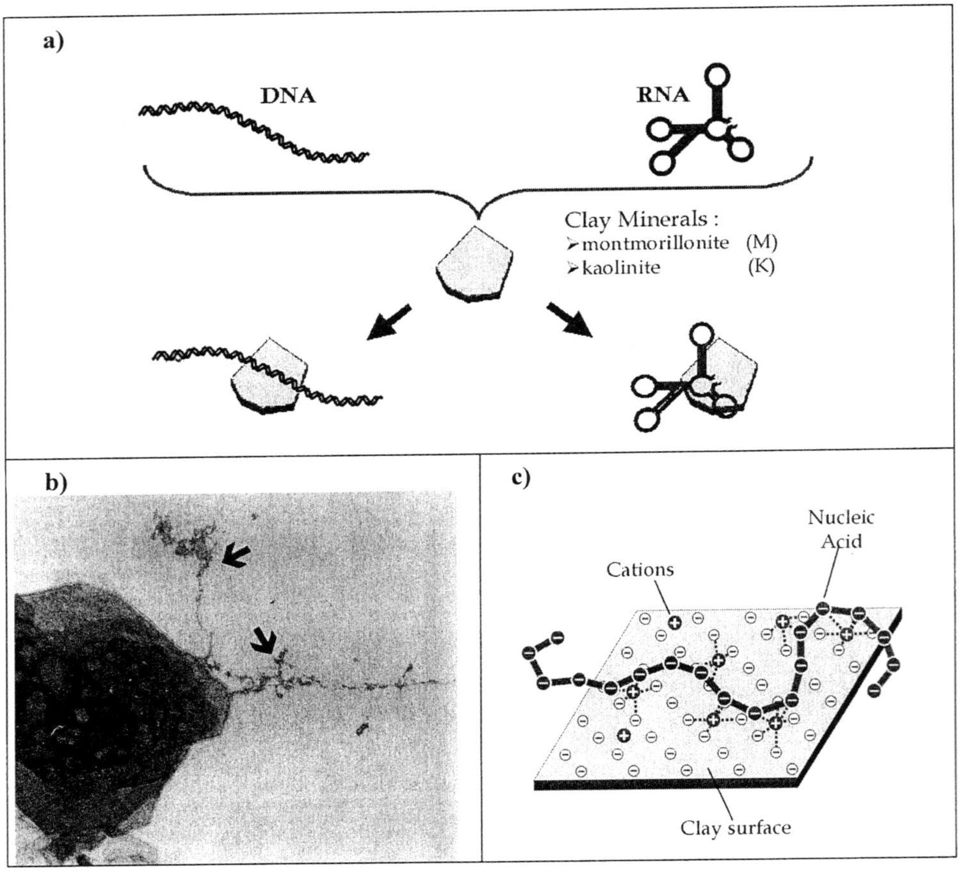

Figure 1 - a) Scheme of preparation of nucleic acid-clay complexes. **b)** TEM micrograph of DNA-K complex (Magnification, X 154 000). Arrows indicates the fibers of DNA. **c)** ."Cation-bridge" adsorption model of nucleic acid molecule on clay surfaces.

With regard to the nature of the interaction between the nucleic acid molecule and the external surface of the clay particle, we investigated the role of different cations (mono- and divalent) in the adsorption process. Our results indicated that an increase in cation concentration produces an increase in the adsorption of nucleic acid, suggesting that the cations act as a "bridge" between the negatively charged DNA and the negatively charged clay surface (Fig. 1.c).

The FT-IR spectra of clay-nucleic acid complexes significantly differ from that of free molecules. For the complexes, there are some bands not directly referable to those of the DNA, which can be ascribed either to the chemical association between the DNA and the clay or to modification of the DNA conformation as a result of its adsorption on the clay (transition B $\rightarrow$ A form). Both possibilities could account for the resistance of the clay-nucleic acid complexes to biotic and abiotic degradation in the environment.

3. Biological properties of nucleic acid-clay complexes

Clay-nucleic acid complexes have specific biological characteristics (Tab. 1). For instance, chromosomal and plasmid DNA bound on M and K are capable of transforming competent bacterial cells for a long period of time (Gallori *et al.*, 1994; Gallori *et al.*, 1998). The adsorption/binding of DNA on clay mineral does not prevent its biological activity, but rather enhances the persistence of the genetic material in the environment. Such persistence would be undetectable in the absence of a specific host; however, when the host is present, the bound DNA can be taken up and expressed. In other words, clay-complexes could act as a "storage" of genetic information in natural habitats. Moreover, clay-bound DNA can be enzymatically replicated and amplified, outside the cellular context, by the polymerase chain reaction (PCR) (Vettori *et al.*, 1996). This supports the hypothesis that in the prebiotic environment, primitive biomolecules could interact with each other and evolve into more complex systems. Recently, the enzymatic replication and amplification of clay-adsorbed RNA molecules was obtained by a reverse transcriptase PCR technique (Franchi *et al.*, 2000). This demonstrates that the passage from a single-strand molecule to a double-strand molecule is possible in an abiotic context.

TABLE 1 - Biological Properties of Nucleic Acid-Clay Complexes

- DNA adsorbed/bound on clay minerals persists in natural habitats and retains the ability to transform bacterial competent cells for long period of time.
- DNA adsorbed/bound on clay minerals is resistant to biotic (i.e. DNases) and abiotic (UV rays) degradation.
- DNA adsorbed/bound on clay minerals can be replicated and amplified by PCR (Polymerase Chain Reaction).
- RNA adsorbed/bound on clay minerals can be retrotranscripted to double strand DNA

4. Conclusions

On the basis of the results obtained to date, we believe that the formation of a close association between prebiotic genetic molecules, *whatever they were*, and mineral surfaces could have been a key step in the preservation of genetic information in primordial habitats, thus allowing the genetic material to express its biological potentiality.

5. References

Bernal, J.D. (1951) *The physical basis of life*, Routledge and Kegan Paul, London.

Ertem, G. and Ferris, J.P. (1996) Synthesis of RNA oligomers on heterogenous templates, *Nature*, **379**, 238-240

Ferris, J.P., Hill A.R., Liu, R. and Orgel, L.E. (1996) Synthesis of long prebiotic oligomers on mineral surfaces, *Nature*, **381**, 59-61.

Franchi, M., Bramanti, E., Morassi Bonzi, L., Orioli, P.L., Vettori, C., and Gallori, E. (1999) Clay-nucleic acid complexes: characteristics and impications for the preservation of genetic mateial in primeval habitats, *Origins Life Evol Biosphere*, **29**, 297-315.

Franchi, M., Vandini, C., and Gallori, E., Enzymatic replication of RNA adsorbed on clay minerals, Abst. Sixth Trieste Conference on Chemical Evolution (Trieste, Italy, 18-22 September 2000), p. 74.

Gallori, E., Bazzicalupo, M., Dal Canto, L., Fani, R., Nannipieri, P., Vettori, C., and Stotzky, G. (1994) Transformation of *Bacillus subtilis* by DNA bound on clayin non-sterile soil, *FEMS Microbiol. Ecol.*, **15**, 119-126

Gallori, E., Franchi, M., Rinaldi, L., and Vettori, C. (1998) Interspecific transformation of *Bacillus subtilis* by clay-bound DNA in non-sterile soil, *Symbiosis*, **25**, 311-322.

Luther, A., Brandsch, R., von Kiedrowski, G. (1998) Surface-promoted replication and exponential amplification of DNA analogues, *Nature*, **396**, 245-248.

Pace, N.R. (1991) Origin of life – Facing up to the physical setting, *Cell*, **65**, 531-533.

Parsons, I., Lee, M.L., and Smith, J.V. (1998) Biochemical evolution. II. Origin of life in tubular microstructures on weathered feldspar surfaces, *Proc. Natl. Acad. Sci.*, **95**, 15173-15176.

Smith, J.V. (1998) Biochemical evolution. I. Polymerization on internal, organophilic silica surfaces of dealuminated zeolites and feldspars, *Proc. Natl. Acad. Sci.*, **95**, 3370-3375.

Stotzky, G., Gallori, E., Khanna, M. (1996) Transformation in soil, in Akkermans, A.D.L., Van Elsas, J.D., and De Brujin, F.J. (eds.), *Molecular Microbial Ecology Manual*, Kluwer Academic Publishers, Dordrecht, pp. 1-28.

Vettori, C., Paffetti, D., Pietramellara, G., Stotzky, G., and Gallori, E. (1996) Amplification of bacterial DNA bound on clay minerals by the random amplified polymorphic DNA (RAPD) technique, *FEMS Microbiol. Ecol.*, 20, 251-260.

CHEMICAL EVOLUTION STUDIES IN ORGANIC COMPOUNDS ADSORBED IN CLAYS

S. RAMOS-BERNAL AND A. NEGRON-MENDOZA

Instituto de Ciencias Nucleares, U.N.A.M.
Circuito Exterior, C. U. 04510, México D.F., México

Abstract

In this study a multiphase system was considered in order to test the hypothesis of Bernal that the clay minerals may act as protectors of organic molecules from the environmental high-energy radiation. Adenine and acetic acid were adsorbed in Na-montmorillonite and exposed to ionizing radiation. The resistance of decomposition of adenine to high-energy radiation changes from total decomposition (in KGy range), to almost no decomposition when it was adsorbed in clay. However, for acetic acid the difference was mainly the selectivity in the way of decomposition. When adsorbed in clay we found decarboxylation and without clay the dimerization reaction was the main way of decomposition.

1. Introduction

A more plausible and realistic way of simulating the environment that could resemble the primitive Earth would require the consideration of multiphase systems. The work done in connection with simulated primitive Earth experiments has been confined to an homogeneous phase rather than heterogeneous one. As far as the experiments in a heterogeneous phase is concerned, the most geologically relevant and abundant surfaces on the primitive Earth are clays.

The original proposal of Bernal in 1951 suggested several roles for the clays.

1) Clay may act as concentrators of biological precursors of molecules. Without such a concentration, most prebiotic scenarios come to a halt because a highly concentrated prebiotic environment may be implausible.
2) It is also possible that condensation reactions may occur inside the clay.
3) Once the molecule is adsorbed into the clay mineral it may act as a protector of such organic molecules from the high energy present in the environment.

Some of these postulates have been extensively studied, for example, the adsorption of important biological molecules such as amino acids, nucleic acid derivatives, etc. However, not much work related to the protection role of clays has been done.

It is known that some organic compounds sorbed in clay minerals are likely to decompose mainly because of the catalytic activity of these minerals. The aim of this

J. Chela-Flores et al. (eds.), First Steps in the Origin of Life in the Universe, 59–63.

work is to get an insight of the behavior of some organic compounds in the presence of Na-montmorillonite. We studied the sorption of adenine as an example of a nucleic acid base, and acetic acid as an example of a carboxylic acid. To observe the role of the clay as a protector agent from the ionizing radiation, the system clay-organic compound was exposed to this high-energy radiation. We compared it to the same system without any irradiation.

Due to radiation the catalytic properties of the clay may change. On the other hand, surface reactions in organic compounds adsorbed into solids can be enhanced if they are irradiated with very penetrating radiation. In this work it is hypothesed that such combinations of interactions enhance the prebiotic evolution of molecules. In this type of systems, the characterization of the solids, and the identification of defects produced by irradiation, may give some clues about how the solid stores energy as well as how does it transfer energy, from the solid to the organic molecules.

To study this protector role of montmorillonite, it is necessary to search the extent of adsorption, the mechanism of adsorption, and especially the place in which the organic compound is binded to the clay. Previous studies showed that for adenine, at acidic pH, the molecule is adsorbed in 95%. In experiments where the interlamellar channel was blocked, it was possible to determine that the adsorption of adenine occurs mainly inside the interlamellar channel. Since at pH 2 the adenine is positively charged, the mechanism of adsorption is by ion interchange with the cations (Na^+ ions) inside the channel. At acidic pH the acetic acid molecule does not present a net charge. The infrared spectra showed that the carboxylic group in acetic acid is present as carboxylate and therefore, the place of binding is the positively charge edges of the clay.

2. Experimental Procedures

2.1. PREPARATION OF SOLUTIONS AND IRRADIATION

Samples without clay. All chemicals used were of the highest purity available. Aqueous solutions of acetic acid 0.8 M, at natural pH, and adenine 10^{-3} M at pH 2 were prepared, employing triple distilled water and in special glass tubes. The samples were deareated with argon.

Samples with clay. The clay used as a substrate to anchor the organic molecules was Na^+-montmorillonite from Crook County, Wyoming U.S.A., (clay mineral standard batch SWY-1). The amount of clay used for each study was calculated in base of its CEC capacity (101 meq/100g of clay). A degassed sample of the Na^+-montmorillonite (100 mg) was weighed in a special glass tube, and the selected organic compound was added (1 ml of acetic acid or 3 ml of adenine). The glass tubes were closed, mixed and shaken at room temperature. After 2 hours the suspensions were centrifuged and the supernatant liquid was separated. The clay samples were extracted with 3 successive aliquots (5 ml each) of 2 M basic solution.

The samples were divided in two groups: a set for blanks with and without clay, and a group that was irradiated.

The samples were irradiated in two ^{60}Co gamma sources facilities. One of these sources is of high intensity (Gammabeam 651 PT). In this source the radiation doses were from 0.01 to 500 KGy. To study the formation of gaseous products, a second low intensity gamma source was used (Gammmacell-200). The radiation doses ranged from 3.7 to 6.2 Gy. After irradiation the samples were analyzed.

2.2 ANALYSIS

Acetic acid solutions. In order to analyze non-volatile products the mixture was centrifuged. A measured amount of the supernatant was evaporated until dryness. Then, methyl esters were prepared according to Negrón-Mendoza and Ponnamperuma, 1976. The analysis was carried out in a GC chromatograph. The column used was a glass column (1.82 m in length and an internal diameter of 4 mm) packed with Silar 7C. For the GC-MS analysis a capillary column of methyl silicon (12 m in length and 0.33 μm of inner diameter) was used. This chromatograph was coupled to a mass spectrometer.

The production of CO_2 was followed as a function of the dose. For this analysis the following procedure was carried: 0.1 g Na-montmorillonite was evacuated in a vacuum line for three hours. Immediately 3 ml of the solution of acetic acid were added to the clay. The system was evacuated for 15 more minutes. The gaseous products formed by the irradiation were extracted employing a Toepler gauge connected to a Gas Chromatograph. The products were identified by their GC retention time. Blank solutions were analyzed, especially for the CO_2 determination.

The clay was recovered from solutions by centrifugation, washed with water, and finally dried at 50 °C. The clay was analyzed by infrared spectroscopy (Perkin Elmer model 500-FT-IR) using bromide potassium disks.

Adenine solutions. The analysis of these samples was done by high-pressure liquid chromatography (HPLC) in a Varian chromatograph model 8055 with a column packed with MCH-10. The detector used was a Varian ultraviolet detector at 260 nm. The mobile phase was a mixture of two solutions. Solution A: a methanol-water mixture (80:10 v/v). Solution B: a pH 2 buffer solution of KH_2PO_4 0.01 M and H_3PO_4 0.035 M. The gradient used was from 94% of solution B down to 60 % of solution B in 12 minutes (Guzmán et al., 2000). After the irradiation, the samples with clay were centrifuged. The supernatant was removed and analyzed by HPLC.

3. Results

Radiolysis of Free Acetic Acid. The irradiation of aqueous acetic acid produced many compounds. The main product obtained was succinic acid (the dimer of acetic acid). The next important compound was tricarballylic acid, which corresponds to the trimer of acetic acid. Other carboxylic acids were also identified, like malonic, carboxysuccinic, β-hydroxy-β-methylglutaric (a linear trimer of acetic acid), 1,2,3-butanetricarboxylic, aconitic, citric and 1,2,3,4-butanetetracarboxylic acids.

The main feature of these series of experiments was the production of the dimer (85 % based in the acid decomposed), as the principal way of decomposition of acetic acid. Other products are explained as secondary reactions from the succinic acid and the decomposition products of the water radiolysis. The target compound decomposition increases as a function of the dose.

Radiolysis of acetic acid-clay complexes. The extent of decomposition of acetic acid with and without clay is about the same, but in presence of clay the number of products identified decreased considerably. The production of the dimer, the main irradiation product, decreases from 85 % (in the system without clay) to 31 % (in systems with clay).

Gaseous products as CO_2 and H_2 were also detected and identified. The production of CO_2 was greater in samples with clay. This formation increases as function of the dose. The source of H_2 is from the radiolysis of water and from the abstraction reactions produced during the radiolysis.

Radiolysis of Free Adenine. The destruction of adenine in terms of radiochemical yield was of 1.1 molecules destroyed for each 100 eV given to the system. This is a low value and suggests the resistance of adenine toward the irradiation in a high radiation field of 1.2 KGy/h (1 Gy = 1J/Kg). Two decomposition products were observed, but then have not been identified.

Irradiation of Adenine-Clay Complexes. The results obtained with the adenine-clay system showed that there was very low decomposition of adenine.

4. Discussion and General Remarks

In terms of adsorption, the most active inorganic constituents of soils is the clay fraction. Clays not only have large surface areas, but also have mineral structures that bear negative charges balanced by exchangeable cations. In addition, clays can bind several organic compounds by van der Waals, ion-dipole, and dipole-dipole interactions. The interaction of organic compounds with clays can be expected to play an important role in chemical evolution studies (Bernal, 1951).

The adsorption process plays an essential part in the roles assigned to clays, particularly in enrichment of organic monomers, specific orientation and arrangement of monomers. The pH at which the adsorption takes place is very important and determines the dominant species of the organic compound. In these series of experiments we used pH 2. In previous studies is reported that adenine is almost completely binded to the clay at acidic pH (Perezgasca et al., 1993). At this pH adenine is positively charged and it is binded to the interlamellar channel, as is confirmed by X-ray spectra. On the contrary, at this pH acetic acid is not dissociated. Yet, the infrared data showed that there is an interaction clay-acid. The acid is as carboxylate bearing a negative charge and it is binded to the positive edges of the clay. Although the maximum adsorption occurs at pH 2, it is important to remark that such pH most probably did not exist in the primitive Earth (Miller and Orgel, 1974. However, the data presented sustain the relevance of microenvironments in origin of life studies.

In an aqueous system the radiation interacts primarily with water molecules. Very reactive species are formed due to this interaction (H, OH, e_{aq} , H_2 and H_2O_2). These radicals attack the organic compound present in the solution in a secondary way, yielding the observed products. In the cases of the irradiation of aqueous solutions of adenine and acetic acid, the extent of decomposition increased with the irradiation doses. At doses of the order of KGy there is a net decomposition of adenine.

For acetic acid in the system without clay the main reactions induced by radiation takes place via the free radicals of the water decomposition. For acetic acid the main reaction is the dimerization. However, in presence of clay, the results showed that the number of products diminished and the generation of CO_2 increased lineally with the radiation dose. Thus, there is a preferential way of decomposition and the decarboxylation reaction is dominant.

There are many works in the literature that deal with the irradiation of aqueous, oxygen-free solution of purines. Still, the radiolysis of this kind of compounds is not well established, and it is poorly understood despite the considerable effort that as been made in this area (Mosqueira, *et al*, 1996).

The complexity of the reactions is considerable and the sites of the reaction with the purine ring are speculative (Cadet and Berger, 1985). The steady state radiolysis of adenine showed a low yield of destruction, regardless of the high reactivity of adenine toward the attack of water decomposition products. This suggested that adenine goes through an overall reconstruction pathway. In the presence of clay, there is a considerable recovery of the adenine, 90%, and can be an example of the role of clay as protector of the organic molecules.

The role of Clay in Radiation Heterogeneous Catalysis. In these experiments, the clay may play a role as an energy moderator of the energy deposited by radiation. Part of the energy adsorbed by the solid decomposes the organic compound adsorbed in it. The mechanism for this energy transfer is not well delineated, although it appears that during that energy transfer process the adsorbent actually participate in the reaction. Thus, the role of clay as protector for the molecules adsorbed in it and exposed to high-energy radiation may be caused by this energy moderator action. It was also observed in the case of acetic acid that the clay directed the reaction along a definitive path rather than several modes of simultaneous decomposition.

5. References

Bernal, J.D. (1951) *The Physical Basis of Life*, Routledge and Kegan Paul, London

Cadet, J., and Berger, M.: (1985) *Int. J. Radiat. Biol.* **47**, 127-143.

Guzmán, A., Ramos-Bernal, S. and Negrón-Mendoza, A. studies (2000) Irradiation of adenine adsorbed in Na-Montmorillonite. Implications to chemical evolution in Astrobiology: Origins from the Big Bang to civilization. Ed. J. Chela-Flores, J. Oro and G. Lemarchand, Kluwer Academic Publisher, p.271-274.

Miller, S.L. and Orgel L. (1974) *The Origins of Life on Earth*, Prentice-Hall, Inc. New Jersey.

Mosqueira, G., Albarrán, G. and Negrón-Mendoza, A. (1996) A Review of conditions affecting the radiolysis due to ^{40}K of nucleic acid bases and their derivatives adsorbed on clay minerals, *Orig. of Life and Evol. Bios.*, **26**, 75-94.

Negrón-Mendoza, A. and Ponnamperuma, C.(1976), Formation of biologically relevant carboxylic acids during the gamma irradiation of acetic acid. *Origins of Life* 7, 191-196.

Perezgasga, L., Negrón-Mendoza, A., De Pablo-Galán, L., Mosqueira, G. (1993) Site of adsorption of purines, pyrimidines and their corresponding derivatives on sodium montmorillonite. *Orig. of Life and Evol. Bios.*, **4**, 47

FROM INANIMATE MACROMOLECULES TO THE ANIMATE PROTOCELL:

In Search of Thermal Protein Phase-Shifting

ARISTOTEL PAPPELIS[1], PETER BAHN[2], RANDALL GRUBBS[3], JOHN BOZZOLA[4], AND PETER COHEN[5].

1. Department of Plant Biology, Southern Illinois University, Carbondale, Illinois 62901 USA (SIUC); 2. Bahn Biotechnology Co., RR2 Box 239A, Mount Vernon, Illinois 62864 USA; 3. Department of History, SIUC; 4. Imaging and Analysis Center, SIUC; and, 5. Department of Chemistry, SIUC.

Abstract. "Glass-state" thermal proteins (TPs) yielded protocells when moistened (at 100 C, instantly on cooling; at 60 C, within 3 sec; at 20 C, within 3 min; and, at 4 C, within 24 h). The TPs were copolymers of: 20 amino acids (AAs) found in "true" proteins, or conjugated TPs (four types of TPs with DOPA, and one type of TP with glucose, lecithin, adenosine monophosphate, ferric ions, or phosphate ions); and, 18 equimolar AAs without aspartic or glutamic acids. The 20 AA copolymer TPs were formed at 180 C (6 h). The 18-AA TPs were formed at 20 C (1 and 5 yr). We infer: the time required to convert TPs (by phase-shifting; TP folding) to protocells is an acellular, "chemical life" process(es). We modified (Italics ours) the NASA definition: Life is a self-sustained chemical system (*acellular, chemical life → protocellular, biological life*) capable of undergoing Darwinian evolution. We infer AAs → TPs in micrometeorites. Melting micrometeoritic ice would convert TPs into protocells. Thus, such micrometeorites could be cosmozoa or panspermia (in the "glass state," these could be called "litho-cosmozoa and litho-panspermia).

1. Introduction

Following their syntheses (self-ordering of AAs), amphiphilic thermal proteins (TPs) can undergo conformational changes in water → (self-assembly) hollow wall-membrane protocellular boundaries, extra-protocellular matrices, and microencapsulated TPs (14-19). These are structural units with multizymic functions: growth, differentiation, and reproduction. TPs were observed (8, 20) (1000 x; TPs "melted" at 60 C on microscope slides, covered with cover glasses) to form animate protocells when moistened (4 to 60 C). Nine TPs were tried at 20, 60, and 100 C and one at 4 C. Our discovery of low temperature (20 C) syntheses of TPs in an 18-AA mixture and the syntheses (1) of many 20-AA-conjugate TPs provided new TPs to extend those findings: 17 TPs at 4 C; 9 at 20 and 60 C; and, 9 after "flash boiling" and cooling). We sustained the null hypothesis that there would be no differences between the two studies (morphologies, diameters, and rates of formation).

2. Materials and Methods

We used two methods to synthesize TPs: 20-AAs with or without substrates for synthesiz-

J. Chela-Flores et al. (eds.), First Steps in the Origin of Life in the Universe, 65–68.

ing conjugate-TPs at 180 C for 6 h (9 TPs; 4 DOPA-TPs; Phos-TP; Fe-TP; glucose-TP; lecithin-TP; and, AMP-TP) (1, 5); and, 18-AAs (equimolar; without aspartic and glutamic acids) aging in glass vials at 20 C for 1 and 5 yr. We visualized protocell formation (8, 20): along margins of "glass-state" TPs ("melted" on microscope slides at 60 C + 1 drop of water at 4, 20, and 60 C); and, after "flash-boiling" TPs. We "recycled" all of the preparations (cooling/reheating protocells and mother liqueur to "flash boiling") and observing each "resurrection" (protocells → TPs → protocells; 4 to 6 replications). Two drops of each preparation were placed on a glass slide, dried at 60 C, and viewed at 500-9000x high vacuum or variable pressure scanning electron microscopy (SEM).

3. Results and Discussion

Protocells were observed (light microscopy) to form along the moistened edges of "glass-state" TPs: within 30 sec at 60 C; within 3-min at 20 C; and, at 4 C, within 24 h. The protocells that formed first were generally small (~ 0.5 mm in diameter) and, at 20 and 60 C, these quickly (within seconds or a few minutes) increased in volume (to ~ 5 mm) as a result of fusing. Occasionally, budding on the larger protocells was observed. These protocells were similar to those produced by "flash-boiling." Following moistening, the TPs formed a white band rich with protocells [similar to white edges in moistened carbonaceous chondrite micrometeorites (13)]. We infer: conformational changes of TPs associated with protocell assembly accounts for the color change: and, the white regions of micrometeoritic material contain TPs and, possibly, protocells. During "flash-boiling" (heating to 100 C), the TPs were solubilized. Protocells formed during cooling (before a microscope slide preparation could be viewed). We refer to the process(es) that enables "glass-stage" TPs (acellular, chemical life) to phase-shift from their conformation(s) to those in water that yield protocells (biological life) the origin-of-life process(es).

The 18-AA mixtures "aging" at 20 C (1 and 5 yr) yielded TPs after "flash boiling." The 1-yr-old mixture (white in color; weakly Biuret positive) exhibited many free protocells and small numbers in aggregates. The 5-yr-old mixture (brownish in color; Biuret positive) exhibited great numbers of free and aggregated protocells. The 18-AA control was white in color and Biuret negative. TPs from 20-AA mixtures were the most strongly Biuret positive and were prolific protocell producers. Protocell preparations from the 1- and 5-yr-old AAs ["flash boiling" → protocells (viewed after moistening, 800-1000x); dried at 60 C, and viewed at 500-9000x SEM] revealed protocells: within a translucent extraprotocellular matrix (light microscopy): and, embedded within an electron opaque extraprotocellular matrix (projecting protocell surfaces visible with SEM). Protocells made from the 5-yr-old mixture were greater in number and larger than those from the 1-yr-old mixture.

One of us (AP) modified the NASA definition of life (10, 12) (**bold** word ours) from "Life is a self-sustained chemical system capable of undergoing Darwinian evolution." to "Life is a self-sustained **cellular** system that undergoes Darwinian evolution." for teaching The History of Biology: Origin of Life (15). This accommodated the need for a functional cellular boundary to classify protocells within the cell theory as proposed by Schleiden and Schwann 1838-1839 (16). Others also were redefining life (12) although without considering the recent discoveries (8, 20) of "lag time" in TP conversion to protocells.

Self-ordering (stereo-electronic) of AAs in TPs (chemical life) is the first "genetic" system and is followed by self-assembly of protocells (molecularly informed biological life) when TPs are moistened (5-6, 8, 14-20). Protocellular metabolism, including non-ribosomal oligo/poly peptide and oligo/poly nucleotide syntheses (along TP template regions), and other cellular capabilities were inferred (5). The definition of "life" is the sticky point!

Oparin took the leap of faith in science: "The origin of life is one of the stages of the Universe evolution." It is "the result of chemical evolution of carbon compounds (7, p. 94)." Gladilin and Suvorov continued: "The origin of life is situated between two transitions that can be observed:" – "chemical evolution" and "biological evolution (p. 103)." What AP did was to "interpolate (p. 103)" between states of cosmogenesis (Universe events) as Gladilin and Suvorov recommended instead of "extrapolating (p.103)" from the time of life known to us today back to microfossils (3.45 billion years old). The cosmogenic continuity pertains to all sites in the Universe where life could have emerged.

Thus, we propose a change in the NASA definition (Italics ours) of life: Life is a self-sustained chemical system (*acellular, chemical life involving conformationally changing TPs → protocellular, biological life*) capable of undergoing Darwinian evolution. To our knowledge, TPs are the only compounds capable of forming protocells. Therefore, TPs are acellular, chemical life. Changing conformation enables TPs to self-assemble into protocells. Whether or not these reactions are self-chaperoned remains to be determined (17, p.130). Could AAs in carbonaceous micrometeorites form TPs in long-term cold reactions or in short-term, heat-driven reactions [as they enter Earth's atmosphere at less than 200 C (3)]? Would melting ice within the micrometeorites induce protocell formation as they fall to Earth's surface? TPs appear to be the stuff that could account for cosmozoa and/or panspermia; "glass-stages" being litho-cosmozoa (formed in Pre-Solar time) or litho-panspermia (formed in Solar time). Carbonaceous micrometeoritic material and other meteoritic sources (Murchison) contain aspartic acid, glutamic acid, glycine, and alanine (2, 9, 13, 21). Pyroglutamic acid, which serves as an N → C polymerization initiator and theoretically aids greatly in limiting products of thermal polycondensation of amino acids, has been detected in a Murchison extract (4). The presence of these amino acids, their precursors, and possible thermal products, is indicated following 6 N HCl hydrolysis: i. e., TPs could have been present in the meteorites. "The presence of abundant carboxy lactams and lactams in carbonaceous chondrites suggests that the amino acids of these meteorites are poised for thermal polymerization (4)."

We conclude: life is a self-sustained chemical system (acellular, chemical life involving conformationally changing TPs → protocellular, biological life) capable of undergoing Darwinian evolution. We infer: These TPs may form in space within carbonaceous meteorites or micrometeorites (= cosmozoa or panspermia), or on planet surfaces and moons (such Titan and Europa) where meteor impacts might eject them into space to land elsewhere in our Solar System "The cyclic amides, given their potential for hydrogen-bonded pair formation, might be considered candidate bases for a primitive sequence coding (4)." Since TPs could be in the "glass state" in cold space, the terms "litho-cosmozoa" and "litho-panspermia are suggested for use to differentiate them from the earlier concepts of cosmozoa and panspermia.

Acknowledgements

Aristotel Pappelis wishes to dedicate his efforts with Sidney W. Fox to the memory of Constantine Zahariades and Gerald Andrews. All of the authors, each in his own way, have been impacted by their contacts with Sidney W. Fox. Our memories of him have inspired us throughout this work. May his memory be eternal.

References

1. Bahn, P. and S. W. Fox. 1996. Thermal proteins +. Chemtech 26: 26 - 29.
2. Brinton, F. B., C. Engrand, D. P. Glavin, J. L. Bada, and M. Maurette. 1998. A search for extraterrestrial amino acids in carbonaceous Antarctic micrometeorites. Origin Life Evol. Biosphere 28: 413-424.
3. Clemett, S. J., X. D. F. Chillier, S. Gillette, R. N. Zare, M. Maurette, C. Engrand, and G, Kurat. 1998. Observation of indigenous polycyclic aromatic hydrocarbons in "giant" carbonaceous Antarctic micrometeorites. Origins Life Evol. Biosphere 28: 425-448.
4. Cooper, G. W. and J. R. Cronin. 1995. Linear and cyclic aliphatic carboxamides of the Murchison meteorite: Hydrolyzable derivatives of amino acids and other carboxylic acids. Geochim. Cosmochim. Acta 59: 1003-1015.
5. Fox, S. W. 1988. *The Emergence of Life: Darwinian Evolution from the Inside*. Basic Books, New York.
6. Fox, S. W., A. Pappelis, and R. Grubbs. 1998. The origin of mind. In *Exobiology; Matter, Energy, and Information in the Origin and Evolution of Life in the Universe,* J. Chela-Flores and F. Raulin, eds., Kluwer Academic Publishers, Dordrecht, pp. 87-92.
7. Gladilin, K. L. and A. N. Suvorov. 1995. General evolutionary process and the origin of life. In *Evolutionary Biochemistry and Related Areas of Physicochemical Biology*, B. F. Poglazov, B. I. Kurganov, M. S. Kritsky, and K. L. Gladilin, eds., Bach Institute of Biochemistry and ANKO, Moscow, pp. 93-104.
8. Grubbs, R., S. W. Fox, A. Pappelis, J. Bozzola, and P.R. Bahn. 1998. Visually retracing the emergence of the evolvable protocell. *Exobiology: Matter, Energy, and Information in the Origin and Evolution of Life in the Universe,* J. Chela-Flores and F. Raulin, eds., Kluwer Academic Publishers, Dordrecht, pp.103-106.
9. Irvine, W. M. 1998. Extraterrestrial organic matter: A review. Origin Life Evol. Biosphere 28: 365-383.
10. Joyce, J.1994. Foreward. In *Origin of Life: The Central Concepts,* D. W. Deamer and G. R. Fleishaker, eds., Jones and Bartlett, Boston, p. xi.
11. Kolesnikov, M. P.1991. Proteinoid microspheres and the process of prebiological photophosphorylation. Origin Life Evol. Biosphere 21: 31-37.
12. Luisi, P. L. 1998. About various definitions of life. Origins Life Evol. Biosphere 28: 613-622.
13. Maurette, M. 1998. Carbonaceous micrometeorites and the origin of life. Origin Life Evol. Biosphere 28: 385-412.
14. Pappelis, A. 1994. Domain Protolife: Link between molecular and biological evolution. American Chemical Society Symposium: Polyamino Acids, the Emergence of Life, an Industrial Applications. p. 6 (Abstr.).
15. Pappelis, A. 1997. *Individualized Learning Program: Study Guide for History of Biology: Origin of Life.* Southern Illinois University at Carbondale, Carbondale, 200 pp.
16. Pappelis, A. and S. W. Fox. 1995. Domain Protolife: The protocell theory. In *Evolutionary Biochemistry and Related Areas of Physicochemical Biology,* B. F. Poglazov, B. I. Kurganov, M. S. Kritsky, and K. L. Gladilin, eds., Bach Institute of Biochemistry and ANKO, Moscow, pp. 151-159.
17. Pappelis, A. and S. W. Fox. 1995. Domain Protolife: Protocells and metaprotocells within thermal protein matrices. In *Chemical Evolution: Structure and Model of the First Cell,* C. Ponnamperuma and J. Chela-Flores, eds., Kluwer Academic Publishers, Dordrecht, pp. 129-132.
18. Pappelis, A. and S. W. Fox. 1996. Thermal Peptides as the initial genetic system. In *Chemical Evolution: Physics of the Origin and Evolution of Life*, J. Chela-Flores and F. Raulin, eds., Kluwer Academic Publishers, Dordrecht pp. 157-165.
19. Pappelis, A., S. W. Fox, and M. Pappagiannis. 1994. Protocells and metaprotocells of the Protolife Kingdom. Transactions Illinois State Academy of Science 86: 57 (Abstr.).
20. Pappelis, A., S. W. Fox, R. Grubbs, and J. Bozzola. 1998. Animate protocells from inanimate thermal proteins: Visualization of the process. In *Exobiology: Matter, Energy, and Information in the Origin and Evolution of Life in the Universe,* J. Chela-Flores and F. Raulin, eds., Kluwer Academic Publishers, Dordrecht, pp. 195-198.
21. Raulin-Cerceau, F., M.-C. Maurel, and J. Schneider. 1998. From panspermia to bioastronomy, the evolution of the hypothesis of universal life. Origins Life Evol. Biosphere 28: 597-612.

HPLC EVIDENCE OF NONRANDOMNESS IN THERMAL PROTEINS

PETER R. BAHN[1] AND ARISTOTEL PAPPELIS[2]

1. Bahn Biotechnology Co., RR2 Box 239A, Mount Vernon, Illinois 62864 USA
2. Department of Plant Biology, Southern Illinois University, Carbondale, Illinois 62901 USA

Abstract. We believe that the first proteins to be synthesized on the Primitive Earth were thermal proteins. Thermal proteins were found by Sidney W. Fox and his colleagues to be nonrandom self-ordered copolyamino acids. HPLC supports this conclusion.

1. Introduction

The general belief among researchers in the origin-of-life field that life originated in hot acidic environments on the Primitive Earth either on volcanic surfaces [1] or at hydrothermal vents [2] supports the contention that the primary dehydration condensing agent which first converted protobiomonomers into protobiopolymers was heat [3,4]. Thermal conversions of saccharides into polysaccharides [5], simple lipids into complex lipids [6], mononucleotides into oligonucleotides [7], and amino acids into thermal proteins [8] have all been accomplished experimentally by the agency of heat.

2. Thermal Proteins and Protein Sequence Space

The thermal copolymerization of amino acids into thermal peptides and thermal proteins has been extensively studied by Sidney W. Fox and his colleagues [9]. This reaction can be symbolized as follows

$$H_3N^+CHR_iCO_2^- \xrightarrow{\Delta} (\text{-}HNCHR_iCO\text{-}HNCHR_iCO\text{-})_x + H_2O$$

where R_i are the various side groups in the 20 naturally occuring amino acids. Both in the solid state and in aqueous solution, amino acids are found in their Zwitterionic form as shown on the left. The reaction is endergonic, having a positive Gibbs Free Energy, as evidenced by the need to add heat to get the reaction to go. An excess of one or both of the dicarboxylic amino acids glutamic acid or aspartic acid, or the diamino acid lysine, is usually required to form a hot but protective liquid melt solvent for copolymerizing the other 17 common amino acids [10]. Thermal proteins, or proteinoids, as Fox often charactrized them, contain considerable numbers of cross-linkages and beta-linkages in their structure [11].

J. Chela-Flores et al. (eds.), First Steps in the Origin of Life in the Universe, 69–72.

However, the first proteins may well have had numerous cross-linkages and beta-linkages, having gotten more linear and regular as they evolved into more modern nucleic acid coded proteins as time went on [12].

Since there are 20 commonly occurring amino acids, in a linear chain poly peptide, the number of possible peptides grows exponentially as peptides of increasing length are considered. That is, the number of possible polypeptides is equal to m^n where m is the number of monomer types and n is the length, in residues, of the polymer chain. Thus, for peptide or protein with a length of 100 amino acid residues, there are 20^{100} different possible protein sequences. This number is inconceivably large. For example, it has been estimated that there are roughly 10^{80} protons and neutrons in the known Universe [13]. Clearly, it can be said that for proteins of moderate length, during the entire course of chemical and biological evolution on the planet Earth, only an insignificant portion of protein sequence space has been or could have been sampled. Modern biological protein sequences could not have been evolved by random sampling of all possible protein sequences. Chemists have been able to generate random batches of polypeptides by utilizing a reaction in which the N-carboxyanhydrides (Leuchs Anhydrides) of the various amino acids are polymerized in an exergonic reaction. The chemistry of such random proteins has been studied quite elegantly by Sela and Katchalsky [14,15]. However, Leuchs polymers have little relation to plausible prebiotic chemistry.

3. HPLC Evidence of Nonrandomness in Thermal Proteins

Although there is already a considerable body of published evidence that thermally polymerized proteins have nonrandom amino acid sequences and nonrandom structures [16-20], some of the best evidence was obtained relatively recently by the late Sidney Fox in 1988 during a sabbatical at the laboratory of the late Prof. Gerhard Braunitzer at the Max Planck Institute for Biochemistry in Martinsried, Germany, and assisted by Dr. Peter Rucknagel [21,22]. The evidence referred to is shown in Figure 1. It is a set of high pressure liquid chromatography (HPLC) separations of two water soluble fractions of thermal peptides prepared by Prof. Fox. In the two chromatograms, time is measured along the abscissa and absorbance is measured along the ordinate. I. The top chromatogram is a separation of thermal polymer of glutamic acid, glycine, and tyrosine, which is symbolized as p(Glu,Gly,Tyr). This polymer was prepared by heating a mixture of 0.74 g Glu, 0.35 g Gly, and 0.90 g Tyr at 180 degrees Centigrade for 6 hours.. II. The bottom chromatogram is a separation of thermal polymer of glutamic acid, alanine, and tyrosine, which is symbolized as p(Glu,Ala,Tyr). This polymer was prepared by heating a mixture of 0.74 g Glu, 0.445 g Ala, and 0.90 g Tyr at 180 degrees Centigrade for 6 hours. The two batches of thermal polymers were made and run through the HPLC under the same conditions. It appears from these HPLCs that thermal peptides can indeed be separated into distinct, homogeneous, and related molecules with different molecular weights. Different families of distinct molecules are generated with different starting mixtures of amino acids. This should not really be that surprising. Amino acids with different side chains can hardly be expected to react identically even under temperatures over 100 degrees Centigrade. The reason that Leuchs Anhydrides do polymerize randomly, in contrast to thermal polymers, is that the energy differences between different Leuchs Anhydride side chain

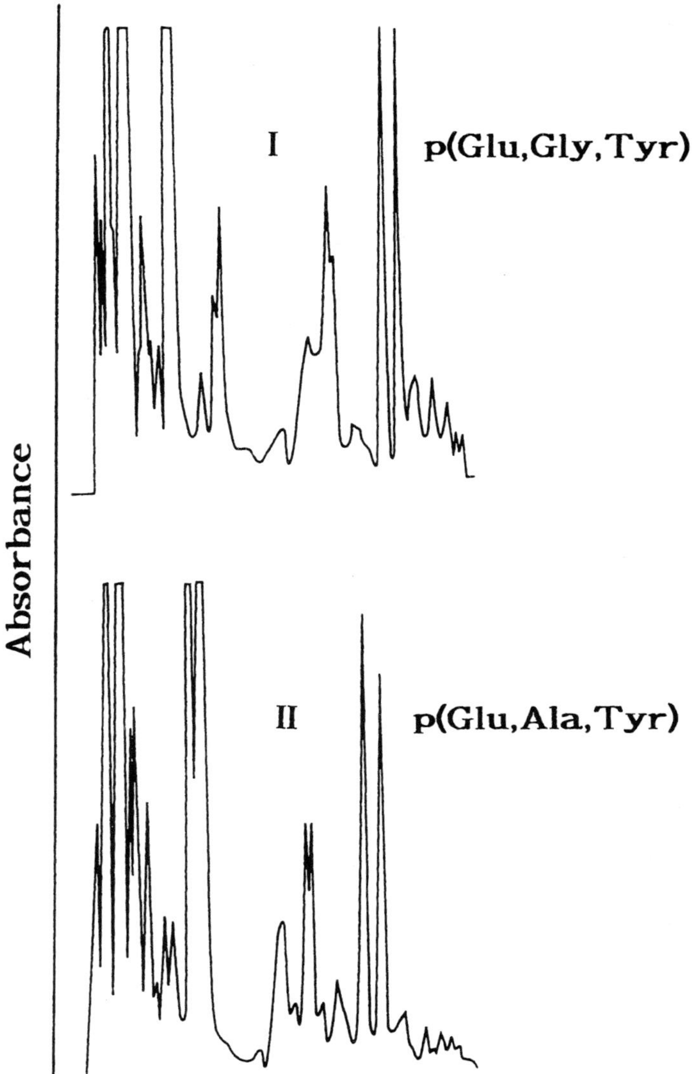

Figure 1. HPLCs of I. p(Glu,Gly,Tyr) and II. p(Glu,Ala,Tyr).

interactions is completely swamped out by the highly energetic identical N-carboxyanhydride group interactions.

4. References

1. Fox, S.W.: Thermal Polymerization of Amino Acids and Production of Formed Microparticles on Lava, *Nature* **201** (1964), 336-337.
2. Walter, M.R. (ed): *Evolution of Hydrothermal Ecosystems on Earth (and Mars?)*, John Wiley & Sons, New York, 1996.
3. Fox, S.W.: Thermal Synthesis of Amino Acids and the Origin of Life, *Geochim. Cosmochim. Acta* **59** (1995), 1213-1214.
4. Imai, E., Honda, H., Hatori, H., Brack, A., and Matsuno, K.: Elongation of Oligopeptides in a Simulated Submarine Hydrothermal System, *Science* **283** (1999), 831-833.
5. Schwartz, A.W.: *The Thermal Polymerization of Ribose*, M.S. Thesis, Florida State University, 1962.
6. Hargreaves, W.R., Mulvihill, S.J., and Deamer, D.W.: Synthesis of Phospholipids and Membranes in Prebiotic Conditions, *Nature* **266** (1977), 78-80.
7. Schwartz, A.W. and Fox, S.W.: Thermal Synthesis of Internucleotide Phosphodiester Linkages, *Biochim. Biophys. Acta* **87** (1964), 696-698.
8. Fox, S.W. and Harada, K.: Thermal Copolymerization of Amino Acids to a Product Resembling Protein, *Science* **128** (1958), 1214.
9. Fox, S.W. and Dose, K.: *Molecular Evolution and the Origin of Life*, Marcel Dekker Inc., New York, 1977.
10. Harada, K. and Fox, S.W.: Characterization of Thermal Polymers of Natural Alpha Amino Acids with Dicarboxylic Amino Acids or Lysine, *Arch. Biochem. Biophys.* **109** (1965), 49-56.
11. Fox, S.W. and Suzuki, F.: Linkages in Thermal Copolymers of Lysine, *BioSystems* **7** (1976), 213-221.
12. Bahn, P.R.: Remnants of the Proteinoid World, *ISSOL Newsletter* **23(3/4)** (1996), 33.
13. Smolin, L.: *The Life of the Cosmos*, 45, Oxford University Press, New York, 1997.
14. Katchalski, E. and Sela, M.: Synthesis and Chemical Properties of Poly-Alpha-Amino Acids, *Adv. Protein Chem.* **13** (1958), 243-492.
15. Sela, M. and Katchalski, E.: Biological Properties of Poly-Alpha-Amino Acids, *Adv. Protein Chem.* **14** (1959), 243-492.
16. Fox, S.W., Harada, K., Woods, K.R., and Windsor, C.R.: Amino Acid Compositions of Proteinoids, *Arch. Biochem. Biophys.* **102** (1963), 439-445.
17. Fox, S.W. and Nakashima, T.: Fractionation and Characterization of an Amidated Thermal 1:1:1 - Proteinoid, *Biochim. Biophys. Acta* **140** (1967), 155-167.
18. Nakashiima, T., Jungck, J.R., Fox, S.W., Lederer, E., and Das, B.C.: A Test for Randomnes in Polypeptides Isolated from a Thermal Polyamino Acid, *Internat. J. Quantum Chem.* **4** (1977), 65-72.
19. Hartmann, J., Brand, M.C., and Dose, K.: Formation of Specific Amino Acid Sequences During Thermal Polymerization of Amino Acids, *BioSystems* **13** (1981), 141-147.
20. Melius, P. and Hubbard, W.: Pyroglutamyl N-Termini of Thermal Polyamino Acids, *BioSystems* **20** (1987), 213-217.
21. Fox, S.W.: The Changing Face of Natural Selection, in Bunyard, P. and Goldsmith, E. (eds): *Gaia and Evolution*, Walderbridge Ecological Centre, Cornwall, England, 1989.
22. Fox, S.W.: My Scientific Discussion of Evolution for the Pope and his Scientists, in *The Harbinger* **XV**, No. 15, Mobile, Alabama (5/27/1997-6/9/1997).

IR SPECTRA OF PROTEIN, THERMAL PROTEIN, THERMAL GLYCOPROTEIN

PETER R. BAHN[1] AND ARISTOTEL PAPPELIS[2]

*1. Bahn Biotechnology Co., RR2 Box 239A,
Mount Vernon, Illinois 62864 USA
2. Department of Plant Biology,
Southern Illinois University, Carbondale,
Illinois 62901 USA*

Abstract. Thermal proteins were made by heating mixtures amino acids at 180 degrees Centigrade for 6 hours, thereby effecting dehydration polymerization. When sugars are added to the amino acid mixtures, thermal glycoproteins are sythesized. IR spectra show that thermal protein has similarities to protein synthesized by the Merrifield technique and that glycan components show up in different parts of the IR spectrum than the protein components in thermal glycoproteins.

1. Introduction

The belief that life may have originated in a hot environment supports the idea that heat may have been a ubiquitous dehydration condensation agent on the Primitive Earth for converting prebiotic sugars into prebiotic polysaccharides and for converting prebiotic amino acids into prebiotic thermal proteins [1,2,3]. Since these two processes could have been acting in concert, it appears that heat could also have produced thermal glycoproteins on the Primitive Earth. Consequently, it is of interest to compare structural similarities of proteins, thermal proteins, and thermal glycoproteins.

2. Proteins, Thermal Proteins, and Thermal Glycoproteins

Proteins are essentially linear polymers of alpha amino acids linked by peptide bonds between the carbonyl carbon of one amino acid and the amino nitrogen of the next succeeding amino acid [4]. Some proteins are cross-linked by disulfide bonds.

Thermal proteins are polyamino acid polymers usually prepared by heating mixtures of amino acids over 100 degrees Centigrade [5]. They have been studied extensively by Sidney W. Fox and his colleagues [6]. An excess of one or both of the dicarboxylic amino acids glutamic acid or aspartic acid, or the diamino acid lysine, is usually required to form a hot but protective liquid melt solvent for copolymerizing the other 17 common amino acids [7]. Thermal proteins, or proteinoids, as Fox often characterized them, contain considerable numbers of cross-linkages and beta-linkages in their structure [8,9].

J. Chela-Flores et al. (eds.), First Steps in the Origin of Life in the Universe, 73–76.

Thermal glycans are polysaccharides prepared by heating sugars over 100 degrees Centigrade. In terms of cooking, the process is generally called caramelization. Ribose is known to undergo thermal polymerization [10]. Glucose also in known to undergo thermal polymerization when heated in the presence of glutamic acid [11]. The glutamic acid cyclizes to pyroglutamic acid which forms a hot protective met for polymerizing the glucose into a polyglucose. Thermal glycans are highly branched polymers.

Thermal glycoproteins are combination polyamino acid polysaccharide polymers prepared by heating sugars and mixtures of amino acids over 100 degrees Centigrade [12].

3. IR Spectra of Protein, Thermal Protein, and Thermal Glycoprotein

IR spectroscopy is a process whereby infrared radiation of differing wavelengths is passed through a chemical substance and absorbances of the incident radiation at various wavelengths are recorded [13,14]. IR absorbances correspond to various symmetric and antisymmetric stretching vibrations, and to various scissoring, rocking, twisting, and wagging bending vibrations in the chemical bonds of the compounds under study. IR spectroscopy of protein structure is not a commonly used technique and this field is in a period of relative infancy [15].

Infrared spectra of proteins usually show absorbance in an amide I region (1600–1700 cm^{-1}) corresponding to the C=O stretching, an amide II region (1500–1600 cm^{-1}) corresponding to the C–N stretching, and an amide III region (1200–1350 cm^{-1}) corresponding to the N–H bending [15]. Infrared spectra of a model protein, a thermal protein, and thermal glycoprotein, and glucose are shown in Figure 1. The IR spectra were taken as KBr pellets of the subject compounds. A. IR spectrum of Alphabetin, a model protein with a molecular weight of 2,396 Daltons. Aphabetin was made by the technique of Merrifield synthesis and it has a sequence from N to C termini of the 20 common amino acids linked in alphabetical order according to the full spelled-out English names of the 20 amino acids. Thus, Aphabetin has the following sequence: N-Ala-Arg-Asn-Asp-Cys-Glu-Gln-Gly-Gly-His-Ile-Leu-Lys-Met-Phe-Pro-Ser-Thr-Trp-Tyr-Val-C. Absorbances in the amide I, II, and III regions are shown in the three main peak areas on the right half of the Aphabetin spectrum. B. IR spectrum of a 2:2:1 Asp:Glu:BNaa (Basic Neutral amino acid equimolar mixture) thermal protein as described previously [12]. C. IR spectrum of a 1:2:2:1 Glucose:Asp:Glu:BNaa thermal glycoprotein as described previously [12]. D. IR spectrum of D(+)Glucose. It is seen that the IR spectrum of thermal protein is quite similar to the IR spectrum of Alphabetin. There are some differences, which is to be expected since as we have already explained, thermal proteins contain features such as beta-linkages and assorted side-chain linkages that are not normally found in modern proteins, for which Alphabetin serves as a useful model. However, the similarity between thermal protein and Alphabetin points out the general overall similarity of modern proteins and thermal proteins. It is also seen that the IR spectrum of the thermal glycoprotein has some similarity to the IR spectrum of thermal protein but that the IR spectrum of the glycoprotein also has some outstanding differences from the IR spectrum of the thermal protein. This is because the IR absorbance bands of the thermal glycan component of the thermal glycoprotein shows up largely at a different frequency range than that of the thermal

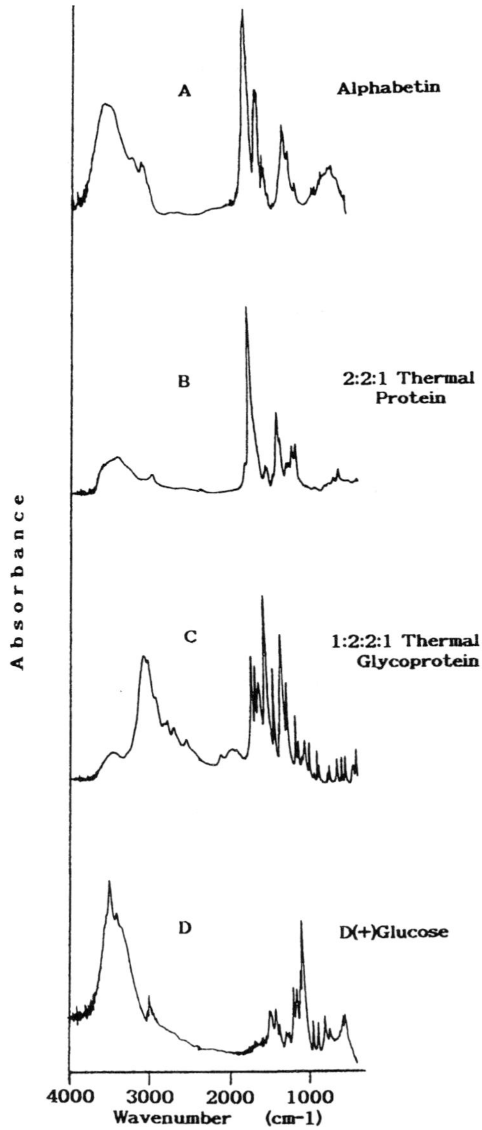

Figure 1. IR Spectra of A. Alphabetin, B. 2:2:1 Thermal Protein, C. 1:2:2:1 Thermal Glycoprotein, and D. D(+)Glucose.

protein component of the thermal glycoprotein. This observation is confirmed by noting that there is not much overlap between the band frequencies of pure glucose with either that of Alphabetin or with the thermal protein. Thus, we note here that IR spectroscopy should provide a very handy way of distinguishing between proteins and glycoproteins, both primitive and modern. In fact, it looks as though the IR spectrum of the thermal glycoprotein is a modified 1:1 combination of the IR spectra of thermal protein and glucose. This establishes that 1:2:2:1 thermal glycoprotein is roughly 50 % glycan and 50 % protein in its two components. Therefore, in addition to IR spectroscopy being able to distinguish between proteins and glycoproteins, IR spectroscopy should also be able to give a rough quantitative estimate of the glycan-to-protein ratio in glycoproteins, both primitive and modern.

4. Acknowledgement

We wish to thank Professor Paul S. Braterman, Department of Chemistry, University of North Texas, Denton, for recording three of the IR spectra in this paper and for helpful commentary regarding the manuscript.

5. References

1. Walter, M.R., (ed): *Evolution of Hydrothermal Ecosystems on Earth (and Mars?),* John Wiley & Sons, New York, 1996.
2. Imai, E., Honda, H., Hatori, H., Brack, A., and Matsuno, K.: Elongation of Oligopeptides in a Simulated Submarine Hydrothermal System, *Science* **283** (1999), 831-833.
3. Fox, S.W.: Thermal Synthesis of Amino Acids and the Origin of Life, *Geochim. Cosmochim. Acta* **59** (1995), 1213-1214.
4. Schulz, G.E. and Schirmer, R.H.: *Principles of Protein Structure,* Springer-Verlag, New York, 1979.
5. Fox, S.W. and Harada, K.: Thermal Copolymerization of Amino Acids to a Product Resembly Protein, *Science* **128** (1958), 1214.
6. Fox, S.W. and Dose, K.: *Molecular Evolution and the Origin of Life,* Marcel Dekker Inc., New York, 1977.
7. Harada, K. and Fox, S.W.: Characterization of Thermal polymers of Natural Alpha Amino Acids with Dicarboxylic Amino Acids or Lysine, *Arch. Biochem. Biophys.* **109** (1965), 49-56.
8. Fox, S.W. and Suzuki, F.: Linkages in Thermal Copolymers of Lysine, *BioSystems* **7** (1976), 213-221.
9. Bahn, P.R.: Remnants of the Proteinoid World, *ISSOL Newsletter* **23(3/4)** (1996), 33.
10. ScHwartz, A.W.: *The Thermal Polymerization of Ribose,* M.S. Thesis, Forida State University, 1962.
11. Fox, S.W. and Bahn, P.R.: Process for Preparing Thermal Carbohydrates, *U.S. Patent Number 4,975,534* (1990).
12. Bahn, P.R. and Fox, S.W.: Thermal Proteins +, *CHEMTECH* **26** (1996), 26-29.
13. Harwood, L.M. and Claridge, T.D.W.: *Introduction to Organic Spectroscopy,* Oxford University Press, New York, 1997.
14. Sorrell, T.N.: *Intrepreting Spectra of Organic Molecules,* University Science Books, Mill Valley, California, 1988.
15. Singh, B.R.: *Infrared Analysis of Peptides and Proteins,* ACS Symposium Series No. 750, American Chemical Society, Washington, D.C., 2000.

PREBIOTIC SELECTION INDUCED BY PERIODIC STARLIGHT

GIORGIO CARERI
Dipartimento di Fisica, Università di Roma "La Sapienza" and INFM
Piazzale A. Moro 2, 00185 Roma, Italy.

Abstract. We consider the transition of the prebiotic polymeric soup into a chemically differentiated and space structured material, this being a crucial step in the origin of animate matter when suitable elements become abundant on the earth. We suggest that Stochastic Resonance (SR) must have been a mechanism by which the periodic warming of the earth's surface may have assisted this transition, by promoting and thus selecting those pathways with reaction time close to earth rotation period. Several points of general relevance to chemical evolution are briefly discussed. This proposal can be tested in the laboratory.

1. Introduction

We consider the evolution of a homogenous prebiotic soup towards a chemically differentiated and space-structured material, this being an early and crucial step of the process ending eventually with a living cell on the earth. From the viewpoint of physics, this step must be considered an ordering transition, since it breaks the symmetry of an homogenous system, while from the viewpoint of biology the increase of information associated to entropy decrease is well in line with the the viewpoint of Szathaamàry and Maynard-Smith (1995), that major evolutionary transitions involved changes in the way information is stored or trasmitted, because this transition, leading from replicating molecules to populations of molecules, must have been relevant to prebiotic evolution. The system in question is often modelled by a dense soup of organic polymers where several auto- or etero-catalytic processes take place and where diffusion processes can be disregarded, as we shall do below in this preliminary paper.

Here we shall suggest that periodic warming of the earth's crust by sunlight may have assisted this step thanks to stochastic resonance (SR), a phenomenon displayed by noisy systems when driven by periodic forcing. As shown below, the main condition required to this end is a time-scale matching between earth revolution period and turnover time of some early catalytic events, to enhance and therefore to select only a few chemical pathways. Even if the SR model proposed below can be applied to polymers of any chemical nature, we shall assume polypeptide chains to have been the main contribution to the prebiotic soup, in order to evaluate the magnitude of some quantities. Actually this assumption, leading to "protein first" in the prebiotic scenario, is widely shared by several authors, as reviewed for instance in Kauffman (1993). And we believe that exactly this lack of early informational polymers required an external factor to allow an ordered growth of the early stuff, a factor we identify in the periodic sunlight.

J. Chela-Flores et al. (eds.), First Steps in the Origin of Life in the Universe, 77–80.

2. Modelling

SR is a phenomenon widely observed in several systems at room temperatures, including biological and geophysical systems as reviewed in Wiesenfeld and Moss (1995), and well studied by statistical physics in models where an activation energy barrier must be crossed under a weak periodic input, as reviewed in Gammaitoni et al (1998). The only requirement is the synchronization of the intrinsic noise of the system with the external periodic forcing, namely a time-scale matching condition between the inverse of the barrier crossing rate and the period of forcing. Moreover, by computer simulation it has been shown that in space extended systems with many degrees of freedom, like assemblies of coupled bistable systems, spontaneous ordering transitions assisted by SR can take place with specific space-patterns.

In an oversimplified model of prebiotic reaction network, one is allowed to assume that each catalytic event took place independently from all others, with a turnover time close to the inverse of its own barrier crossing rate. Therefore we disregard the difference in binding and unbinding process exhibited by different chemical pathways. In this way we can simplify the complexity of the different reactions, by focusing only on the barrier crossing rates, which are the quantities of concern in SR studies. And it must be added that in these studies a bistable energy profile is often assumed for the potential energy, a quite suitable model for the activation energy profile versus the reaction coordinate leading from substrate to product in chemical kinetics, as the one considered here.

In a prebiotic context, the above mentioned time-scale condition requires that the turnover time of catalytic events occurring in the soup be a factor 10^{-7} lower than the turnover time observed today in enzymes (typically about 1ms in acqueous solution). This factor is conceivable, in view of the lower efficiency of prebiotic catalysts and of the lower reaction rates observed in nearly anhydrous conditions in present enzymes (about 10^{-3} or lower , as reviewed in Rupley and Careri (1991). We note in passing that nearly anhydrous conditions were certainly needed in peptide synthesis, since this reaction is thermodynamically favoured by dehydration.

If the above briefly outlined model is accepted, one must conclude that in the wide turnover time spectrum exhibited by the polymer soup, only turnover events close to circadian time must have been enhanced by SR, and thus selected from the remaining messy stuff. This disordered polymeric stuff become eventually food for the few sun-assisted pathways, in a prebiotic Darwinian-type process. The symmetry breaking character of this SR assisted transition emerges in the turnover time spectrum, nearly white before and nearly monocromatic after this transition.

In order to observe this proposed selection, one further condition to be satisfied is that the periodic temperature change of the earth crust must have been sufficient for SR-assisted crossing in the polymerization reaction. One can see that this condition is satisfied if one assumes a temperature change of only 50K (involving an energy of 100 cal mol^{-1}), and an energy barrier of 20 Kcal mol^{-1} (about 10 times the free energy for one pepetide bond). These quite safe and well conservative assumptions yield a value 5.10^{-3} for the ratio of the two energies involved. This value is close than that evaluated in the problem of periodically recurrent ice ages, induced by the different solar energy influx in the 10^5yr modulation period of earth orbital eccentricity, as considered by Benzi *et al.* (1981) in their early paper where the concept of SR was originally put forward using a bistable energy profile.

The above oversimplified model can be improved in several ways. For instance, instead of an assembly of independent particles oscillating at the bottom of their potential well, one can consider a loop of coupled oscillators where the product of one catalytic event enters as substrate for a nearby event, as widely observed today in biochemical cycles. Or else, one can include other geothermal phenomena occurring on the earth's surface, by considering aperiodic driving forces superimposed to circadian sun's radiation periodicity. Even the Gaussian noise of the single oscillator can be replaced by colored or 1/f noise, as displayed by real systems. All these improvements have been already considered in the literature of SR studies, as mentioned by Gammaitoni *et al.* (1998). Moreover, the conventional SR considered above can be characterized by information theory tools, for instace following the recent work by Goychuk and Hanggi (2000). These and other original improvements will be considered elsewere by this author.

3. Discussion

On the grounds of the model outlined above we suggest a simplified prebiotic scenario of the earth surface were separate regions of poorly hydrated polypeptide stuff were progressively evolving from a wide distribution of polymer sizes and reaction times towards an essentially single chemical pathway centered on the circadian time, and growing by assimilation of other non assisted pathways. This picture is attractive for prebiotic chemical evolution from several viewpoints, but here we shall only consider the following two points.

i) The symmetry breaking induced by periodic sun light thanks to SR is a fact merely grounded on statistical physics, and as such quite independent from the chemical nature of the polymeric system involved. Therefore this point should be accepted by current theories of the origin of life, as a further specification of environmental conditions. For instance in a recent theory proposed by Lifson (1998), life originated from an unspecified autocatalyst which replicated itself and was forced to evolve by natural selection determined by the environment, and chemical selection was the first step in the transition from inanimate to animate matter. The model proposed here is perfectly consistent with this theory, and with "protein first" theories reviewed by Kauffman (1993) and by Lifson (1998) as well.

ii) A central role in the model proposed here is played by the selected polymerization induced by thermal cycling. This effect comes from SR theory, but it can be observed in the laboratory if a suitable matching between the turnover time of a suitable catalyst and thermal cycling is achieved. Besides the intrinsic interest on this experiment from the viewpoint of polymer kinetics, one can reverse this argument and use this experiment to distinguish between random drift of polymerization and emergence of coordinated regularities in some chemical complexities, capable to persist and to grow by metabolic self regulation. For this search , the best candidates seems to be a mixture of glicine, alanine, valine and aspartic acid, since these amino acids are supposed to have been the more abundantly formed in prebiotic synyhesis. And it is worth mentioning that, according to Ito et al. (1986) this interesting mixture after hydration-dehydration cycles by microwave heating up to 80°C, displayed resulting polymers with well defined secondary structures and catalytic activities.

4. Concluding remarks

In this short report we have not considered the obvious extention of SR modelling to already proposed photochemical processes induced by ultraviolet component of prebiotic solar energy on suitable chromophores. Nor we have discussed that the prebiotic SR selection considered here can be the ancestor of the still unexplained circadian clock which functions to synchronize our organism with its environment. Before to say that this circadian rhythm is a kind of kinetic relict of the prebiotic turnover, one must be sure that other recently proposed mechanisms grounded on genetic code or on the limiting cycle are not succesful.

The above outlined model is quite different from other models proposed for the origin of biological order, grounded on statistical physics as well and mentioned by Kauffman (1993), where order spontaneously emerges at a threshold in an interacting self-replicating homogenous medium. Since SR effects by periodic warming of the earth crust cannot be ignored, there is a need for an improved new theory where the previous theoretical attempts should be extended to include the occurrence of SR and the likely presence of etherogenous catalytic events as well.

Finally, the generality of SR phenomenon suggests that the SR model shoud not be limited to the earth surface only. Since in this Conference several people are exploring the possibility that life exists elsewere in the universe, it seems advisable to suggest that priority is given to investigation of regions periodically warmed by starlight. Otherwise even if a polymeric soup will be detected, that soup may not be able to evolve toward a differentiated material and later to a living cell. Again, the so called " biological order ", namely that dual order of structures and processes that reciprocally condition each other as discussed in Careri (1984), and which is displayed by all forms of life on the earth, cannot be ignored.

5. References

Benzi, R., Parisi,G., Sutera,A., and Vulpiani, A., (1981) Stochastic resonance in climatic change,*Thellus* **13** 10-16.

Careri,G., (19847 *Order and disorder in matter*, Benjamin/Cumming Publ. Menlo Park Cal.

Gammaitoni,L., Hanggi,P., Jung, P., and Marchesoni,F., (1998) Stochastic Resonance, *Rev.Mod.Phys.* **70** 224-287.

Goychuk,I., and Hanggi,P., (2000) Stochastic resonance in ion channels characterized by information theory, *Phys .Rev. E* **61** 4272-4280.

Kauffman,S.A., (1993) *The origins of order,* Oxford Univ.Press, Oxford.

Lifson,S., (1997) On the crucial steps in the origin of animate matter, *J. Mol . Evol.* **44** 1-8.

Ito,M., Handa,N., and Yanagama,H., (1986) Architecture of models for prebiotic synthesis of proteins: the structure and function of polypeptides synthetized in a fluctuating system. *Origin of life* **16** 494-495

Rupley, J.A., and Careri, G., (1991) Protein hydration and function, *Ad. Prot. Chem.* **41** 37-172.

Szathmàry,E., and Maynard-Smith, J., (1995) The evolutionary transitions , *Nature* **374** 227_232

Wiesendeld,K., and Moss,F., Stochastic resonance and the benefit of noise: from ice ages to crayfissh and SQUIDS , *Nature* **373** 33-36

A model system for studying the role of phosphoryl group in energy transfer system

BO QU, WENQING WANG AND JILAN WU

Department of Technical Physics, Peking University, Beijing 100871, China

Abstract: The discovery of alkyl phosphonic acids in the Murchison meteorite suggests the possibility of delivery of these water soluble, phosphorus containing molecules by meteorites to the early Earth. Phosphonic acids may have been precursors to the first nucleic acids [1]. We propose the model of energy transfer system as the following steps: (1) alkyl hydrocarbon absorbed far UV energy. (2) R-P=O group accepted energy from excited alkanes. (3) The P-C bond in phosphonate may be broken to form phosphate and corresponding acid in the presence of oxygen.

1. Introduction

Light energy is one of the significant energy resources in the origin of life. The identification of a suite of alkyl phosphonic acids in Murchison meteorite [2] has created a new perspective of the supplies of organic phosphorus for the early stages of chemical evolution. Because the C-P bond is considerably more stable to hydrolysis than the C-O-P ester bond [3], phosphonic acids would persist in the prebiotic environment even under extreme conditions. Alkyl phosphonic acids can be regarded as being related to orthophosphorus acid, via substitution of the P-H grouping by a P-C bond. We study the dependence of energy transfer rate on the molecular structure of the organophosphoryl compounds in cyclohexane solution by pulse radiolysis method. We argue that the R-P=O group in phosphonic acid may play an important role in an instant accepting energy from excited alkanes which absorbed far UV energy efficiently.

2. Results and Discussion

Cyclohexane (CH) is considered as a model hydrocarbon existed in primordial atmosphere. Benzophenone (BP) is used as a probe. The absorption peak with λ_{max}= 560 nm is assigned to the ketyl radicals (BH) of benzophonone, which

J. Chela-Flores et al. (eds.), First Steps in the Origin of Life in the Universe, 81–85.

needs more than 10 μs after pulse irradiation to reach its maximum value. $R_3P{=}O$ is used instead of R $(OH)_2P{=}O$ since it is easy to make sure which bond is broken to form corresponding acid in the final product analysis. Tributyl phosphine oxide (TBPO), trioctyl phosphine oxide (TOPO), tributyl phosphate (TBP) and trioctyl phosphate (TOP) with BP-CH system were selected to study the energy transfer mechanism.

2.1 ENERGY TRANSFER MECHANISM

There was existed two-fold scavenge effect—TBP, TBPO, TOP and TOPO accepting excitation energy from CH* and scavenging proton from $[C_6H_{12}^+ \ldots e^-]$ geminate pair. The process of energy transfer existed as follows:

$$CH^* \rightarrow \text{products} \tag{1}$$

$$CH^* + S \rightarrow S^* + CH \tag{2}$$

$$S^* + BP \rightarrow S + BP^* \rightarrow \rightarrow BH \tag{3}$$

Here S and S* represent TBPO or TOP and its excited state. The kinetic equation can be expressed as Stern-Volmer form:

$$\frac{1}{G_{BH}} = \frac{1}{G_{s^*}} = \frac{1}{G_0}\left(1+\frac{k_1}{k_2}\cdot\frac{1}{[S]}\right) \tag{4}$$

where G_0 and G_s* are G values of CH* and S*. It agrees well with the linear parts.

Taking k_1=1.0_$10^9 M^{-1}s^{-1}$[4], the values of k_2 are obtained in the order: TOPO > TBPO > TOP > TBP, as list in Table 1.

TABLE 1. The rate constants of energy transfer between CH^* and the phosphoryl compounds

solute	TBP#	TBPO	TOP	TOPO#
K_2 ($M^{-1}.s^{-1}$)	4.8×10^9	5.2×10^{12}	8.8×10^{11}	5.8×10^{12}

These terms are recalculated, after dose correction, from Yikui Du[4].

From Table 1, it can be seen that the rate constant of excitation energy transfer is very large and the link between C-P makes the excitation energy transfer much easier than C-O-P linkage.

It was found that when the concentration of solutes increased, the molecules of solutes diffuse into spur and react with geminate pair, which is different to react outside of the spur. Therefore Stern-Volmer equation is no longer reasonable. A proton transfer process, which destroys geminate pairs in spur was proposed.

$$[C_6H_{12}^{+\cdot} \ldots e^-] + (C_4H_9)_3PO \longrightarrow C_6H_{11}^{\cdot} + C_4H_{10}^{+}\text{-}\overset{\overset{\large O}{\|}}{P}(C_4H_9)_2 + e^- \qquad (5)$$

$$\frac{1}{\Delta G - G_{fi}} = \frac{1}{G_{gi}}\{1 + (\alpha[S])^{-1/2}\} \qquad (6)$$

The process fits in with WAS equation:

Here ΔG is the yield of ketyl radical formed after capturing proton by TBPO or TOP at definite concentration. G_{gi} is the initial yield of geminate pairs for cyclohexane and G_{fi} is the yield of free ion pairs, taking G_{fi}=0.1 [4]. $\lambda = k/\alpha = 2\times10^{11} s^{-1}$ keeps a constant for CH [5].

Plotting $(\Delta G - G_{fi})^{-1}$ vs. $[S]^{-1/2}$, we get a good linear relationship. The values of the scavenging rate constant k and α were obtained, as list in Table 2.

TABLE 2. The values of α and scavenging rate constant k

solute	TBP#	TBPO	TOP	TOPO#
α (M^{-1})	6.3	60	27	199
κ ($M^{-1}.s^{-1}$)	1.3×10^{12}	1.2×10^{13}	5.4×10^{12}	4.0×10^{13}

These terms are recalculated from previous work[4].

The ratios of scavenging rate constants were:

$$k_{TOPO}/k_{TOP} = k_{TBPO} / k_{TBP} = 7\sim9$$

$$k_{TOPO} / k_{TBPO} = k_{TOP} / k_{TBP} = 3.4\sim4.2$$

It can be seen that the energy transfer would take place with very large rate constants, which means that phosphory molecules might accept ionization energy via the environmental excited alkanes easily.

Interestingly, when phosphoryl g roup reacts with geminate pair, tetravalent radical might be formed. It may be an active center for radical combination to form new compounds:

$$\geqslant P{=}O + [C_6H_{12}^{+}\ e^{-}] \longrightarrow (\geqslant \dot{P}{-}OH)^{*} + C_6H_{11}^{\cdot} \qquad (7)$$

2.2 A POSSIBLE PATHWAY FOR PHOSPHINE OXIDE EVOLVING TO CORRESPONDING PHOSPHATE

The γ irradiation experiments in TOPO-cyclohexane solution saturated with air were carried out and two kinds of radiolytic products were separated. One was analyzed to be phosphonic acid, and the other was an oligomer of TOPO, phosphonic acid and its ester. It might be just the way via which phosphonic compounds evolving to phosphoric acid-based nucleotide analogues in life.

3. Conclusion

Phosphoryl compound can accept ionization energy easily from excited alkanes via energy transfer and proton transfer process. The dependence of energy transfer rate on molecular structure shows that the P-C bond (the direct linking P and C atoms), high polarity of phosphoryl group and long alkyl chain make the excitation transfer and proton transfer easier. Our model study demonstrate that alkyl hydrocarbon absorbed far UV energy via a number of pathways of energy transfer and create a new perspective of an energy transformer for phosphonic acid –based nucleotide evolving to phosphoric acid-based nucleotide analogues in life.

4. References

1. Bunch T.H., Chang S. (1980), "Carbonaceous Chondrites II. Carbonaceous Chondrite Phyllosilicates and Light Element Geochemistry as Indicators of Parent Body Processes and Surface Conditions", Geochim. Acta, 44, 1543-1577

2. Cooper G. W., Onwo W.M., et al. (1992), "Alkyl Phosphonic Acids and Sulfonic Acids in the Murchison Meteorite", Geochim. Cosmochim. Acta, 56, 4109-4115

3. Schwartz, Alan W. (1996), "Exogenous Phosphorus Compounds in Early Chemical Evolution", ISSOL' 96 8th ISSOL Meeting, 11th International Conference on the Origins of Life Abstracts, 39

4. Yikui Du, Jilan Wu, et al. (1999), "A Pulse Radiolysis Study on Energy Transfer Mechanism in Trioctyl Phosphineoxide-Cyclohexane and Tributylphosphate-Cyclohexane Systems by Using Benzophenone as a Probe", Radiat. Phys. Chem, 54, 455-461

5. Katsumura Y., Tabata Y., et al. (1982), "The Formation Process of the Excited State of Cycloalkane Liquid Using Picosecond Pulse Radiolysis", Radiat. Phys. Chem., 19 (4), 267-276

DEMONSTRATION OF THE PVED IN PHASE TRANSITION OF D-/L-ALANINE & VALINE

WENQING WANG, YAN GONG, WANQING CHEN

Department of Technical Physics, Peking University, Beijing 100871, China

Abstract: The parity-violating electroweak neutral current (mediated by the Z^0 boson) has been suggested as the main physical force underlying the observed homochirality. We show experimentally that the parity violating in crystals of alanine and valine enantiomer, induced by the electroweak interaction of the conduction electrons with the nuclei of the lattice, can give rise to an enantiomeric PVED in a second order λ phase transition driven by the electron spin reversal of α hydrogen of the asymmetric center under critical temperature. This effect may have played a role in the origin of the homochirality.

1. Introduction

Andrea Szabo-Nagy & Lajos Keszthelyi (1999)[1] reported PVED was observed in the crystallization process of tris(1,2-ethanediamine) cobalt(III) and tris (1,2-ethanediamine) iridium(III) racemic mixtures. It is based on natural circular dichromism, that is, a difference in absorption coefficient of the two enantiomers for CPL. G.L.J.A.Rikken & E. Raupach (2000)[2] used the chiral Cr(III) tris-oxalato complex with unpolarized light in a magnetic field parallel to the irradiation direction, a small excess of one enantiomer is produced, and vice versa on reversing the magnetic field direction. Since the PVED is proportional to the sixth power of the atomic number, one should expect a considerable enhancement of PNC effects in heavy atoms $_{27}$Co, $_{77}$Ir and $_{24}$Cr. Proteins in life are only constituted of L-α amino acids. The hydrogen of αC is the asymmetry center between enantiomers. Even though the atomic number of hydrogen equal to unity, since the simplicity of the hydrogen only one electron around the nucleus, the behavior of αC-H was selected in present paper as the main fragment for studying PVED.

As for the Z force, left and right handed electrons have Z charges of opposite signs and approximately equal magnitudes. The difference in sign causes right handed electrons to be attracted to the nucleus by the Z force and left handed ones to be repelled. The weak force acts in an opposite way on left and right

J. Chela-Flores et al. (eds.), First Steps in the Origin of Life in the Universe, 87–90.

handed electrons, so that the mirror reflections are subtly different—the L-enantiomer has a lower energy than the D-enantiomer. From the calculation of Mason, Tranter and MacDermott [3,4] the theoretical value of PVED is $\approx 1\times 10^{-17}$ kT for the amino acid. The PVED of alanine and valine is 3.0×10^{-19} eV/molecule and 6.2×10^{-19} eV/molecule respectively. Owing to the extremely small Compton wavelength of the Z^0 boson of mass M (λ = h/MC $\approx 2.5 \times 10^{-19}$ m), the parity violating weak interaction occurs only when the electron is localized inside the nucleus. Spin is crucial in determining how a particle behaves.

In a hydrogen atom, alligning the spins of the proton and the electron increases the atom's total spin from zero to one, and its mass by increasing 6.0×10^{-15} GeV[5].We have noticed that in Rikken's work, the spin forbidden transition from the ground state $^4A_{2g}$ to the excited ^2E_g state shows fairly large values for g_{NCD} and g_{MCD}, and expect a substantial g_{MchA} (e.$e_{MchA} = g_{MchA}/2$) at the transition.

2. Experimental

In previous work we found a novel phase transition of D-/L-alanine and valine at 270K by the method of specific heat measurement[6]. The specific heat jump of valine enantiomer was the λ type and released differential specific heat ΔCp =50 mJ/g K =6.2×10^{-5} eV/molecule K (M =117.15) at critical temperature. The slope difference of two curves of D- and L-valine reflected various phase transition mechanism and the ΔCp value approached to the released energy (6.0×10^{-6} eV) by electron spin reversal between D-valine and L-valine.

DC magnetization measurements of D-/L-alanine and D-valine crystals were carried out by SQUID magnetometer (QUANTUM Design, MPMS-5) with differential sensitivity 1E-8 EMU to 1 Tesla at a field of 0.1 T [6] 1T and –1T from 5K to 300K with the emphasis on 200 to 300K(Fig.1). Magnetization measurements have shown alanine and valine enantiomers to be diamagnetic, since they have even numbers of electrons which form closed, magnetically neutral shells. A magnetic drop occurs for D-alanine and D-valine that is coincident with T_c of specific heat measurement. Due to the parity violating of Z force, the electrons of α-H in D-alanine (D-valine) attempt to orbit the external DC magnetic field lines, causing spin up electrons to orient down and produce the susceptibility χ_p drop. Owing to the electron spin reversal of H-αC in D-valine, it might be associated with a magnetic transition possibly from paramagnetism to antiferromagnetism in the crystal lattice of D-alanine and D-valine.

Solid state ^{1}H NMR spectra of D-alanine powders was measured as a function of temperature by Bruker DRX 300WB (AVANCE) with a 300 MHz, 89mm magnet. The chemical shift anisotropy of the nuclei was averaged out by magic-angle spinning. All the ^{1}H-MAS spectra were externally referenced to tetramethylsilane with the chemical shift value of the methyl resonance assigned as 0 ppm. Fig.2 showed the temperature dependence with the ^{1}H MAS spectra of a polycrystalline D-alanine by single pulse excitation with a delay time of 4.5 μs at a spinning speed of 5 kHz with 256 scans. The spectrum of D-alanine showed three main peaks at 4.772, 3.409 and 0.910 ppm at 290K respectively. The signals were assigned to be the amino protons, α proton and the methyl protons going from low to high magnetic field and the chemical shift values were coincident with those observed in the ^{1}H echo-MAS spectra[7]. The peak of ^{1}H-αC showed an obvious upshielding under the decrease of the temperature. It would be explained that the shielding effect of metal hydrogen was higher than the hydrogen of α carbon. The integral peak ratio of D-alanine was further proved the amplification prediction in the λ phase transition.

3. Conclusion

Due to the smallness of the Z force, many scientists have kept wondering its effect on the chemical properties of molecules. We argue that the PVED may occur not only for a system in its chiral equilibrium geometry, but also for a transition state. Magnetic fields are common in the cosmos. Unpolarized light as driving force under a parallel or antiparallel magnetic fields could produce one enantiomer excess in space. Engel & Macko [8] discovered the excess L-amino acids in the Murchison meteorite, represented the definite identification of exochirality. We assume that homochirality was in existence during the latter stages of the prebiotic era. Alanine and valine being the essential biomolecules, were presented prior to the prebiotic-biotic transition. Our experiments showed a significant PVED effect which was amplified at T_c, caused by the electron spin asymmetry of α hydrogen in condensed state of alanine and valine.

4. References

1. Andrea Szabo-Nagy and Lajos Kesthelyi (1999) "Demonstration of the parity-violating energy difference between enantiomers" *Proc. Natl. Acad. Sci. USA* **96,** 4252-4255.

2. G. L. J. A. Rikken and E. Raupach (2000) " Enantioselective magnetochiral photochemistry" *Lett. to Nature*, **405,** 932-935.

3. Mason , S. F. And Tranter G. E. (1984) " The parity-violating energy difference between enantiomeric molecules", *Molecular Physics* **53,** 1091-1111.

4. MacDermott, A.J. (1995) "Electroweak enantioselective and the origin of life" *Origins of Life and Evolution of the Biosphere,* **25,** 191-199.

5. Klaus Rith & Andreas Schafer (1999) " The mystery of nucleon spin" *Sci. American,* **July** , 58-63.

6. Wenqing Wang et al., (2000) " Parity violation of electroweak force in phase transitions of single crystals of D- and L-alanine and Valine" *J. Biol. Phys.* **26,** 51-65.

7. Jianzhi Hu et al., (1996) "High resolution ^{1}H spectra of powdered solids observed by Hahn Echo Pulse Sequence with Magic-angle spinning" *Solid State Nuclear Magnetic resonance* **6,** 85-94.

8. Engel, M. H. and Macko, S.A. (1997) " Isotopic evidence for extraterrestrial nonracemic amino acids in the Murchison meteorite, *Lett. to Nature* **389,** 265-268.

TWO PUNCTUATION SYSTEMS IN THE GENETIC CODE

ROMEU CARDOSO GUIMARÃES

Dept. Biologia Geral, Inst. Ciências Biológicas, Univ. Federal Minas Gerais, Belo Horizonte MG 31270.901 Brazil; romeucg@icb.ufmg.br

Abstract. In a schema for evolution with (a) early fixation of aminoacyl-tRNA synthetase (aRS) class 2 attributions coherent with homogeneous (Ho; RR, YY) principal dinucleotides (pDiN; anticodonic central3' base): the hydropathic outliers - S, P, G - plus D, N, K, F, and (c) late fixation of aRS1 coherent with mixed pDiN (Mx; RY, YR; I, M, C, W, R, Y, Q), the intermediate stage (b) corresponding to deviants (aRS2/Mx, aRS1/Ho) plus V: (a) early proteins were enriched in coils and turns, stabilizer amino acids concentrated at the N-end, destabilizers at the C-end, a polarity characterizing non-specific punctuation; (c) specific punctuation extended chains with fMet. Initiation pDiN were slipped, an 'inversion' analogous to mRNA capping. The intercalated wobble position at formation of the initiation tRNA couple was eliminated. Termination codons had their tRNAs deleted; their presence conflicted with initiation, competing with fMet-tRNA for initiation codons, through pairings involving the slipped pDiN of initiation and the standard pDiN of termination.

1. Introduction

Organization of the code based on pDiN sectors and quadrants (Guimarães, 1998) noted that punctuation boxes belonged in the aRS1/Mx sector, forming apparently 'inverted' pDiN pairs. An explanation for such 'inversions' is now presented in an unified consistent mechanism, and amino acid contribution to protein stabilization is introduced.

2. Materials and Methods

Nucleic acid complementarity divided the code in a succession of halves. **(a)** A *sector* of the Ho pDiN was separated from that of Mx. Extreme hydropathies belonged in the Ho, intermediate in the Mx. **(b)** *Quadrants* of sectors were united through complementarity (RR:YY; RY:YR; TABLE I). **(c)** *Axes*, composed by boxes of (*h*) homodinucleotides and of pDiN built from (*p*) complementary bases, each divided into the symmetrical *cores* (G &/or C-only boxes) and *tips* (A &/or U-only, corners). **(d)** *Non-axial* pairs of boxes (central R: central Y; [Ho] GA:CU, AG:UC; [Mx] AC:UG, GU:CA).

J. Chela-Flores et al. (eds.), First Steps in the Origin of Life in the Universe, 91–94.

3. Results and Discussion

TABLE I. Complementarity of central bases is maintained. Anticodonic: 3' of pDiN are complementary to each other (c) but pairing with the 5' (w) base of the partner. Alternative initiation codons showed strict conservation of pairing through their 3rd position with the 1st of Met anticodon, besides the central position of both (except ThrACG). So, the slipped pDiN of initiation [standard] has the variable position 'inverted'.

tRNA:tRNA		ELONGATION	INITIATION
5'w central 3'c	***mRNA***	5' central 3'w	5' central 3' [N U] G
3'c central 5'w	***: tRNA***	3' central 5'w	3' central 5' [U A] C

3.1. *NON-SPECIFIC PUNCTUATION*

Regularities of sequences related to protein half-lives (the N-end rule; TABLE II) correlated with the pDiN organization: (**a**) *stabilizers*: primary (**1**), resident at N-end (M,V,G,P); secondary (**2**), internal, most effective at N-end + 1 (S,A,T); (**b**) *destabilizers*: weak (**3**; C,E,Q,D,N); intermediate (**4**; H,I); strong (**5**), utilized by conjugation to proteins (K,R); strong, resident at the N-end (**6**; F,L,Y,W).

TABLE II. The N-end and C-end rules. Degrees of contribution to protein destabilization (numbered) adapted from Varshavsky (1992). Preferred+, avoided- at protein C-ends from Arkov *et al.* (1995). Stabilizers concentrated in the hydropathic outliers, central G, 3'C, RY. Amino acids of the C-end rule filled the Ho axis (strong destabilizers at tips UU+K, AA+F, L; strong stabilizers at core GG-P, CC-G,), hydropathic outliers (GA+S) and central G (GC+A, GU-T), the latter containing the 2 boxes of the Mx sector and secondary stabilizers.

RR	Phe+ 6, Leu+ 6	Ser+ 2	Cys 3, Trp 6	Tyr 6	**YR**
	Leu 6	Pro- 1	Arg 5	His 4, Gln 3	
	Val 1	Ala+ 2	Gly- 1	Asp 3, Glu 3	
RY	Ile 4, Met 1	Thr- 2		Asn 3, Lys+ 5	**YY**

Stabilizing amino acids were concentrated at N-ends (Varshavsky, 1992) of long-lived proteins and the C-end rule (our nomenclature) complemented precisely those data. None of the authors commented on the relevance of the findings for the code organization. We consider stabilizing properties idiosyncratic to amino acids, properties of proteolytic systems and amino acid distribution in sequences following them.

N-end and C-end rules were consistent in indicating strict involvement of aRS2 (but the L hexacodonic expansion) in non-specific punctuation, and of the Ho axis in the C-end rule, in a core/tips symmetry. Other central G attributions (S,A,T) were inconsistent.

Nagel and Doolittle's (1991) proposal is reinforced, that aRS2 built an early translation system, adding that the polar distribution of stabilizer and destabilizer amino acids constructed a punctuation mechanism. Main consequence should be concentrating function and stability at the 3' end, destabilizers pushed to the C-end.

3.2. *SPECIFIC PUNCTUATION*

Main event in initiation was adoption of the slipped pDiN (5' C central A) for fMet-tRNA, the standard one (central A3' U) remaining dedicated to elongation. Initiation is the bi-component reaction for building the 1st peptide bond. Inversion here is analogous to mRNA capping and eliminated the wobble pairing across the two 1st tRNAs.

Components of punctuation differ from those composing other paired boxes in the sense that interferences became qualitative, a not tolerated confrontation of opposites, bound to occur anywhere initiation was established. Initiation is an extension of the peptide N-end with fMet and termination the deletion of the charging systems for the C-end +1 amino acids, corresponding to termination codons and analogous to suppressors. The slippage mechanism of initiation was a cause for the deletions of tRNAs corresponding to termination codons, through pairing conflicts.

Relevant sources of conflicts (TABLE III) should: (a) obey the physiologic configuration, pDiN of the initiation partner being slipped, that of termination standard [Ac, Ba1, Cc, Da1]; (b) match the 3 termination sites in the code; (c) present at least a standard base pair formed with the central base of the initiation partner.

TABLE III. Conflicts due to mutual sequestration of initiation and termination codons or anticodons. Dinucleotide (b, c) pairings had one nucleotide dislocated to the staggered configurations. Standard [pDiN], slipped pDiN. Standard base pair, p; G:U pair, w; X, main (UA), x, secondary (CA) termination anticodons or correspondent codons. Termination signs from variant codes (Osawa, 1995; footnote) also tested.

A) Crossed-combination pairings of initiation codons with termination anticodons

	Triplet pairings: (a) Both standard or slipped [5' central] 3'w [3' central] 5'w		Doublet pairings: (b) Initiation standard, termination slipped [5' central] 3'w 3' central 5'w		(c) Initiation slipped, termination standard 5' central 3'w [3' central] 5'w	
	5' Val [GU]G	Met [AU]G	Val [GU]G	Met [AU]G	Val [GU]G	Met [AU]G
x 3' ACU					p, p	p, p
X AUU					p, w	p, w
X AUC					p, w	p, w

B) Pairings of initiation anticodon with termination anticodons in tRNA pools

	(a1) initiation slipped, termination standard, or (a2) *vice versa* 5'w [central 3'] [3' central] 5'w		(b) both slipped 5'w central 3' 3' central 5'w	(c) both standard 5'w [central 3'] [3' central] 5'w
	5' fMet	C[AU]	C[AU]	C[AU]
x 3' ACU			/, p	
X AUU		/, p, /	/, p	
X AUC		/, p, /		

C) Crossed-combination pairings of initiation anticodon with termination codons (subheadings as in **A**)

	5'w [central 3'] 3'w [central 5']		5'w [central 3'] 3' central 5'w	5'w central 3' 3'w [central 5']
	5' fMet	C[AU]	C[AU]	C[AU]
x 3' AGU				p, p
X AAU				/, p
X GAU				/, p

D) Pairings of initiation codons with termination codons of mRNA (subheadings as in **B**)

	[5' central] 3'w 3'w [central 5']		5' central 3'w 3'w central 5'		[5' central] 3'w 3'w [central 5']	
	5' Val [GU]G	Met [AU]G	Val [GU]G	Met [AU]G	Val [GU]G	Met [AU]G
x 3' AGU			p, /	p, /		
X AAU	/, p, w	/, p, w	p, /	p, /		
X GAU	/, p, w	/, p, w				

(**A**, **C**) Pairings depicted in (c), with x or X, valid for the whole respective boxes. Standard pairings involving the central base of the initiation partner detected in (b) with Cys, Tyr and His. (**B**, **D**) Pairing depicted in (b) with x also shown by ArgUCY, and with XAUU also by Gln. Pairings depicted in (a) with X also shown by Tyr, His and Gln.

Matches explaining the deletions of 'termination' tRNAs were the Ac combination, with dislocated doublet pairings, termination anticodons sequestering initiation codons and competing with the initiation anticodon. Such pairings may not occur in present day physiology but it is indicated that they should at times of fixation of the code.

Similar results in the reverse combinations (Cc), termination codons being blocked by the initiation anticodon and x with stronger pairings than X, could be related to the observations of variant codes, where transfers of termination codons to amino acid coding were more frequently of x than of X. These were detected in restricted genomes, with benefits of avoiding punctuation conflicts and obtaining another amino acid code.

Pairings in tRNA pools (Ba1, in the physiologic configuration, and Bb) would not, by themselves, explain the coherent development of termination signs but could have combined with those in Ac, with reinforcement. Between the 2 X signs, the most conflictive combination was on the UAA codon, consistent with the finding of its transfer to Tyr in a variant code and the rationale in the preceding paragraph.

Conflicts indicated in Da1 and Db, in situations of 'circularized' mRNAs, with head-to-tail interactions (Parsch *et al.*, 1998), would be solved through development of sequence contexts where such pairings would not be favored.

Non-specific punctuation is at work in present day proteins, from which data were obtained to infer on the early ones. Met is expendable from the N-end of proteins, being removed from a great proportion of them after translation, especially when the N-end +1 amino acids are the secondary stabilizers (see Varshavsky, 1992); its stabilization property was necessary for it being the initiator but not for maintenance of stability.

The 7 aRS2/Ho attributions contained the 6 most preferred in coils and turns (G, P, S, D, N, K; Creighton, 1993), in line with Jurka and Smith's (1987) proposal for the early constitution of proteins.

Acknowledgments

Discussions with Carlos H. Costa Moreira; support from FAPEMIG, FAPESP, CNPq.

4. References

Arkov, A.L., Korolev, S.V. and Kisselev, L.L. (1995) Non random distribution of sense codons prior to stop codons in human genes, in B.F. Poglasov, B.I. Kurganov, M.S. Kritsky and K.L. Gladilin (eds.), *Evolutionary biochemistry and related areas of physicochemical biology*, Bach Inst. Biochem. and ANKO, Moscow, pp. 341-349.

Creighton, T.E. (1993) *Proteins: Structures and Molecular Properties*, Freeman, New York.

Guimarães, R.C. (1998) Genetic code: hydropathic, dinucleotide type and aminoacyl-tRNA synthetase class organization, in J. Chela-Flores and F. Raulin (eds.), *Exobiology-Matter, Energy and Information in the Origin and Evolution of Life in the Universe*, Kluwer Acad. Publ., Dordrecht, pp. 157-160.

Jurka, J. and Smith, T.F. (1987) -turn-driven early evolution: the genetic code and biosynthetic pathways, *J. Mol. Evol.* **25**, 15-19.

Nagel, G.M. and Doolittle, R.F. (1991) Evolution and relatedness in two aminoacyl-tRNA synthetase families, *Proc. Nat. Acad. Sci. USA* **88**, 8121-8125.

Osawa, S. (1995) *Evolution of the Genetic Code*, Oxford Univ. Press, Oxford UK.

Parsch, J., Stephan, W. and Tanda, S. (1998) Long-range base pairing in *Drosophila* and human mRNA sequences, *Mol. Biol. Evol.* **15**, 820-826.

Varshavsky, A. (1992) The N-end rule, *Cell* **69**, 725-735.

STUDIES ON METAL FERROCYANIDES AS PREBIOTIC CATALYST

KAMALUDDIN

Department of Chemistry
University of Roorkee, Roorkee – 247667, India

Processes in chemical evolution must have involved several catalysts multifunctional in nature. Our previous work has demonstrated that metal ferrocyanides posses strong adsorption affinity towards amino acids and nucleotides. It is assumed that during the course of chemical evolution, cyanide could have formed some metal ferrocyanides. As these metal ferrocyanides are mostly insoluble in water, and they are therefore, considered to have settled at the bottom of the primeval ocean or at sea shore. We propose that metal cyanogen complexes might have acted as an important prebiotic catalyst.

A series of metal ferrocyanides of general formula $M_2[Fe(CN)_6].xH_2O$, where M= Cr, Mn, Co, Ni, Cu, Zn and Cd were prepared through a method reported by Kourim *et al.* [1]. Characterization of the metal ferrocyanides was carried out by elemental analysis and on the basis of i.r. spectra. The purity of metal ferrocyanides was checked by comparing X-ray diffraction data of the complexes. Surface area measurements of metal ferrocyanides were carried out. Specific surface areas of Cu-, Zn-, and Ni- ferrocyanides were found as 444.7, 412.4, and 302.1 (m^2/g), respectively. Surface area of Cr-, Mn-, Co-, and Cd- ferrocyanides were comparatively less and below 100 m^2/g range [2]. Interaction of different metal ferrocyanides with a class of organic compounds were studied and the results found are categorized as follows:

1. Interaction of Metal Ferrocyanides with Amino Acids

Interaction of representative of three classes of amino acids viz., aspartic, histidine, and glycine as acidic, basic, and neutral, respectively was studied with Cu- and Zn- ferrocyanides [3]. Adsorption studies were carried out at different pH with the help of buffers. Histidine and glycine showed increased adsorption with increase in pH from acidic to neutral where as aspartic acid was adsorbed only above pH 7.0. Higher adsorption of amino acids at pH above their isoelectric points suggested that adsorption is taking place through their anionic forms. At pH 7.0 adsorption of the three amino acids studied was in the range 50-80% on both Zn- and Cu- ferrocyanides. Adsorption of histidine was found maximum in comparison to glycine or aspartic acid.

J. Chela-Flores et al. (eds.), First Steps in the Origin of Life in the Universe, 95–97.

2. Interaction of Copper and Zinc Ferrocyanides with Ribose and 2′-Deoxyribose -5′- Monophosphate Nucleotides

Adsorption of different purines and pyrimidines ribose and 2′-deoxy ribose nucleotides was studied [4] on copper and zinc ferrocyanides at pH 7.0. Higher adsorption (70-90%) of both ribose and 2′-deoxy ribose nucleotides of purines (5′- AMP, 5′- GMP, 2′- d- 5′- AMP, and 2′- d- 5′- GMP) in comparison to (20-70%) that of pyrimidine nucleotides (5′- UMP and 5′- TMP) on both copper and zinc ferrocyanides was explained probably due to additional binding site in the imidazole ring of purines. Adsorption of the nucleotides on copper and zinc ferrocyanides followed Langmuir adsorption isotherm and values of X_m and K_L obtained for the nucleotides also confirmed the above observations.

3. Interaction of Metal Ferrocyanides with Aromatic Amines

Interaction of one each of the three categories of aromatic amines viz., p-toliudine (more basic), aniline (basic), and p-chloroaniline (less basic) has been studied with a number of metal ferrocyanides [5-6] namely, Mn-, Co-, Ni-, Zn-, and Cd-ferrocyanides. It was observed that Mn-ferrocyanide rapidly reacted with all the three amines forming colored oligomers on its surface.

Percent binding of p-toluidine (56-95%) was found highest in comparison with aniline (44-55%) and p-chloroaniline (16-31%) on Co-, Ni-, and Zn-ferrocyanides. Cadmium ferrocyanide showed least adsorption affinity (17-28%) for the above three aromatic amines. Relative adsorption of amines was explained as proportional to their relative basicities. The adsorption data obtained were found to fit Laugmuir adsorption isotherm. Langmuir constants viz., X_m and K_L were determined.

Reaction of aromatic amines with manganese ferrocyanide was found to be appreciable in alkaline medium at pH about 8. In acidic medium no reaction took place. Red brown colored products were deposited on the surface of manganese ferrocyanide, which were extracted with benzene. Mass spectral analysis of the products showed formation of tetramer, trimer and a dimer due to reaction of aniline, p-toluidine and p-chloroaniline, respectively.

4. Interaction of Metal Ferrocyanides with Aminopyridines

Studies on interaction of metal ferrocyanides with aminopyridines were done [7-9]. It was observed that Co-, Ni- Cu-, and Zn-ferrocyanides showed strong adsorption affinity towards all the three aminopyridines namely, 2-amino, 3-amino, and 4-aminopyridines.

On the other hand all the three aminopyridines showed chemical reaction with Mn-ferrocyanide and gave colored products. Co-, Ni-, Cu-, and Zn-ferrocyanide at neutral pH and 25^{0}C showed 32-99% adsorption range for all the three aminopyridines whereas Ni-ferrocyanide showed least adsorption affinity in the 15-18% range.

Manganese ferrocyanide on reaction with aminopyridines gives yellow colored product. Reaction of 3-aminopyridine was found very fast as compared to reaction with 2-amino and 4-aminopyrindines. Mass spectral studies showed formation of a dimer azopyridines for all the three aminopyridines studied.

In view of the above studies, it can be suggested that metal ferrocyanides once formed in the prebiotic environment might have interacted with important biomolecules in order to protect them from degredation and have catalysed their oligomerization.

5. References

1. Kourim,V.,Rais,J., and Million,B.:Exchange Properties of Complex Cyanides-I: Ion Exchange of Caesium on Ferrocyanides ,*Inorg. Nucl. Chem.* **26** (1964), 1111-1115.
2. Alam, T.: Studies on Metal Ferrocyanides (II) as Prebiotic catalyst, *Ph. D. Thesis*, University of Roorkee, Roorkee (1998),35.
3. Kamaluddin, Nath, M., Deopujari, S.W., and Sharma, A.: Role of Transition Metal Ferrocyanides (II) in Chemical Evolution, *Origins of Life and Evolution of the Biosphere* **20** (1990), 259-268.
4. Kamaluddin, Nath, M., and Sharma, A.: Role of Metal Ferrocyanides in Chemical Evolution: Adsorption of Ribose and 2′- Deoxy ribose 5′- Nucleotides on Metal Ferrocyanides, *Origins of Life and Evolution of the Biosphere* **24** (1994), 469-477.
5. Alam, T., Tarannum, H., Viladkar, S., and Kamaluddin: Oxidation of Aniline and its Derivatives by Manganese Ferrocyanide, *Oxidn. Communications* **22**(4) (1999), 599-609.
6. Alam, T., Tarannum, H., Kumar, M.N.V.R. and Kamaluddin: Adsorption and Oxidation of Aromatic Amines by Metal Hexacyanoferrates (II), *Talanta* **51** (2000), 1097-1105.
7. Alam, T. and Kamaluddin : Interaction of Aminopyridines with Metal Hexacyano- ferrates (II), *Bull.Chem.Soc. Japan* **72**(8) (1999), 1697-1703.
8. Alam, T. and Kamaluddin: Interaction of 2-Amino, 3-Amino, and 4-Aminopyridines with Nickel and Cobalt Ferrocyanides, *Colloid and Surfaces* **162** (2000), 89-97.
9. Alam, T., Tarannum, H., Kumar, N., and Kamaluddin: Interaction of 2-Amino, 3- Amino, and 4-Aminopyridines with Chromium and Manganese Ferrocyanides, *Colloid & Interface Sciences* **224** (2000), 133-139.

STABILITY OF RIBONUCLEIC ACID IN PROTECTIVE ENVIRONMENTS OF ALKANES $\geq n$-C_{18} - RESULTS FROM EXPERIMENTS IN LABORATORY

VICENTE MARCANO[1], PEDRO BENITEZ[1] , LUIS FAJARDO[1] AND ERNESTO PALACIOS-PRÜ[1,2] *
[1] *Electron Microscopy Center, Evolutionary Biology and Chemistry Laboratory, University of the Andes, P. O. Box 163, Mérida, Venezuela.* prupal@ula.ve
[2] *Internacional Institute of Advanced Studies, USB, Caracas, Venezuela.*

1. Introduction

It is generally agreed that the life have originated between 3500-3800 Myr ago (Orgel 1998). Before 3600 Myr, the oceans would probably have been repeatedly vaporized by large impacts, which could sterilize the planet eliminating small and large molecules and life itself (Maher & Stevenson 1988, Sleep et al. 1989, Zahnle & Sleep 1997, Lyons & Vasavada 1999). However, reconstruction of the last heavy bombardment indicates that large impacts up to ~ 100-km in diameter occurred still 3600 Myr ago and generated excessive temperatures on Earth able to melt most of the planet's mass (Lyons & Vasavada 1999). Though it is difficult to suppose that an accumulation of organic molecules could be possible before of this age, it would be feasible only during the intervals among each impactor fall (Marcano et al. 2000a), which would have surface temperatures between 85-110°C generated by CO_2 greenhouse effects (Kasting 1993), volcanic activity and lower Earth's rotation rate (~ 14 hours).

It is frequently stated that modern replicative systems based on RNA and DNA suggest an abundance of these molecules and of its precursors during the early Earth (RNA world). However, the RNA world or even the pre-RNA world under conditions ≥ 85°C are difficult to conceive due to the high instability of these molecules (White 1984, Miller & Bada 1988, Lindahl 1993, Bada et al. 1995, Forterre et al. 1995, Levy & Miller 1998, Shapiro 1999). For instance, at pH 7 and 100°C, the half-lives of the nucleobases are: A = 1 yr, G = 0.8 yr, U = 12 yr, T = 56 yr and C = 19 days, whereas at 250°C, they are between 1 and 35 min, and at 350°C between 2 and 15 seconds. Even RNA molecules at pH 7 give a half life of 2 min at 250°C for the hydrolysis of each phosphodiester bond, at 350°C the half life is 4 seconds. The hydrolysis observed for the N-glycosyl bond in adenosine is 1.3 min at 250°C. Finally, the half life of ribose is 73 minutes at pH 7 and 100°C. However, RNA molecules had have other factors inhibiting their survival on the early Earth, such as photodecomposition by UV-radiations in the UVC and UVB regions and probably alterations by electron excitation caused by electric

* To whom correspondence should be mailed.

J. Chela-Flores et al. (eds.), First Steps in the Origin of Life in the Universe, 99–102.

discharges, viz. excessive ion and free-radical production. We show here based in experimental data that the existence of thermostable *n*-alkane $\geq C_{18}$ environments on the Earth´s surface (< 3600 Myr) (Marcano et al., 2000a) could offer various advantages for the survival of RNA molecules such as occur with other biomolecules (Marcano et al. 2000b, c). The first advantage is represented by their capacity to the protection against rapid hydrolysis in high temperatures generated by CO_2 greenhouse effect or by small impacts; second, such coats may avoid or reduce the oxidative decomposition of inmersed RNA molecules generated by hydrolysis, electrolysis or photolysis of several gases; third, absorbances of UVC radiations by heavy *n*-alkanes allow the protection against photolysis, and fourth, due to the low dielectric constant (2.0-2.5 x 10^6 cycles) and conductivity (~ 0.5 μs/cm) of these alkanes, this layer might serve as an insulator thereby protecting the synthesis and accumulation of RNA molecules from alterations by electron excitation caused by excess of ions from electric discharges.

2. Experimental section

Quantification of RNA molecules (NAQ) was determined using a Gene-Quant *pro* RNA/DNA Calculator model Amersham Pharmacia Biotech. FTIR absorbance measurements were made by a Perkin-Elmer Fourier transform IR spectrometer. Products not soluble in organic solvents were studied by fluorescence microscopy. This last procedure was found to be especially useful for the study of their surface and density. Experiments were carried out in a 2.85-liter glass (Pyrex) reactor containing 100 ml of mineral oil and 500 mg of RNA from Torula yeast (Sigma). Heavy mineral oil was used (d = 0.85 g/cm^3, Sigma) as a model of primordial alkanes > *n*-C_{18}. Simulating the absence of UV atmospheric shielding, the reactor was exposed to UVC-radiations (253.7 nm lamp, 2200 μW cm^{-2}) in a dark room. The experiment was carried out simultaneously under spark discharges (continuous arc) obtained from a high-frequency Tesla coil (50 c/s, 120 V input; 30 kv, 0.35 amp, 10.5 KW output, Electro Technic Products, model BD 10) for 15-45 minutes. Temperatures ranges chosen were between 100-250°C. When heated, the content of the flask reached 100% of their final temperature within ~ 5-10 min. This was measured by a thermocouple inserted into of the flask. One type of prebiotic-like atmosphere was selected for these experiments having $H_2O + CO_2$ in equimolar concentrations and pressures ranging between ~ 0.6-0.8 bars. All the raw products were dispersed in Milli-Q water, stirred, separated from the mineral oil by chloroform, decanted, and filtered through Whatman No. 1.

3. Results and Discussion

Kinetics of the decomposition of the RNAs indicates rate constants k = 15.33 mg min^{-1} at 250°C, k = 10.8 mg min^{-1} at 200°C, k = 2,91 mg min^{-1} at 150°, and k = 0.25 mg min^{-1} at 100°C, whereas activation energy is E_a = 3,42 x 10^4 J.degree.mg, according to Arrhenius´s equation (Figure 1). These data reveal a half life for RNA of 50 min at

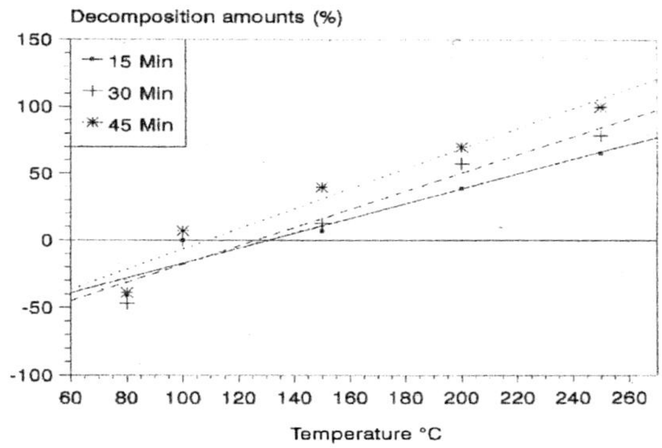

Figure 1. Temperature profile for the decomposition of RNA under protected conditions by heavy *n*-alkanes.

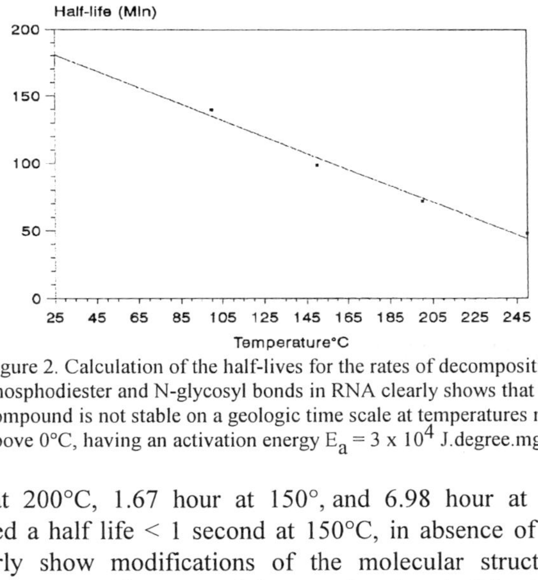

Figure 2. Calculation of the half-lives for the rates of decomposition of phosphodiester and N-glycosyl bonds in RNA clearly shows that this compound is not stable on a geologic time scale at temperatures much above 0°C, having an activation energy $E_a = 3 \times 10^4$ J.degree.mg.

250°C, 1.28 hour at 200°C, 1.67 hour at 150°, and 6.98 hour at 100°C (Figure 2). Control data showed a half life < 1 second at 150°C, in absence of mineral oil. FTIR spectral data clearly show modifications of the molecular structure of the RNA, indicating that the decomposition rate of these molecules shall be increased at higher temperatures and at more prolongated times. Lower absorbances about 1215 cm^{-1} revealed loss of the N-glycosyl bonds whereas lower absorbances about 1000 cm^{-1} showed loss of the phosphodiester bonds ($\nu_{C\ O}$ vibrations). The sites in the RNA molecule where would occur the longer damages (Lindahl, 1993) are showed in the figure 3. However, FTIR spectral evidences of hydrolytic attack and oxidative damages were not seen.

The presence of an amphiphilic component in the heterogenous products was determined by surface tension measurements in aqueous media by microscopy, at 20°C and 1-bar. Superficial tension values have ranges between 1.94×10^6 and 5.82×10^5 dynes cm^2 of pressure. This component shows spheroidal shape, dark brown and solid

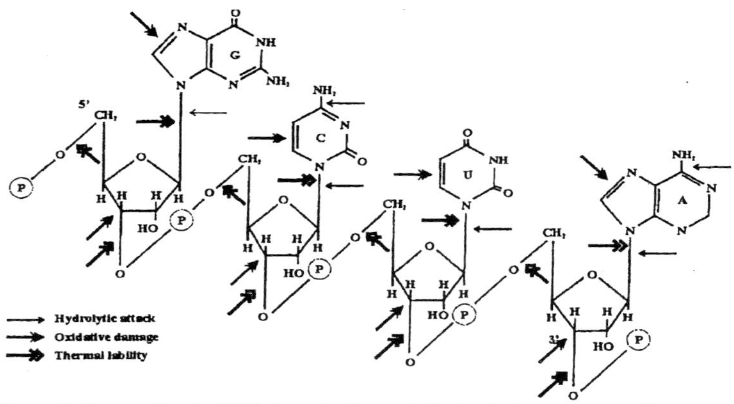

Figure 3. Target sites for RNA decay in the early Earth. A short segment of one strand of the RNA is shown with the four common bases. Sites susceptible to hydrolytic attack, oxidative damage, and thermal lability are indicated by arrows.

appearance, uniformity of size (3-5 μm in diameter). Residues from the simulations dispersed in aqueous media for microscopy and examined by fluorescence revealed negative UV excitation and absence of accretion.

In conclusion, although the half-lives of the RNA molecules were slightly improved under protected conditions by *n*-alkane $\geq C_{18}$ (viz. 1.67 hour vs < 1 sec at 150°C), the thermal lability of the N-glycosyl and of the phosphodiester bonds could prevented the formation and accumulation of primitive RNA and other RNA-like polymers on the warm early Earth´s surface. Therefore, it is more feasible that the RNA world began when the Earth became cooler.

4. References

Bada, J. L., Miller, S. L. and Zhao, M.: 1995, *Origins of Life Evol. Biosphere* **25**, 111.

Forterre, P., Confalonieri, F., Charbonnier, F. and Duguet, M.: 1995, *Origins of Life Evol. Biosphere* **25**, 235.

Kasting, J. F.: 1993, *Science* **259**, 920.

Levy, M. and Miller, S. L.: 1998, *Proc. Natl. Acad. Sci. USA* **95**, 7933.

Lindahl, T.: 1993, *Nature* **362**, 709.

Lyons, J. and Vasavada, A. R.: 1999, *Origins of Life Evol. Biosphere* **29**, 123.

Maher, K. A. and Stevenson, D. J.: 1988, *Nature* **331**, 612.

Marcano, V, Benitez, P. and Palacios-Prü E.: 2000a, Accumulation of alkanes $\geq n$-C_{18} in the early Earth, in *Astrobiology: Origins from the Big Bang to the Civilization*, J. Chela-Flores, G. Lemarchand and J. Oró (eds.), Kluwer Academic Publishers, The Netherlands.

Marcano, V, Benitez, P. and Palacios-Prü E.: 2000b, Advantages of the alkanes $\geq n$-C_{18} as protectors for the synthesis and survival of critical biomolecules in the early Earth, in *Astrobiology: Origins from the Big Bang to the Civilization*, J. Chela-Flores, G. Lemarchand and J. Oró (eds.), Kluwer Academic Publishers, The Netherlands.

Marcano, V, Benitez, P. and Palacios-Prü E.: 2000c, An experimental approach to production of peptide-like compounds in the early terrestrial planets. *Planetary and Space Science* **50** (in press).

Miller, S. L. and Bada, J. L.: 1988, *Nature* **334**, 609.

Orgel, L.: 1998, *Origins of Life Evol. Biosphere* **28**, 91.

Shapiro, R.: 1999, *Proc. Natl. Acad. Sci.* **96**, 4396.

Sleep, N. H., Zahnle, K. J., Kasting, J. F. and Morowitz, H. J.: 1989, *Nature* **342**, 139.

White, R. H. 1984, *Nature* **310**, 430.

Zahnle, K. J. and Sleep, N. H.: 1997, in *Comets and the Origin and Evolution of Life*, Thomas, P. J. Chyba, C. F. and McKay, C. P. (eds.), Springer-Verlag, New York, pp. 175-208.

IS BORIC ACID THE MISSING LINK IN PREBIOTIC CHEMISTRY?

Benoît PRIEUR

Ecole Normale Supérieure, Laboratoire de chimie, CNRS 8642,
24 rue Lhomond, F-75231 PARIS Cedex 05, FRANCE
email: benoit.prieur@theoriginoflife.org
website: www.theoriginoflife.org

1. Introduction

Boric acid has always been neglected in prebiotic chemistry because it is not a major component of biological macromolecules. Here we argue for a potential catalytic role of boric acid in peptide and nucleic acid synthesis. Today boron is an important component of sea water (0.4 mM). Its concentration at prebiotic times was probably higher. In water, boron appears mainly in the form of boric acid or its derivatives (esters and anhydrides).

2. Boric acid and sugars

In 1874 Vignon was the first to describe that borates form esters with cis-diols such as ribose (1,2,3). The stability of a given ester depends on the carbohydrate and is highest at alkaline pH (4,5). Borate esterification can change the conformation and rotation of carbohydrates (6). Furthermore, boric acid catalyzes epimerization of arabinose to ribose in the presence of molybdate salts (7). When complexed, ribose is primarily in furanose form while in the absence of borate, only 20% of ribose adopts this form (Figure 1) (4,5). Complexation with boric acid protects carbohydrates against degradation and caramelization (8,9). For example, when a solution of ribose in water is heated at 100°C under an argon atmosphere, a black color develops in a few minutes due to aldehyde polymerization (10). We have observed that in the presence of excess boric acid, this color develops only slowly, appearing after 48 h.

J. Chela-Flores et al. (eds.), First Steps in the Origin of Life in the Universe, 103–106.

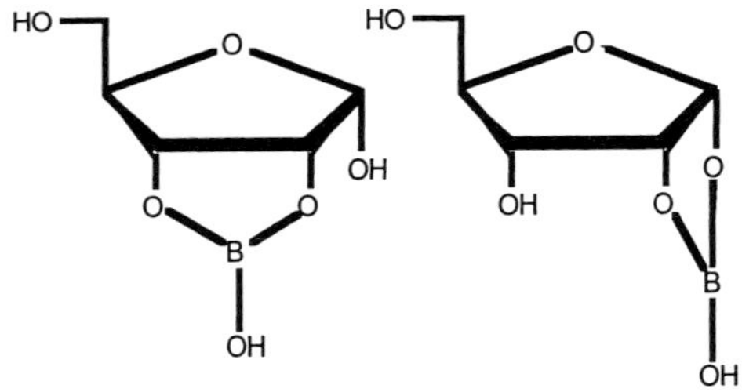

Figure 1: Complexation of boric acid on ribose

3. Synthesis of ribose in the presence of boric acid

First described by Boutlerow in 1861, the formose reaction consists of condensation of formaldehyde (aldolization) in aqueous alkaline solution (11). In these drastic conditions the synthesized sugars are fragile and can be quickly destroyed. We have studied the activity of boric acid in a condensation reaction of formaldehyde in acidic medium. Our preliminary findings indicate that it does not appear possible to obtain monosaccharides in detectable quantities by reacting formaldehyde with boric acid (3.5 M/l) (100°C, 4 weeks, pH 6.5). Other experimental conditions must be tested, particularly Kiliani-Fischer sugar synthesis in acidic medium.

4. Adenylation reactions in the presence of boric acid

We have succeeded in performing adenylation of ribose with tetraborate sodium salts. We have not yet found a way to perform adenylation of ribose in acidic conditions.

5. Boric acid and amino acids

The interaction between the carboxylic group and boric acid as catalyst was first described by Cottle and Young in 1955 in the industrial preparation of amides in organic solvents (12). We have studied the effects of boric acid on polymerization of glycine in water (Figure 2). Our preliminary results show synthesis of some cyclic and linear oligopeptides in water.

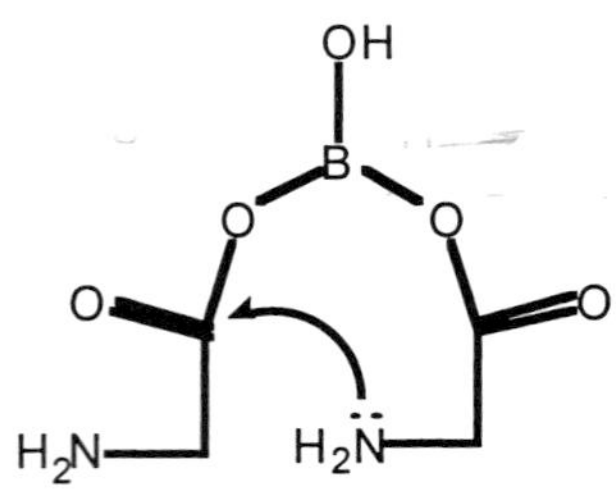

Figure 2: Activation of the carboxylic moiety of glycine by boric acid

6. Boric acid and polyphosphates

In water, boric acid can also form an anhydride bond with bases, in the form RO^- (amino acid, phosphate, ...). The interaction of boric acid with inorganic phosphate is certainly the most interesting to explain the formation of polyphosphates by exchange of an anhydride bond, because we know that it is very difficult to condense phosphate to pyrophosphate in mild conditions (Figure 3) (13).

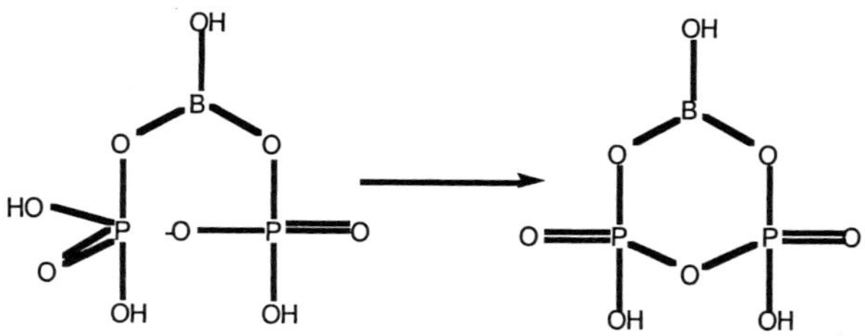

Figure 3: Boric acid could allow phosphate to concentrate and polymerize

7. Boric acid, sugar, nucleoside and phosphate

The question of sugar and nucleoside phosphorylation is entirely open. There are no known conditions which allow phosphorylation of sugars and nucleosides. Complexation of sugar and nucleoside with boric acid has a distinct advantage because it could allow the approach and fixation of a phosphate group to the sugar. The catalytic role of boric acid in the phosphorylation of sugar has a second advantage: it might allow polymerization of phosphoribose to polyphosphoribose.

8. Boric acid and aminoacyl-RNA

In the current aminoacyl-tRNA, the amino acid is linked to the 2' or 3' extremity by an ester bond. The prebiotic synthesis of this component is unrealistic. On the other hand, the formation of a 2',3'-boro-aminoacyl RNA is very easy, especially if the RNA has a physico-chemical environment adapted to the amino acid (Figure 4).

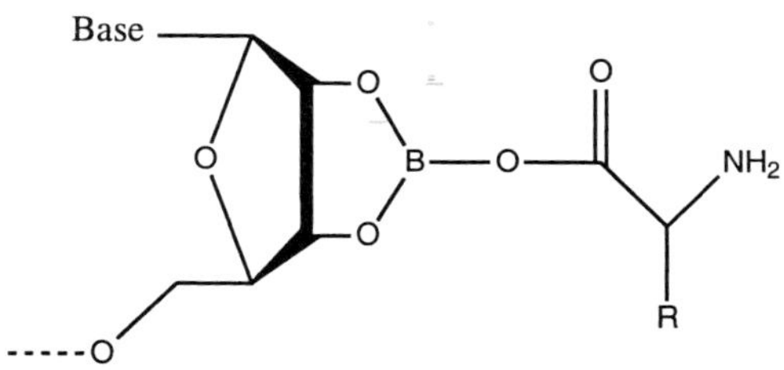

Figure 4: A bond between amino acid and RNA can be formed spontaneously with boric acid

9. Conclusions

Boric acid was an abundant element in the prebiotic ocean. It is also an indestructible molecule. Its catalytic properties make otherwise impossible reactions appear possible. It might very well be the missing link in prebiotic chemistry.

10. References

(1) Vignon, L., (1874), Compt. Rend., 78, 148
(2) Böeseken, J., (1949), Adv. Carbohydr. Chem., 12, 81
(3) Angyal, S.J. and McHugh, D.J., (1957), J. Chem. Soc., 1423
(4) Chapelle, S. and Verchère, J.F., (1989), Carbohydrate Research, 191, 63-70
(5) Verchère, J.F. and Sauvage, J.P., (1988), Bull. Soc. Chim., 2, 263-266
(6) Böeseken, J., (1944), Chem. Zentr., I, 546-547
(7) Hiroshi, H., Takashi, T. and Masahiko, H., (1985), Japanese Patent 61,212,592
(8) Levy, M. and Doisy, E.A., (1929), J. Biol. Chem., 77, 733-751
(9) Niculescu, M., (1941), Z. Anal. Chem. 122, 335-344
(10) Larralde, R., Robertson, P. M. and Miller, S., (1995), P.N.A.S. USA, 92,8158-60
(11) Boutlerow, M.A., (1861) C.R. Acad. Sci., 53, 145
(12) Cottle, D.L. and Young, D.W., (1955), U.S. Patent 2,711,415
(13) Beck, A., Lohrmann, R. and Orgel, L., (1967), Science, 157, 952

PHOSPHORAMIDE-BONDED POLYNUCLEOBASEPHOSPHATES AS RIBOSELESS PRECURSORS OF RNA ?

GILBERT TURIAN
Laboratoire de Microbiologie et Bioénergétique, Université de Genève, 20, boulevard d'Yvoy, CH 1211 Genève 4, Suisse

To set in motion the evolutionary organization required self-reproducing information storage admittedly assumed by the "RNA world". However, because of the scarce prebiotic availability of ribose, this role of RNA would have been preceded by sugarless prenucleic polymers of a "pre-RNA world" (Picirilli, 1995; Brack, 1998).

Parallely to other precursor systems such as peptide nucleic acids (PNAs), we have proposed a deterministic coding scenario (Turian, 1996, 1998a,b) in which prebiotically available nucleobases would have stereo-chemically recognized amino acids of fortuitously formed peptides and "pinched" them as proximal anticodon doublets (C-C for glycine, C-G for alanine, etc.) parallely locked by phosphoramide, P~N bondings of the creatine phosphate type on polyphosphates, thereby validating their primal encoding.

1. Phosphoramide Locking of Nucleobases on Polyphosphates

First experimental evidence for possible bonding of purine/pyrimidine nucleobases to phosphate groups of oligo(tri)phosphates was obtained by their incubation in slightly alkaline water and the measured UV hypochromaticity reversed by acid pH (see 2) of their putative phosphoramidate products (Turian and Schoenenberger, 1997). Further incubations and assays by ^{31}P-NMR were conducted in the presence of the condensing agent EDC, a water-soluble derivative of carbodiimide, tautomer of prebiotic cyanamide, known to catalyze the dehydrating change from the linear triphosphate to the cyclic trimetaphosphate (Keele and Miller, 1995), a ring structure which could thus provide the high energy bonds required to drive the phosphoramidations of both imidazole and nucleobases. This was confirmed by the sharp increase of the –21.5 ppm peak signaling trimetaphosphate, in the additional presence of these nitrogen bases opening its ring for their phosphoramide binding, thereby displacing the chemical equilibrium in favor of additional cyclizations (Turian *et al.*, 1998).

Further experimental evidence of the role of trimetaphosphate as phosphoryl donor was provided by its Mg^{2+}-enhanced nucleophilic attack either by imidazole (i) or each of the 4 purine (Pu) or pyrimidine (Py) nucleobases (Turian *et al.*, 1999). Such cycle opening of the trimetaphosphate (Fig. 1a) was signalized by the decrease of its single

J. Chela-Flores et al. (eds.), First Steps in the Origin of Life in the Universe, 107–110.

resonance peak paralleled by the appearance of the two peaks characteristic of the linear triphosphate consequently bearing the nucleophilic P~N bonded nucleobase. The enhancing effect of Mg^{2+} ions on the cycle opening by all nitrogen bases could be ascribed to their shielding of the electrically-charged phosphate groups (Westheimer, 1987) as well as to their cooperation in the nucleophilic effect finalized by the hooking of the bases on phosphate groups.

Final polymerization of the dibasehexaphosphates (coding doublets) by heat treatment and splicing of 1 Pi on 3 detected by NMR (Fig. 1 *left*; Turian *et al.*, unpublished results) would further lead to linear polydibasetetraphosphates insuring lining-up of the nucleobases on polyphosphate chains. The informational tape thereby produced could then be decoded by acylation-amination into new template peptides bonded at the expense of the transitorily broken P~N bonds in coupled, self-reinforcing cycles of prenucleic replication-translation (model in Turian, 2000).

2. Acidic Opening of the "Energy-rich" Phosphoramide Bonds

Phosphoramides - the simplest example being amidophosphoric acid $H_2N\text{-}PO_3H_2$ - are stable in alkaline solution, but labile at acid pH values as well illustrated by phosphoguanidines such as creatine phosphate which dissociates at low pH as:

$$=\text{C-NH-P}^+\text{-O}^- \rightarrow \text{-C=NH}^+ \leftrightarrow \text{P}^+\text{-O}^-$$

In neutral to slightly alkaline media, it is the inhibition of the normal resonance associated with the symmetry of the guanidine ion which confers them the "energy-rich" property of their P-N bond (average 7.0 Kcal mol^{-1}). Oppositely, in low pH media, their resonance form, with the positive charge adjacent to the positively charged phosphorus ion, leads to their hydrolysis by repulsion of their similar electric charges.

In nucleobases as in imidazole (Vogel, 1984), the amino groups are protonated in acidic conditions when the pH of their solvent is lowered to values close to their pKas (from 4.5 for C to 3.5 for U through 4.2 for A). Such so-called prototropic tautomerism is accompanied by movements of protons from one ring nitrogen atom to another one in pyrimidines as in purines (Fig. 15 in Jeffrey and Saenger, 1994). Here also that acid-induced reprotonation leads to the repulsion of the NH^+ group to the electrophilic, positive phosphate acceptor and thus to the hydrolytic opening of the P~N bond (Fig. 1b).

3. C_1-C_5 Ribosylation-bridging of Opened Phosphoramide Bonds

Ribose as such was difficult to synthesize prebiotically (ref. in Brack, 1998). Its first precursor, formaldehyde, can oligomerize by the formose reaction to a mixture of sugars from which only low concentrations of D-ribose could be resolved (Joyce, 1989).

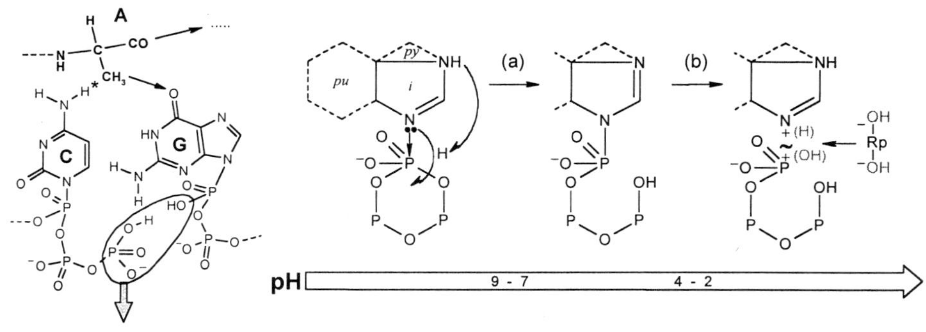

Figure 1. Model of a polyphosphate-built pre-RNA, *left* with coding nucleobases (C+G for A, etc...) locked on decyclized triphosphates by P~N bonds (a), acid-opened (b) and thus powered (~) to intercalate hydroxylated riboprecursors (Rp).

To add to these difficulties, bases and ribose molecules must then come together to react and make nucleosides, a difficult process requiring heating (see Schwartz, 1998). Hence, our proposal that the process of intercalation of ribose precursors would be powered by the high energy released at the opening of the phosphoramide bond and its further bridging electrically driven by the bipolar attraction of the negatively charged hydroxyl groups of the riboprecursors according to the following scale of increased complexity:

- C_1 methylene glycol [$CH_2(OH)_2$], a product of CH_2O hydrolysis in rainwater (Kasting and Brown, 1998), could provide the shortest bridge between a base and a phosphate group (Fig. 1) which could thus be considered as the prototype of a nucleotide (protoribonucleotide);
- C_2 glycolaldehyde (Pitsch *et al.*, 1995) when condensed by two (= C_4) on the already phosphorylated C_1 of methylene glycol could close the pentose cycle of a ribonucleoside;
- C_3 glyceraldehyde provides with its terminal C the phosphorylation site as C_5 of ribose after its condensation with glycolaldehyde, thereby closing the pentose cycle. Another C_3, glycerol, proposed to replace ribose in simpler informational molecules (see Brack, 1998) can be condensed with 1 formaldehyde into a C_4 of the acyclonucleoside (Joyce, 1989; Orgel, 1992) or with 2 CH_2O to complete the ribose cycle by providing the necessary C_5 for phosphorylation of the nucleoside.

The straight *bipolar* bridging of the P~N bonds of polynucleobasephosphates by riboprecursors complexifying into ribose might have been an advantageous short-cut to the progressive, *monopolar* completion of first formed either ribonucleosides or ribosephosphates in the chemical evolution to the ribonucleotides of the RNA world.

Acknowledgments for technical help by A. Fehr and A. Cattaneo (Microbiologie), Dr. E. Rivara-Minten (Chimie pharmaceutique) and Dr. P.-Y. Morgantini (Chimie physique).

4. References

Brack, A. (ed.) (1998) *The Molecular Origins of Life*, Cambridge University Press, Cambridge, U.K.

Joyce, G.F. (1989) RNA evolution and the origin of life, *Nature* **338,** 217-224.

Jeffrey, G.A. and Saenger, W. (1994) *Hydrogen Bonding in Biological Structures*, Springer-Verlag, Berlin, Heidelberg.

Keefe, A.D. and Miller, S.L. (1995) Are polyphosphates or phosphate esters prebiotic reagents? *J. Molecular Evolution* **41,** 693-702.

Kasting, J.F. and Brown, L.L. (1998) The early atmosphere as a source of biogenic compounds, in A. Brack (ed.), *The Molecular Origins of Life*, Cambridge University Press, Cambridge, U.K., pp. 35-56.

Orgel, L.E. (1992) Molecular replication, *Nature* **358,** 203-209.

Picirilli, J.A. (1995) RNA seeks its maker, *Nature* **376,** 548-549.

Pitsch, S., Krishnamurthy, R., Bolli, M., Wendeborn, S., Holzer, A., Minton, M., Lesueur, C., Schlönvogt, I., Jaun, B. and Eschenmoser, A. (1995) Pyranosyl-RNA('p-RNA'): Base-pairing selectivity and potential to replicate, *Helvetica Chimica Acta* **78,** 1621-1635.

Schwartz, A.W. (1998) Origins of the RNA world in A. Brack (ed.), *The Molecular Origins of Life*, Cambridge University Press, Cambridge, U.K., pp. 237-254.

Turian, G. (1996) Polarity at onset of genetic coding. I. Bipolar bondings in the two-step takeover of peptide templates by prenucleic-ribonucleic acids, *Archs Sci. Genève* **49,** 213-227.

Turian, G. (1997) Polarity at onset of genetic coding. II. Primary recognition of amino acids by base doublets of prenucleic sugarless polymers secondarily taken over by ribonucleic acids, *Archs Sci. Genève* **50,** 95-104.

Turian, G. (1998a) Polarity at onset of genetic coding. III. Prenucleic stereochemical interactions between amino acids and nucleobases doublets, *Arch Sci. Genève* **51,** 147-152.

Turian, G. (1998b) Origin of life. I. Recurrent riddles about its genetic coding, *Archs Sci. Genève* **51,** 311-323.

Turian, G. (2000) Model of prenucleic replication cyclically coupling encoding to decoding of peptide templates, *Archs Sci. Genève* **53 (3),** *in press.*

Turian, G. and Schönenberger-Solà, I. (1997) Spectral evidence for phosphoramide bondings between nucleobases and tripolyphosphate possibly generative of prenucleic polybasephosphate chains, *Archs Sci. Genève* **50,** 145-152.

Turian, G., Rivara-Minten, E. and Cattaneo, A. (1998) Similar ^{31}P-NMR signals emitted by imidazole and the nucleobases presumably phosphoramido-bonded to tri(meta)phosphates, *Archs Sci. Genève* **51,** 187-193.

Turian, G., Rivara-Minten, E. and Cattaneo, A. (1999) Further ^{31}P-NMR evidence of phosphoramide bonding of nucleobases by Mg^{2+} enhanced nucleophilic attack on cyclic triphosphate, *Archs Sci. Genève* **52,** 209-216.

Vogel, H.J. (1984) ^{31}P-NMR studies on phosphoproteins, in *Phosphorus-^{31}P NMR Principles and Application*, Academic Press, Inc, New York, p. 105-153.

Westheimer, F.H. (1987) Why Nature chose phosphates, *Science* **235,** 1173-1177.

Section 3:

Possible sources of life's origins

COMETARY MOLECULES & LIFE'S ORIGIN

JOAN ORÓ
Department of Biology and Biochemistry
University of Houston
Houston TX 77204-5934, USA

1. Introduction

The present paper discusses the problem of the origin of life on Earth. It is based on previous experimental work concerning the prebiological synthesis of biochemical molecules, such as adenine from hydrogen cyanide (Oró, 1960) and the subsequent hypothesis developed by the author on the role that cometary matter may have had on the formation of biochemical molecules on the primitive Earth. It was suggested (Oró, 1961) that the capture of comets by the primitive Earth provided the molecular precursors necessary for the eventual appearance of life on our planet. A major paper by A. Delsemme (2000) has been published recently that explains in detail the cometary origin of the biosphere, and provides support for our initial experimental and theoretical studies (Oró, 1960, 1961). The present paper is in agreement with Delsemme (2000) major conclusions, and we believe that the work from both laboratories provides a reasonably complete physical and chemical insight into the difficult problem of the origin of life on the primitive Earth.

2. The Biogenic Elements

It is humbling to realize that most of the so called observable universe is made by about 98% hydrogen and helium, primarily generated in the BIG BANG primordial explosion of the cosmos 15 thousand million years ago. Of the remaining 2% of matter, approximately 1% is made by five of the biogenic elements (C, N, O, S, P) in addition to hydrogen. The remaining 1% is constituted by all the other heavier chemical elements up until uranium and beyond. The key points are that the elements necessary for life constitute a substantial portion of the observable universe, and that they are continuously formed in stars and diffuse to the interstellar medium.

3. The Interstellar Medium – The Universe is Organic

From an organic chemist point of view it is important to recognize that the nuclides of carbon and of the other biogenic elements are quite stable at the very high temperatures of stellar nuclear synthesis, and that as soon as the circumstellar temperatures decrease they acquire their complementary set of electrons and are transformed into reactive radicals ready to interact with other similar reactive biogenic element radicals to generate all kinds of organic molecules and other compounds. It is

J. Chela-Flores et al. (eds.), First Steps in the Origin of Life in the Universe, 113–120.

of interest for organic chemistry and life that the bulk of the molecules of the interstellar medium are organic or made of biogenic elements. It is not surprising therefore that the first five neutral molecules detected in interstellar space were made of biogenic elements. These first five molecules were: ammonia, water, formaldehyde, hydrogen cyanide and cyanacetylene. Furthermore, more than 75% of the molecules, are organic. Indeed, from the point of view of a molecular chemist, one could say, as we have emphasized in the past, that chemically the universe is essentially organic.

4. Formation of the Earth-Moon System

From cosmochemical and geochemical studies it is well known that the gravitational collapse of the proto-solar nebula gave rise to the formation of the Solar System, with the Sun, the rocky terrestrial planets, the gaseous giant planets, as well as the satellites, asteroids, comets and meteorites. It is also known, from radioactive decay dating of meteorites that the age of the Solar System is approximately 4,550 million years old.

There have been several theories for the formation of the Earth-Moon system (Oró et al., 1990). But it has not been until the past two decades that a correct theory has been developed. According to this theory, the Earth-Moon system has been the result of a collision of a cosmic body, the size of Mars, with the proto-Earth. Thanks to the work of William Hartmann, and Cameron and colleagues (e.g. Cameron and Benz, 1991), it is known that the impact of the collision melted the surface and part of the mantle of the Earth, and that most of the colliding's body-matter was ejected into orbit and eventually coalesced to form our natural satellite, the Moon. In this way the dynamic aspects, such as the angular momentum of the Moon, as well as the similarities and small differences in the chemical composition between the Moon and the Earth's mantle, are explained. The results of the analysis of the lunar samples from the Apollo program are in line with this interpretation. As a consequence of this massive collision no volatiles remained on the Earth and on the Moon (Delsemme, 2000).

5. Cometary Matter Captured by the Primitive Earth

Where did all the water and other volatiles of the terrestrial biosphere come from? The simple answer is from comets (Oró, 1961; Delsemme, 2000). Based on Urey's calculations on the probability of cometary collisions with the Earth, I arrived at the conclusion (Oró, 1961) that our planet, must have acquired up to 10^{18} g, that is to say, the weight of the biosphere, of cometary matter during its early phases of accretion. In a revision of this work with A. Lazcano and others we arrived at the value of 10^{23} g. It is of interest that independent calculations by other scientists produced similar quantitative values (Oró et al., 1992). The value of $1.0 \times 10^{24\text{-}25}$g is that by Delsemme who considers that all the water of the oceans is of cometary origin. This has now been explained in detail in two recent papers (Delsemme, 1998, 2000).

The fact that not long ago some astronomers found a discrepancy in the deuterium content D/H of comets Halley, Hyakutake, and Hale-Bopp with that of ocean water, is explained by the fact that these are comets that come from the Oort's cloud, highly enriched in deuterium in accordance with the low temperatures prevailing in the interstellar space region where they originally came from. In contrast, the comets that brought the water to the primitive Earth, were comets from the Jupiter's region, which had essentially the lower deuterium value in agreement with the D/H content of the terrestrial ocean waters (Delsemme, 1998, 2000).

6. Molecular Composition of Comets

Aside from the isotopic composition, the molecular composition of comets may be slightly different for different comets. However, it should be similar to the extent that the source of their molecules is the same, the interstellar medium (Greenberg and Hage, 1990). This notwithstanding, the proto-Solar nebula may have been processed to different degrees depending on the history of the comets during the formation of the Solar System. In short, we would have Oort cloud comets, Kuiper belt comets, and Jovian comets, the latter during the early formation of the Solar System. The volatile molecules detected in several comets by microwave spectroscopy correspond to some of the molecules detected in the interstellar medium, for instance ammonia, water, formaldehyde, hydrogen cyanide, cyanacetylene, etc. Some of the organic molecules detected by mass spectrometry in the dust of Halley's comet with the VEGA I and GIOTTO space probes (Kissel and Krueger, 1987), include adenine, xanthine and other biochemical molecules.

7. Cometary Molecules as Precursors of Life

This is only an attempt to conceptually simplify the problem of the origin of life, so that one can see that only a few cosmic or cometary molecules are needed to produce the three major groups of biochemical compounds: a) Amino acids and peptides; b)Purines, pyrimidines, monosaccharides, nucleotides and their polymers, c) Fatty acids, phospholipids and membrane structures. In a more condensed graphical version, the corresponding chemical reactions, can be seen in **Fig. 1**

Fig 1. Prebiotic synthesis from cometary molecules.

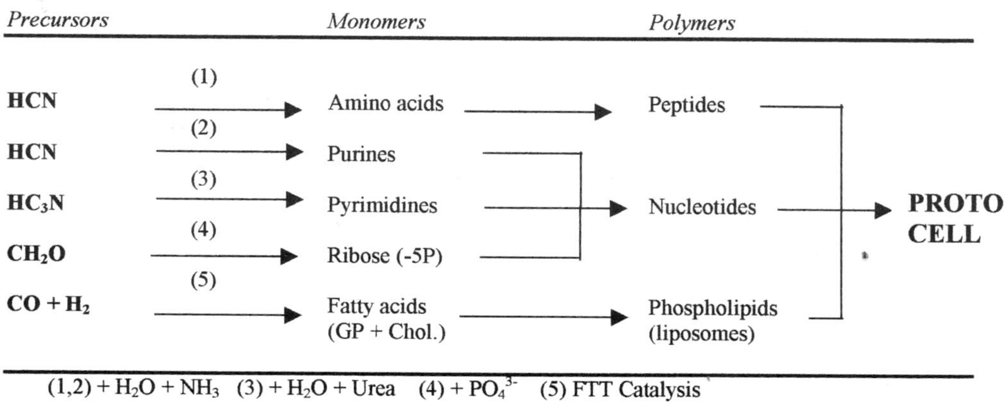

This condensed version of the problem is only an introduction to the detailed reactions that are required for the prebiotic chemical evolution processes, which presumably took place on the primitive Earth, and preceded the appearance of life on our planet.

8. Prebiotic Formation of Amino Acids

One of the first attempts to synthesize biochemical compounds under possible primitive Earth conditions was carried out in 1951 in the laboratory of Melvin Calvin (1965). However, they only obtained small amounts of formic acid and formaldehyde from the reduction of carbon dioxide in aqueous solution by ionizing radiation. Obviously, as pointed out by Urey one could not expect to generate important biochemical compounds from a non-reducing atmosphere of CO_2. The classic experiment of Stanley Miller (1953) on the formation of amino acids in a reducing atmosphere of H_2, CH_4, NH_3 and H_2O vapor by the action of electric discharges under possible primitive Earth conditions is well known, for it was the first successful experiment testing the Oparin-Haldane hypothesis of chemical evolution on the origin of life. Even though such process of amino acid syntesis probably occurred on the primitive Earth, if the above cometary hypothesis is considered (Oró, 1961; Delsemme, 2000) the primitive Earth, already contained many amino acids and other biochemical compounds when electric discharge processes made their contribution to prebiotic synthesis. Both types of syntheses could have provided a continuous source of amino acids and other compounds necessary for the appearance of life on Earth. At any rate, if the amino acids found in the Murchison meteorite are considered as a probable example of the ones formed from cometary sources on the primitive Earth, a significant fraction of the biological amino acids would have been available for the synthesis of proteins and other compounds.

9. Prebiotic Formation of Purines

The formation of adenine from ammonium cyanide or from HCN in aqueous ammonia solution is well known (Oró, 1960). The simplest description of the reaction involves 5 molecules of HCN which condense sequentially (forming a dimer, a trimer, etc) to generate one molecule of adenine which indeed can be properly called pentameric hydrogen cyanide. (5 HCN $\rightarrow$ $H_5C_5N_5$). During the past forty years the original work has been confirmed and extended in several laboratories (Lowe et al., 1963; Ferris and Orgel, 1965). All the purines have been synthesized (adenine, guanine, xanthine, and hypoxanthine) and several other biochemical compounds are formed in the process depending on the conditions of the reaction (Oró, 1965). Furthermore, in the absence of liquid water, that is with HCN and NH_3 alone, adenine is formed in 15% and higher yields. (Wakamatsu et al., 1966). More recently the formation of adenine, glycine and other amino acids, as well as, guanine has been observed to occur at very low temperatures (Levy et al., 1999, 2000). These authors suggest the possibility of formation of these compounds in Europa. It is therefore not surprising that Kissel and Krueger (1987) detected the presence of adenine, xanthine, as well as, purine in the dust of comet Halley. Furthermore, adenine formation may also occur in a collapsing molecular cloud (Chakrabarty and Chakrabarty, 2000). These findings provide additional support for the role of comets in the formation of biochemical compounds on the primitive Earth (Oró, 1961, Delsemme, 2000).

10. Formation of Pyrimidines

The pyrimidines, uracil and cytosine are easily synthesized by evaporation of a solution of cyanacetaldehyde (Robertson and Miller, 1995). This compound is in turn obtained from cyanacetylene and cyanate, which were used in earlier experiments (Ferris et al., 1968). Uracil is also obtained by the hydrohysis of cytosine. Thymine was obtained in our laboratory by the methylation of uracil with formaldehyde (Stephen-Sherwood, et al., 1971). It is of interest to point out that the most stable pyrimidine, namely, uracil, has also been detected in the Murchison meteorite and in comet Halley.

11. Formation of Monosaccharides

The aldol condensation of formaldehyde to monosaccharides is well known. The formation of the different products, among them ribose have been discussed by several authors (e.g. Oró, 1965). It is of interest from a cosmic point of view that formaldehyde was the first organic molecule detected in interstellar space, and that comet Halley apparently had large amounts of this compound and was detected by several authors in the form of the normal polymer polyoxymethylene (Huebner W.H., 1987). Therefore when cometary material was captured by the primitive Earth there should have been plenty of formaldehyde to react to form monosaccharides and other biochemical compounds. The formation of 2-deoxyribose was carried in our laboratory, through the aldol condensation of glyceraldehyde with acetaldehyde. As we have seen acetaldehyde is present in comet Halley and glyceraldehyde can be obtained by aldol condensation of formaldehyde. The two isomeric 2- deoxypentoses were formed: 2-deoxyxylose and 2-deoxyribose. The best catalyst was found to be simple NH_4OH and the reaction could be carried out at room temperature for over one month, and 2-deoxyribose could be obtained in yields as high as 15% (Oró, 1965).

12. Formation of Nucleosides and Nucleotides

The synthesis of purine nucleosides in small yields is not difficult. Thus, when a mixture of purine bases are heated with ribose and salts of sea water (which contain magnesium chloride, a catalyst for this reaction) small but reasonable yields of different purine nucleosides are obtained, namely, 4,5% β-inosine, 2% β-adenosine, 3,2% β-guanosine and 1,8% β-xanthosine (Fuller et al., 1972). This is what would be expected during the evaporation of coastal lagoons of the primitive Earth's oceanic cometary waters. The synthesis of pyrimidine nucleosides and nucleotides is more difficult, because it requires that once the α-anomers are formed they be rearranged to the natural isomers (β-anomers) with the help of irradiation with ultraviolet light (Sanchez and Orgel, 1970). The phosphorylation of nucleosides can be carried out with orthophosphates by dry heating at relatively low temperatures (below 100°C) in the presence of urea and ammonium chloride, as shown by Lohrmann and Orgel (1971). Minor modifications of this method produce reasonable yields of 5'-nucleosidemonophosphates minimizing the formation of 2':3' cyclic phosphate derivatives, as shown recently by Reimann and Zubay (1999). This was done with the four nucleosides, uridine, cytidine, inosine and adenosine. In summary, the phosphorylation of nucleosides involve reactions using moderate temperatures and ambient humidity simulating the conditions resulting from the evaporation of primeral lakes from cometary waters.

13. High Energy Phosphates and Coenzymes

The synthesis of pyrophospate under possible primitive Earth conditions has been shown by Miller and Parris (1969). It occurs by reaction of orthophosphate with cyanate on the surface of hydroxylapatite, presumably forming carbamyl phosphate which reacts with another phosphate molecule to generate pyrophosphate. As is known, phospate has been detected in fluffy interplanetary dust particles (IDP'S) of probable cometary origin (Macia et al., 1997). Once inorganic pyrophosphate is available, transphosphorylations can take place. In this way AMP has been further transphosphorylated to ADP and ATP (Miller and Parris, 1969). Furthermore the synthesis of all deoxyribomononucleosides mono-, di- and triphosphates has been accomplished in our laboratory by simple evaporation in aqueous solutions in the presence of inorganic phosphate and cyanamide. Under similar prebiotic conditions a number of simple phosphate containing coenzymes can also be synthesized. For instance, in addition to ADP and ATP, the coenzymes UDPG, CDP-choline, CDP-ethanolamine and other coenzymes have also been obtained in our laboratory (Oró, 1994).

14. Synthesis of Polynucleotides and Polypeptides

The formation of biochemical monomers presumably took place during the first phase of prebiotic evolution. The conditions for the formation of biopolymers had to be somewhat milder and at any rate more specific. In the presence of condensing agents, or activating compounds, it has been possible to synthesize oligopeptides and oligonuccleotides. In our laboratory we have been able to synthesize oligodeoxyribonucleotides up to 8 nucleotides long and polyglycines of up to 15 monomers long by means of the condensing agent cyanamide (Stephen-Sherwood et al., 1974). The synthesis of polynucleotides has been studied particularly in the laboratories of Orgel and of Ferris, using activated nucleotides. For instance, Ferris and coworkers (Ferris, 1998) have shown that a common clay mineral, montmorillonite, is an excellent catalyst for the polymerization of 5'-phosphorimidazolides of the nucleoside bases. Polymers up to 40 to 50 units long have been obtained in this way.

15. Replication of Polynucleotides and the RNA World

In Orgel's laboratory they have synthesized polymers of GMP, (PolyG) starting with a template of PolyC (Inoue and Orgel, 1982). The monomers of GMP were

activated with 2-Me Imidazolide. In this way PolyG is obtained as the template of PolyC. However, the reaction only works with the isomeric D-mononucleotides, not with racemic mixtures of D- and L-mononucleotides.

The well known work on ribozymes by Cech and Altmann, and the Gilbert notion of an RNA only world elevated the importance of RNA over DNA. The idea of an RNA world first, is possibly a breakthrough on the theory of the origin of life that may solve some of the problems in prebiotic evolution. The fact that a single molecule may embody the three major characteristics of life: encoding information, catalysis and replication, appears to be a simplifying principle. An optimist view is that of "The molecular biologist dream" of Joyce and Orgel (1999). Simply stated this says that the primitive Earth was loaded with a pool of activated ribonucleotides waiting to be polymerized into RNA. But, as we pointed out, there are problems with the replication of RNA. An alternative possible solution to the RNA world problem is that of chemical coevolution of ribonucleotide oligomers and catalytic peptide protoenzymes as discussed by Lahav (1993). Another important suggestion is that of peptide nucleic acid (PNA) as a model structure for the primordial genetic material. This would be a pre-RNA world as proposed by Nielsen (1993) and more recently elaborated by Miller and coworkers (2001). See also Eschenmoser's views in Orgel (2000, 2001).

The prebiological synthesis of phospholipids and the formation of liposomes for the encapsulation of replicating and catalytic biomolecules, in order to generate a primordial protocell is discussed in a paper from our laboratory (Oró, 1994). It is hoped that in the next century significant advances will be made in our understanding of the origin of life on Earth (Schopf, 2001).

References

Calvin, M. (1969) *Chemical Evolution*. Oxford Univ. Press, New York, 278 pp.

Cameron, A.G.W. and Benz, W. (1991) The origin of the Moon and the single impact hypothesis IV, *Icarus,* **92**, 204-216.

Chakrabarty, S.K. and Chakrabarty, S. (2000) Adenine abundance in a collapsing molecular cloud. *Indian J. Phys.* **2**, 97-99.

Delsemme, A. (1998) The deuterium enrichment observed in recent comets is consistent with the cometary origen of sea water, *Planetary and Space Science,* **47**, 125-131.

Delsemme, A. (2000) Cometary Origin of the Biosphere, *Icarus,* **146**, 313-325.

Ferris, J.P. (1998) Catalyzed RNA synthesis for the RNA world. In Brack, A. (Ed.) *Molecular origins of life: Assembling pieces of the puzzle*, Cambridge Univ. Press, pp 255-268.

Ferris, J.P. and Orgel, L.E. (1965) Aminomalononitrile and 4-amino-5-cyanoimidazole in hydrogen cyanide polymerization and adenine synthesis. *J. Am. Chem. Soc.* **87**, 4976-1977.

Ferris, J.P., Sanchez, R.A. and Orgel, L.E. (1968) Synthesis of pyrimidines from cyanoacetylene and cyanate. *J. Mol. Biol.* **33**, 693.

Fuller, W.D., Sanchez, R.A. and Orgel, L.E. (1972) Solid state synthesis of purine nucleosides. *J. Mol. Evol.* **1**, 249.

Greenberg, J. M. and Hage, J. I. (1990) From Interestellar dust to comets: A unification of observational constraints, *Ap. J.*, **361**, 260-274.

Huebner, W.E. (1987) First polymer in space identified in Comet Halley. *Science,* **237**, 628-630.

Inoue, G.F., and Orgel, L.E. (1982) Oligomerization of (guanosine 5' phosphor) -2-methylimidazolide on polyC: an RNA polymerase model. *J. Mol. Biol.,* **162**, 201-207.

Joyce, G.F. and Orgel, L.E. (1999) Prospects for understanding the origin of the RNA world. In Gesteland R.F., Ceck T.R. and Atkins, J.F. (Eds.). *The RNA world. The Nature of Modern RNA suggests a prebiotic RNA*. Cold Spring Harbor Laboratory Press, pp 49-77.
Kissel, J. and Krueger , F.R. (1987) The organic component in dust from comet Halley as measured by the PUMA mass spectrometer on board Vega I, *Nature,* **326**, 755-760.
Lahav, N. (1993) The RNA world and co-evolution hypotheses and the origin of life: Implications research, strategies and perspectives. *Origins Life Evol. Biosph.,* **23**, 329-344.
Levy, M., Miller, S.L. and Oró, J. (1999) Production of guanine from NH_4CN polymerizations. *J. Mol. Evol.* **49**, 165-168.
Levy, M., Miller, S.L., Brinton, K. and Bada, J.L. (2000) Prebiotic synthesis of adenine and amino acids under Europa-like conditions *Icarus,* **145**, 609-613.
Lohrmann, R. and Orgel, L.E. (1971) Urea-inorganic phosphate mixtures as prebiotic phosphorylating agents. *Science,* **171**, 490-494.
Lowe, C.U., Rees, M.W. and Markham, R. (1963) Synthesis of complex organic compounds from simple precursors: formation of amino acids, amino acid polymers, fatty acids and purines from ammonium cyanide, *Nature,* **199**, 219-222.
Macià, E., Hernandez, V. and Oró, J. (1997) Primary sources of phosphorus and phosphates in chemical evolution. *Origins Life Evol. Biosph.,* **27**, 459-480.
Miller, S.L. (1953) Production of aminoacids under possible primitive earth conditions. *Science,* **117**, 528-529.
Miller, S.L. and Parris, M. (1964) Synthesis of pyrophosphates under primitive Earth conditions, *Nature,* **204**, 1248-1250.
Miller, S.L. et al., (2000) Peptide Nucleic Acid rather than RNA may have been the first genetic molecule *(Presented, sixth Trieste Conference on Chemical Evolution, september 18-20).*
Nielsen, P.E. (1993) Peptide Nucleic Acid (PNA): A model structure for the primordial genetic material? *Origins Life Evol. Biosph.,* **23**, 323-327.
Orgel, L. (2000) A Simpler Nucleic Acid. *Science,* **290,** 1306-1307.
Orgel, L. (2001) The origin of biological information. In: *The Emergence of Life, Proceedings of the Gold Medal Simposium* (Shcopf, J. W., ed.), *In press.*
Oró, J. (1960) Synthesis of adenine from ammonium cyanide, *Biochem. Biophys. Res. Commun.* **2**, 407-412.
Oró, J. (1961) Comets and the formation of biochemical compounds on the primitive Earth, *Nature,* **190**, 389-390.
Oró, J. (1965) Stages and mechanisms of prebiological organic synthesis. In *The Origins of Prebiological Systems* (S.W.Fox, ed.), pp 137-171, Academic Press, New York, New York.
Oró, J. (1994) Chemical synthesis of lipids and the origin of life. *J. Biol. Physics,* **20**, 135-147.
Oró, J., Miller, S.L. and Lazcano, A. (1990) The origin and early evolution of life on Earth. *Ann. Rev. Earth and Planet. Sci.,* **18**, 317-356.
Oró. J., Mills, T. and Lazcano, A. (1992) Comets and the formation of biochemical compounds on the primitive Earth-A review, *Origins Life Evol. Biosph.,* **21**, 267-277.
Reimann, R. and Zubay, G. (1999) Nucleoside phosphorylation: A feasible step in the prebiotic pathway to RNA. *Origins Life Evol. Biosph.,* **29**, 229-247.
Robertson, M.P. and Miller, S.L. (1995) An efficient prebiotic synthesis of cytosine and uracil, *Nature,* **375**, 772-774.
Sanchez, R.A. and Orgel, L.E. (1970) Synthesis and photoanomerization of pyrimidine nucleosides. *J. Mol. Biol.,* **47**, 531.
Schopf J.W. (2000) Solution to Darwin's dilemma: Discovery of the missing precambriam record of life *(Lecture presented at the Sixth Trieste Conference on Chemical Evolution, sept. 18-20).*
Stephen-Sherwood, E., Odom, D.G. and Oró, J. (1974) The prebiotic synthesis of deoxythymidine oligonucleotides, *J. Mol. Evol.,* **3**, 323-330.
Stephen-Sherwood, E., Oró, J. and Kimball, A.P. (1971) Thymine: A possible prebiotic synthesis. *Science,* **173**, 446-447.
Wakamatsu, et al. (1966) Synthesis of adenine by oligomerization of hydrogen cyanide, *J. Org. Chem.,* **31**, 2035-2036.

WHEN AND HOW QUICKLY DID LIFE EMERGE ON EARTH?

J. MAYO GREENBERG

The Raymond and Beverly Sackler Laboratory for Astrophysics at Leiden Observatory, P.O. Box 9513, 2300 RA Leiden, The Netherlands
e-mail: greenber@strw.LeidenUniv.nl

Abstract
Necessary conditions for early and rapid chemical evolution of prebiotic organics in comet dust to the first life forms are discussed.

1. Introduction

Besides the question of the chemistry of life's origins there are three other questions which may be posed: "How soon after the major impact era on earth did life appear?" or "Could it have started even before?" and "How long did it take for the prebiotic chemical constituents to organize themselves into some sort of metabolizing, reproducing life form?". In effect one asks whether a certain level of stable, not necessarily static conditions, is required to allow prebiotic chemistry to produce life. This may be one of the reasons why the origin of life in hydrothermal vents has emerged as a viable theory. First of all it occurs in regions which are shielded from external impacts and secondly it seems to provide a cyclic –and in that sense – stable dynamic environment for prebiotic chemistry. The alternative suggestion that prebiotic chemistry started from exogeneous sources such as comets provides the kind of material in which sufficiently directed and stable conditions to initiate life quickly and early offers a number of advantages. In section 2, I will summarize the reasons we believe life originated as early as it could have, given the early geological evidence. In section 3 I will give some arguments showing how comets could have delivered the seeds of life. In section 4, I will indicate plausible reasons why the seeds could have initiated life in short times consistent with the prevailing instabilities at the earth's surface.

2. When did life begin?

An answer to this question is still hidden in the fact that the conditions on the earth are now and have always been subjected to changes and weathering which turns over and modifies the surface because of tectonics and the atmosphere. This means first of all there are limitations to the dating of rocks. This has been amply discussed elsewhere in this volume by Stephen Moorbath. The oldest detected rocks are no older than about 3.6 – 3.8 billion years. On the other hand the findings of fossils also seem to be limited by

J. Chela-Flores et al. (eds.), First Steps in the Origin of Life in the Universe, 121–130.

this same dating problem. If one follows the idea that the amount of fossil evidence seems to decrease in the same proportion as the amount of datable rocks one winds up with an ambiguity in the time for life's emergence. If the fall-off in the amount of fossils and the amount of datable rocks is such that the ratio is a constant, the evidence for the earliest fossils places no strict limit on the earliest life (see Fig. 1). I present this not as a proof that life originated before the major impact era ended 3.8 billion years ago but rather to suggest that, if life did originate on the earth's surface (read primeval ocean) then it either could have happened many times even before conditions became less hostile, or that, if it started only after the heavy bombardment ended it succeeded very quickly. Even in the immediate post impact era, one major impact may have occurred every 6000 years.

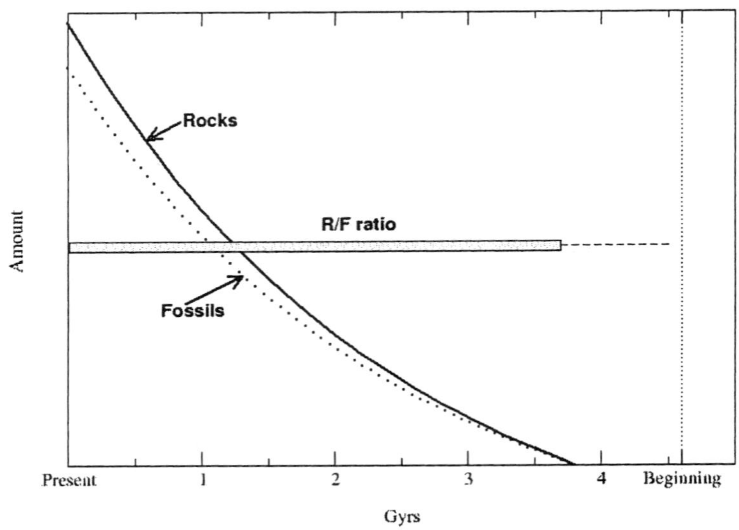

Fig. 1 Schematic of the decay curves for the amounts of datable rocks and of fossils found as a function of time. The ratio of these amounts appears to be roughly constant with an earliest time of ~3.8 Gyr. The dashed curve extends the ratio to earlier times and indicates the possibility that the earliest life on earth may predate the earliest fossil findings.

3. Comets as the bearers of the seeds of life.

A) THE CHEMISTRY OF INTERSTELLAR DUST AND COMETS.

The prebiotic chemistry contained in comets is now becoming increasingly recognized as having originated in the interstellar dust. The details of the chemical evolution of the interstellar particles are most directly traced via the presence of frozen mantles on these particles. The resemblance of these frozen ices to the comet coma molecules is shown in Table 1.

Table 1: Comparison of the abundances of ices in the interstellar medium (towards IRS9) and of cometary volatiles (at -1 AU). Cometary abundances are from Table 7 in Crovisier (1999). ISM ice abundances are taken from a compilation of Schutte (1999) from various sources.

Species	Interstellar ices	Cometary volatiles	
H_2O	=100	=100	
CO	10--40	2-20	
CH_3OH	5	1-7	
CO_2	10	2-6	
H_2CO	2-6 tentative	0.05--4	
HCOOH	3 tentative	~ 0.1	
CH_4	1--2	0.7	
other hydrocarbons	??	~ 1	C_2H_2, C_2H_6
NH_3	<10	0.5	
O_3	≤ 2	??	
OCN^-	≤ 0.5-2	0.2	nitriles+ HNCO
OCS	0.2	0.4	OCS + CS
SO_2	??	~ 0.1	
H_2	≥ 1	??	
N_2	??	??	
O_2	??	??	

There are clearly important simple prebiotic molecules seen here, such as formaldehyde, and various CN groups represented by these volatile components. However, as an additional product of the chemical processes occurring in and on dust grains, there is substantial evidence for the presence of much more complex organic molecules on the dust grains. These have been variously referred to as organic refractories or as "yellow stuff". The direct astronomical evidence for such molecules is limited mainly to the observations of 3.4 ?m absorption features corresponding to CH stretches in CH_2 and CH_3 groups in aliphatic hydrocarbons (Butchart et al.., 1986; Pendleton et al.., 1994). However, laboratory simulations, including space simulations, have provided a plausible basis for believing that there is a wide variety of prebiotic molecules manufactured in space (Greenberg et al.., 1972). In the Leiden laboratory we have produced many such samples (Schutte, 1988) of which an example is shown in Table 2.

Table 2. GC-MS and HPLC analysis of the products resulting from the simulation of interstellar grain photolysis (taken from Briggs et al., 1992). These products were obtained from the photolysis of H_2O:CO:NH_3 = 5:5:1. The amounts are the average of ten analyses

Compound	Amount
Glycolic acid ($HOCH_2CO_2H$)	30.0
3-hydroxy-propionic acid ($HOCH_2CH_2CO_2H$)	1.0
Formamidine ($HCNH(NH_2)$)	0.1
2-hydroxy-acetamide ($HOCH_2CONH_2$)	18
Hexamethylene tetramine ($(CH_2)_6NH_4$)	1.25
Urea (NH_2CONH_2)	2.5
Biuret ($NH_2CONHCONH_2$)	0.07
Oxamic acid (NH_2COCO_2)	0.1
Ethanolamine ($HOCH_2CH_2NH_2$)	0.5
Glycerol ($HOCH_2CH(OH)CH_2OH$)	6.25
Glycine ($NH_2CH_2CO_2H$)	0.4
Oxamide ($NH_2COCONH_2$)	0.4
Glyceric acid ($HOCH_2CH(OH)CO_2H$)	7.5
Glyceramide ($HOCH_2CH(OH)CONH_2$)	18.7
Ethylene glycol* ($HOCH_2CH_2OH$)	6.2
Serine@ ($HCOCH_2CN(NH_2)COOH$)	tr
Alanine@ ($CH_3CH(NH_2)COOH$)	tr

* It was not possible to determine whether the ethylene glycol observed using ^{13}CO contains ^{13}C.
@ Obtained by J.R. Cronin (unpublished).

Table 2 gives the results of the GC-MS and HPLC analysis of one of our interstellar analog produced organics following ultraviolet processing of interstellar ice mantles (Agarwal et al.., 1985; Briggs et al.., 1992). This is work which is ongoing and should be taken as illustrative rather than final. Nevertheless in broad outline we can clearly see how important interstellar prebiotic chemistry can be. More recently we have subjected such material – which we refer to as first generation organics – to further ultraviolet processing and the material not only turned brown but exhibited a whole new range of molecules categorized as polycyclic aromatic hydrocarbons (PAH's) with aliphatic side groups (Greenberg et al., 2000). A schematic of the evolution of dust with this highly irradiated mantle material (dust in diffuse clouds) to the dust as it may evolve in a protosolar collapsing molecular cloud is indicated in Fig. 1 of Greenberg and Li (1998). What is presumed to occur to the dust in the cloud of increasing density is that first volatile molecules from the gas stick and are photoprocessed leading to an inner layer of "yellow stuff" organics and a final unphotolyzed layer of ices including all the gas phase produced molecules form an outermost layer. In addition, there are known to be much more finely divided dust particles in space and these will accrete along with the ices in the final stages of accretion (Li and Greenberg, 1997)..

Ultimately, we believe these fully accreted dust grains whose mean size (diameter) is of the order of a few tenths of a micron will coagulate in the outer regions of the protosolar nebula to form cometesimals and finally comets. The current close correlation between the dust from icy molecules and the comet coma and molecules leads us to believe that all the above processes take place at temperatures low enough to preclude significant evaporation of the dust before coagulation into comets (Mumma, Weissman, Stern, 1993; Bockelée-Morvan et al., 2000; Crovisier, 1999). The implication of this is that certainly the interior constituents of the grains – the complex organics – have been preserved as mantles on the dust. Confirmation of the preservation has had its most direct observation via the mass spectrometry of the dust of comet Halley which exhibited apparently equal amounts of the organic elements O, C, N with the rocky elements Si, Mg, Fe (Kissel et al., 1986a, 1986b; Jessberger and Kissel, 1991)

B) THE FLUFFY NATURE OF COMETS AND COMET DUST.

When the interstellar dust coagulated to form comets it was impossible for it to become fully packed. This would require either high pressure compression, as occurs from gravitational forces in the interior of a much more massive body, or by evaporation and filling of the interstices. Both of these are excluded by the nature of comets. What is clearly indicated is that comet dust; i.e., that material which remains following evaporation of the volatile components by solar heating after coming off the comet surface, is extremely loosely packed. In fact, quantitative calculations show that its porosity is $P \approx 0.95$ or 95% empty space (Greenberg and Hage, 1990). Reinserting the volatile fraction on the individual dust components brings the porosity down to about $P = 0.8$.

C) COMET DUST DELIVERY OF PREBIOTIC ORGANIC TO THE EARTH.

When a comet approached the early earth the main difference from the present epoch was that the atmosphere was not only different in composition – it was predominantly CO_2 - it was also most likely as much as 10 times the present level (Kasting, 1993). One of the implications of this kind of atmosphere is that it was not conductive to the production of precursor organic molecules as was assumed in the Miller experiment (Miller, 1953). This already suggests the need for an exogeneous source of prebiotic molecules. However, another implication is that should a low density fragment of a comet impact such an atmosphere it could undergo rapid braking and, because of its high surface to volume, it would not be greatly heated in its descent to the earth's surface. This has been quantitatively discussed by Basiuk et al. (1999) where it has been shown that tenth micron particles (or small aggregates of such particles) would not be heated enough to substantially modify such molecules as amino acids. Thus when a comet impacted the earth's atmosphere, the smaller pieces which were initially ablated could survive the impact velocity chemically "intact". Furthermore, there is theoretical justification for the fact that even before impacting the earth's atmosphere, the comet might have (if its size is ≥0.5 km) split, as Shoemaker/Levy did in the neighborhood of Jupiter, and shed much of its material as dust (Rickman and Greenberg, 1998; Davidsson 1999) The bottom line here is that a substantial fraction of the prebiotic chemistry can survive in comet dust which lands on the earth.

D) ADVANTAGES OF COMET DUST.

The advantages offered by comet dust can be summarized by the expression "The right stuff at the right time". Insofar as timing, implicitly, comet dust was raining continuously and in great abundance on the earth ever since the earth was born. Most of this material was also being destroyed by the heavy bombardment but as the heavy bombardment began to decay the periods of relative quiet became longer and longer. At some point, the conditions became favourable long enough for life to be created. How long is long enough? Here the intrinsic properties of comet dust play a role. As already noted, comet dust already contains a large fraction of prebiotic molecules. And furthermore because it is packaged in fluffy aggregates of tenth micron particles, the smaller comet dust will have largely preserved these organics intact during descent through the earth's atmosphere. We are led to believe that by say 4.4 billion years ago there was a substantial amount of liquid water on the earth's surface (Mojzsis et al., 2001; Wilde et al., 2001) in which the comet dust would have landed. What we then picture is that, even if only 1/10 % of an impacting comet survived as such comet dust, there would have been present in the ocean as many as 10^{25} such fluffy grains of about 5-10 ?m diameter. While it is still being argued how much of the earth's water originated in comets, it is hard to believe that such a large input of material to the earth's surface material as implied by the major impact era would not have contributed a substantial fraction. During the 500 million year period of massive bombardment the equivalent of 20 million comets hit so that the current ocean mass of $2\ 10^{24}$ g is about the same order of magnitude as the water brought by comets (Chyba, 1990). It is not unreasonable to expect that some of this survived. The fact that the D/H ratio of the present ocean is less than that in current interstellar matter but still greater than the cosmic abundance ratio implies that some fraction originated in the interstellar chemically fractionated forms (Greenberg, 1986). If the comets contributed H_2O to the earth then it is likely that the other volatile components of the comet could have also been contained in the comet "rain"; e.g., such molecules as formamide and formic acid would have been present. We note here that one comet of 5 km radius (comparable with comet Halley) contains about $10^{16} - 10^{17}$ grams of water. Given the survival of the 10^{25} comet dust grains immersed in the rich ocean soup – not pure water but water bearing its own dissolved prebiotic molecules – presents a very strong presumptive basis for the origin of life. A recent discussion of possible pathways for the synthesis of purine, adenine, cytosine and 4(3H) – pyrimedinone from formamide suggests a series of conditions which are coincidentally met by the "comet dust in an early ocean" (Saladino et al., 2001).

Complex molecules required to initiate life on earth very likely came from outer space rather than having been formed on the earth (Chyba et al., 1990; Greenberg and Mendoza-Gómez, 1992). In any case, the abundance of complex organics in interstellar dust as inferred from infrared spectra and from laboratory analog experiments is certainly an immense reservoir of prebiotic molecules. There is reason to believe that the interstellar dust in the presolar nebula had an excess of lefthanded amino acids (Cronin and Pizzarello, 1997; Engel and Macko, 1997). It is these molecules-as a major fraction of the interstellar dust- which makes up about 20% of the mass of a comet (Greenberg, 1982, 1998). Favorable factors provided by comet dust to the initiation of life involve four factors: (1) high porosity, (2) inorganic (mineral) surfaces for catalysis, and of course, (3) the presence of preformed chiral organics, (4) an ocean rich in small soluble prebiotic organics. We envisage that these comet dust fragments may fulfill the requirements suggested by Krueger and Kissel (1989) for chemical thermodynamics to start molecular self-organization. Each comet dust particle consisting of a porous aggregate of submicron silicate cores with organic refractory mantles sitting within a water bath (the local "ocean") containing nutrients satisfies the conditions for non-equilibrium thermodynamics as in Nicolis and Prigogine (1977). (See Greenberg, 2000)

The discussion by Krueger and Kissel (1989) points out particularly to the chemical composition of the organics predicted by Greenberg (e.g., Briggs et al. 1992) - nitrogen as a hetreroatom and as reactive precursors of purines and pyrimidines-possibly cytosine. The oxygen is present in aldehydes and carboxylic acids, carbon oxides and of course, water. Most of the hydrocarbons are unsaturated and the aldehydes may polymerize to form sugars given the right conditions.

4. The Rosetta mission and the origin of the solar system

The most pristine relics of the ealrly solar system are comets. From the first space missions to comet Halley we already became aware of their complex chemistry and their resemblance to an aggregate of interstellar dust (Greenberg, 1986). Both the US NASA and the the European ESA have planned missions to comets and asteroids which will reveal even more details about their composition. It is the goal of the ESA Rosetta mission to explore the nucleus of a comet (P/Wirtanen) in detail from close distance over an extended period of time not only remotely from the Orbiter, but also by means of a Lander on the surface of the nucleus. (Verdant and Schwehm, 1998). What were the conditions in the solar system when comets were formed? How was the collapsing cloud material affected by heating and turbulence? One of the aspects of comets that will tell us about these conditions is the density. Remarkably, as much as we have learned about comets there are still debates about their density. The Rosetta orbiter will both see the comet and it will be able to measure its mass from its orbit. Thus, for the first time we will know whether comets are as fluffy as has been predicted. (Greenberg, Hage, 1990; Li, Greenberg, 1999).This information is critical to appreciate the initial aggregation processes which occurred in the pre-solar nebula. The orbiter will follow the comet from the time it is about 4.5AU from the sun until about 1AU. During this time, the COSIMA (Kissel, 1999) instrument will capture the dust and analyze it for both its organic and inorganic constituents, including isotopic abundances. Far from the sun the cometary particles are expected to contain many of the ices remaining from the original interstellar dust. Farther in, the remaining constituents will be the complex organic and the rocky/metallic components.

The molecular composition of the organic phase of the solid cometary particles will be used to establish its exobiological relevance as possible organic precursors to the origin of life. The COSAC experiment on the lander (Rosenbauer et al., 1999) is also aimed at the in situ investigation of the matter of a cometary nucleus with respect to its chemical and isotopic composition. In addition it will measure its chirality (handedness) of the organics to answer the question of whether the left handedness of the amino acids of life on earth originated from the interstellar dust which aggregated to form comets. Not only will the chemistry and density be measured but also the structure. The grain morphology will be studied with an atomic force microscope with nm resolution. This should tell us something about the size of the individual postulated interstellar dust particles. Considering what will be achieved in the next 20 years I would venture the prediction that within another couple of decades we will have the capability to 'mine' pristine comet nucleus material and bring it back to the earth for laboratory analysis. If comets are indeed made of interstellar dust, we will have in our hands the ultimate form of molecular cloud dust resembling the dust around low mass protostars - 4.6 billion years ago. We will know the material out of which the solar system was formed.

What might we anticipate being accomplished in the next millenium? When we retrieve material from a comet intact (and we will someday) and bring it back to earth one thing to try is to distribute small pieces (5-10?m) in appropriate water baths and follow their chemical evolution. We can expect that newer experiments techniques will have developed to the point at which this will be possible. I do not predict that this will make life but we can expect to find whether these processes of directed chemical evolution occur and at what rate. In a way, this is one phase of the COSIMA project where dust from the comet will land on a water saturated surface before being subject to mass spectroscopy.

Thought is already being directed by Krueger, Kissel and the author to simulating the porous structure of comet dust in which laboratory organics are imbedded and placing these structures in water. It is just a matter of providing such a model to determine how effectively the chemical evolution is directed.

5. Acknowledgements

I thank Jochen Kissel and Franz Krueger for many stimulating discussions.

6. References

Agarwal, V.K., Schutte, W., Greenberg, J.M., Ferris, J.P., Briggs, R., S. Connor, C.P.E.M. Van de Bult, Baas, F. Photochemical reactions in interstellar grains, photolysis of CO, NH_3 and H_2O. Origins of Life 16 (1985) 21-40.

Basiuk, V.A., Douda, J., Navarro-Gonzalez, R., Trasnport of extraterrestrial biomolecules to the earth: problem of thermal stability, Adv. Space Res. 24 (1999) 505-514

Bockelee-Morvan, D., Lis, D.C., Wink, J.E., et al., New molecules found in comet C/1995 01 (Hale-Bopp); investigating the link between cometary and interstellar material, Astron. Astrophys. 353 (2000) 1101-1114

Briggs, R., Ertem, G., Ferris, J.P., Greenberg, J.M., McCain, P.J., Mendoza-Gómez, C.X., Schutte, W., Comet Halley as an aggregate of interstellar dust and further e vidence for the photochemical formation of organics in the interstellar medium, Origins of Life and Evolution of the Biosphere 22 (1992) 287-307

Butchart, I., McFadzean, A.D., Whittet, D.C.B., Geballe, T.R., Greenberg, J.M. Three micron spectroscopy of the galactic centre source IRS 7. Astron. Astrophys. 154 (1986) L5-L7

Chyba, C.F., Impact delivery and erosion of planetary oceans in the early inner Solar System, Nature 343 (1990) 129-133.

Chyba, C.F., Thomas, P.J., Brookshaw, L., Sagan, C., Cometary delivery of organic molecules to the early earth, Science 249 (1990) 366-373.

Cronin, J.R., Pizzarello, S., Enantiomeric analysis of chiral organic compounds from the Murchison meteorite, Science 275 (1997) 951.

Crovisier, J., Solids and volatiles in comets: from nuclei to cometary atmospheres. In: Formation and Evolution of Solids in Space, J.M. Greenberg, A. Li, Eds. (Kluwer, Dordrecht, Netherlands, 1999), pp.389-426.

Davidsson, B.J.R., Tidal splitting and rotational breakup of solid spheres, Icarus 142 (1999) 525-535

Engel, M.H., Macko, S.A., Isotopic evidence for extraterrestrial non-racemic amino acids in the Murchison meteorite, Nature 389 (1997) 265-268

Fegley, B. Chemistry of the solar nebula. In: The Chemistry of Life's Origins, eds. J.M. Greenberg, C.X. Mendoza-Gómez, V. Pirronello. Kluwer, 1993, p. 75-147

Greenberg, J.M., What are comets made of - a model based on interstellar dust, in: Comets (ed. L.L. Wilkening), Univ. of Arizona press, 1982, p. 131-163.

Greenberg, J.M. Evidence for the pristine nature of comet Halley. In: The comet nucleus sample return mission, Proc. workshop Canterbury, UK (15-17 July 1986), SP-249 (1986) p. 47-55

Greenberg, J.M. The chemical composition of comets and possible contribution to planet composition and evolution. In: The galaxy and the solar system (eds. R. Smoluchowski, J.N. Bahcall and M.S. Matthews), Un. of Arizona Press, 1986, p. 103-115.

Greenberg, J.M., Hage, J.I. From interstellar dust to comets: a unification of observational constraints. Astrophys. J. 361 (1990) 260-274

Greenberg, J.M., Mendoza-Gómez, C.X. The seeding of life by comets. Adv. Space Research 12 (1992) nr. 4, p. 169-180.

Greenberg, J.M., Making a comet nucleus, Astron. Astrophys. 330 (1998) 375-380,

Greenberg, J.M., The secrets of stardust, Scient.Amer. 283 (2000) Dec. p. 46-51.

Greenberg, J.M., Gillette, J.S., Munoz-Caro, G.M., Mahajan, T.B., Zare, R.N., Li, A.,Schutte, W.A, De Groot, M., Mendoza-Gomez, C. Ultraviolet photoprocessing of interstellar dust mantles as a source of polycyclic aromatic hydrocarbons and other conjugated molecules. Ap.J. Lett. 531 (2000) L71-L73.

Jessberger, E.K., Kissel, J., Chemical properties of cometary dust and a note on carbon isotopes, in: Comets in the Post-Halley Era (eds. R. Newburn, M. Neugebauer, J. Rahe), Kluwer, Dordrecht, 1991, p. 1075-1092.

Kasting, J.F., "Early evolution of the atmosphere and ocean", in: The chemistry of life's origins (Greenberg, J.M., Mendoza-Gómez, C.X., and Pirronello, V., eds.), Dordrecht, Kluwer, p. 149-176, 1993.

Kissel, J., Brownlee, D.E., Büchler, K. et al., Composition of comet Halley dust particles from Giotto observations, Nature 321 (1986) 336-337

Kissel, J., Sagdeev, R.Z., Bertaux, J.L. et al., Composition of comet Halley dust particles from Vega observations, Nature 321 (1986) 280-282

Kissel, J., In situ measurements of evolved solids in space with emphasis on cometary particles. In: Formation and Evolution of Solids in Space (eds. J.M. Greenberg, A. Li), Kluwer, Dordrecht, 1999 p.427.

Krueger, F.R., Kissel, J., Biogenesis of cometary grains - thermodynamic aspects of self-organization, Origins of Life etc. 19 (1989) 87-93..

Li, A., Greenberg, J.M. A unified model of interstellar dust. Astron. Astrophys. 323 (1997) 566-584.

Li, A., Greenberg, J.M. The distributed CO in comet Halley. Space Sci. Rev. 90 (1999) 69-74

Miller, S.L., A production of amino acids under possible primitive Earth conditions, Science 117 (1953) 528-529

Mojzsis, S.J., Harrison, T.M., Pidgeon, R.T., Oxygen-isotope evidence from ancient zircons for liquid water at the earth's surface 4,300 Myr ago, Nature 409 (2001) 178-180.

Mumma, M.J., Weissman, P.R., Stern, S.A., Comets and the origin of the solar system: Reading the Rosetta stone, in E.H. Levy, J.I. Lunine and M.S. Matthews (eds), Planets and Protostars III, Univ. of Arizona Press, Tucson, 1993, 1177-1252.
Nicolis, G., Prigogine, I., Self-organization in nonequilibrium systems, Wiley, 1977.
Pendleton, Y.J., Sandford, S.A., Allamandola, L.J. et al., Astroph. J. 437 (1994) 683
Rickman, H., Greenberg, J.M., Tidal splitting of comets in earth's vicinity, Highlights of Astronomy (IAU) (J. Andersen, ed.) 11A (1998) 262-265.
Rosenbauer, H., Foselier, S.A., Chielmetti, A. et al., Adv. Space Res. 1999
Saladino, R., Crestin, C., Costanzo, G., et al., A possible prebiotic synthesis of purine, adrenine, cytosine and 4(3H)-pyrimidinone from formamide: implications for the origin of life. Bioorg. Med. Chemistry, 2001.
Schutte, W.A. The evolution of interstellar organic grain mantles. Thesis, Leiden, 17 February 1988.
Schutte, W.A. Laboratory simulation of processes in interstellar ices. In: Formation and evolution of solids in space. Greenberg, J.M., Li, A. (eds.). Kluwer, 1999, p. 177-201.
Verdant, M., Schwehm, G.H. The international Rosetta mission. ESA Bull. 93 (1998)
Wilde, S.A., Valley, J.W., Peck, W.H. et al.,Evidence from detrital zircons for the existence of continental crust and oceans on the earth 4.4 Gyr ago, Nature 409 (2001) 175-178

COSMOCHEMISTRY IN THE EARLY UNIVERSE

DENIS PUY
Institute of Theoretical Physics, University of Zurich
Winterthurerstrasse, 190 - 8057 Zurich (Switzerland)
and
PAUL SCHERRER
Laboratory for Astrophysics, 5232 Villigen,
(Switzerland)

Abstract. At early times the Universe was filled up with an extremely dense and hot gas. Due to the expansion it cooled below the binding energies of atoms which led to the formation of the first nuclei. In the physical environment of the post-recombination period of hydrogen, molecules such as H_2, HD and LiH can be formed. The proto-object formation, resulting from the growth of linear density fluctuations in the early Universe, can have an important impact on the chemical state of the Universe. Hence it can be enriched with metals, and thus lead to the formation of the first pre-biotic molecules.
In this contribution, I will present some scheme for the formation of primordial molecules and discuss the consequence of the formation of first stars on the existence of possible primordial pre-biotic.

1. Introduction

The conventional model of nucleosynthesis is the process by which chemical elements and their isotopes are formed, the heavy elements (carbon, oxygen, nitrogen and heavier ones) are thought to be the result of thermonuclear burning in stars, and especially the relatively rare stars which become supernovae [1], [2].
Standard big bang nucleosynthesis generated just a few elements: hydrogen, deuterium, helium and lithium, traces of beryllium and boron. The recombination of this primeval plasma led progressively to form the neutral species at $\sim 10^9$ yr after the big bang. Little is known about the history of the Universe corresponding to 10^6-10^9 years (or to redshifts of $z \sim 1000 - 5$)

J. Chela-Flores et al. (eds.), First Steps in the Origin of Life in the Universe, 131–134.

after the big bang, and only recently astrophysicists begun to investigate seriously the post-recombination era and particularly the problem of the formation of the first molecules [3], [4], [5].
It is usually supposed that the first objects (first stars or galaxies) could be formed from a gas cloud by the gravitational attraction. The clouds contract adiabatically and some kind of *cryogen* is necessary to form bound systems. Since the primeval gas does not contain heavy elements like carbon, primordial molecules such as H_2, HD and LiH are good candidates to lead to an efficient cooling [6], [7]. In this context primordial molecules can play an important role in the pregalactic gas particularly on the thermochemical stability [8], [9].
In this contribution, I will present in Sect. 2 the formation of primordial molecules. Then in Sect. 3, I will describe how these molecules can influence on the formation of the first objects. In this context, I will discuss the important influence of these proto-objects on the second generation of molecular formation, which will be important for the formation of biomolecules. I will describe, in Sect. 4, some possible outlooks of this study.

2. Formation of primordial molecules

At the end of the recombination period of hydrogen, the ions became progressively neutralized in the order of their ionization potentials. With the creation of neutral species, molecular ions can initiate the molecular formation. The chemistry of the early Universe is the chemistry of the elements hydrogen and its isotope deuterium, helium, lithium and its isotopic forms. The ongoing physical reactions are immense after the recombination of hydrogen, the main processes are collisional (ionization, radiative recombination, attachement...) and radiative due to the presence of the cosmic microwave background radiation (photoionization, photodissociation, photodetachement) [4], [5].
Two main routes lead to H_2:

$$H^- + H \longrightarrow H_2 + e^- \quad \text{and} \quad H_2^+ + H \longrightarrow H_2 + H^+, \tag{1}$$

when HD is formed through H_2 [10] and LiH from LiH^+ [11]

$$H_2 + D^+ \longrightarrow HD + H^+ \ , \quad LiH^+ + H \longrightarrow LiH + H^+. \tag{2}$$

The reaction rates depend on the temperature. Thus the temperature and density evolution equations must be solved simultaneously, which needs in turn the simultaneous determination of molecular cooling and heating rates [4], [5]. From the initial abundances given by the standard big bang nucleosynthesis [12], the integration of the chemical network lead to the following abundances of the main molecules at $z \sim 5$:

e^-/H	H_2/H	HD/H	LiH/H
$\sim 3 \times 10^{-4}$	$\sim 10^{-6}$	1.2×10^{-9}	$\sim 7 \times 10^{-20}$

3. First stars and cosmochemistry

The first objects formed after the molecular formation had no metals heavier than lithium. The role of molecular cooling in the fragmentation of primordial gas clouds is important [13]. The excitation of the rotational levels of H_2 and HD induce a cooling which can trigger thermal instability in the collapsing cloud [6], [9]. Thus, primordial molecules play a crucial role on the evolution of the first objects, through the process of fragmentation. The first objects can be, in some cases, massive stars [14].
Thus, as soon as stellar processes occur, proto-objects can lead through SN explosions to the formation of other molecules species, such as CO, CI or HCN. They in turn are important sources of contamination of the medium, and thus can offer different ways of early pre-biotic molecular formation. Chakrabarti showed that a significant amount of adenine, a DNA base, may be produced during molecular cloud collapse [15], through the chain reaction (HCN addition):

$$HCN + HCN \longrightarrow ... + HCN... \longrightarrow H_5C_5N_5 \text{ (adenine)}. \qquad (3)$$

Actually, the direct detection of the prebiotic molecules is not easy. For example, the tentative of detection of glycine were negative or below the confusion limit [16].

4. Outlooks

We have seen that through the formation and the evolution of the first objects, it is useful to consider DNA bases that have been produced at an early stage of the Universe. A large lepton asymmetry in the standard big bang nucleosynthesis could lead to the formation of a non-negligible number of heavier elements [17], and thus a strong incidence on the cosmochemistry. For example, the existence of primordial carbon or nitrogen can produce primordial bio-molecules or pre-biotic molecules such as amino acids, sugars, adenine just after the cosmological recombination period.
Thus, the formation of DNA bases could happen in the early history of the Universe. Barrow pointed out the link between the biochemical timescales for the evolution of life and the astrophysical timescales that determine the time required to create an environment supported by a stable hydrogen burning star [18]. In the context of pre-biotic molecular formation in

the early Universe, it is not necessary to establish this natural link between the stars and the prebiotic chemistry. Pre-biotic molecules could have contamined the first objects and planets from the beginning. Nevertheless these molecules could have been destroyed during the formation of planetary disks. Comets carrying away these molecules are thus interesting sites of protection and potential sources of contamination for planetary systems. Thus, life bases could be appeared earlier than what we think. In this context the emergence of extraterrestrial intelligence should be not exceedingly rare (see [19], [20], [21])

Acknowledgments

I would like to thank Prof. Chela-Flores for the organization of this 6th Trieste Conference, Lukas Grenacher and Flavia Puy-Svara for useful discussions and comments. Part of this work was supported by the *Tomalla Foundation* and by the Swiss National Science Foundation.

References

1. Signore M., Puy D. 1999, New Ast. Rev. 43, 185
2. Signore M., Puy D. 2000, New Ast. Rev. accepted, astro-ph/0010634
3. Lepp S., Shull M. 1984 280, 465
4. Puy D., et al. 1993, A& A **267**, 337
5. Galli D., Palla F. 1998 A&A 335, 403
6. Puy D., Signore M. 1996, A&A 305, 371
7. Puy D., Signore M. 1997, New Ast. 2, 299
8. Puy D., Signore M. 1998, New Ast. 3, 27
9. Puy D., Signore M. 1998, New Ast. 3, 247
10. Palla F., Galli D., Silk J. 19995 ApJ 451, 44
11. Stancil P., Lepp S., Dalgarno A. 1996 ApJ 458, 401
12. Sarkar S. 1996 Rep. Prog. Phys. 59, 1493
13. Silk J. 1977 ApJ 211, 638
14. Tegmark M. et al. 1997 ApJ 474, 1
15. Chakrabarti S., Chakrabarti S.K. 2000 A&A 354, L6
16. Combes F., Nguyen-Q-Rieu, Wlodarczark 1996 A& A 308, 618
17. Dubrovich V., Dolgov A. 2000 private communication
18. Barrow J., proceedings the Varenna Conference *The Origin of Intelligent Life in the Universe* 1998, astro-ph/9811461
19. Carter B. 1983 Philos. Trans. R. Soc. London A 310, 347
20. Barrow J., Tipler F. 1986, The Anthropic Cosmological Principle (Oxford:Clarendon)
21. Livio M. 1999 ApJ 511, 429

ABSORPTION AGAINST THE COSMIC 2.7 K BACKGROUND

A technique for detection of asymmetric top molecules in cosmic objects

SURESH CHANDRA[1], W.H. KEGEL[2], A. K. SHARMA[1]
[1] *School of Physical Sciences, S.R.T.M. University, Nanded 431 606, India; Email: sch@iucaa.ernet.in*
[2] *Institut für Theoretische Physik der J.W. Goethe Universität, D-60054 Frankfurt am Main, Germany; Email: kegel@astro.uni-frankfurt.de*

In cool cosmic clouds, where non-linear molecules are formed, kinetic temperature in the cloud may be insufficient to excite some of these molecules to higher rotational energy levels, and therefore, they are not capable to generate their emission spectra. Hence, the detection of such molecules through emission spectra becomes infeasible. However, absorption of radiations, coming from the background source(s), corresponding to some transition(s) in these molecules, may be possible, as the low energy levels of the molecule are always populated.

Observation of an interstellar line in absorption against the cosmic 2.7 K background is an unusual phenomenon. The intensity, I_ν, of a line generated in an interstellar cloud, with homogeneous excitation conditions, is given by

$$I_\nu - I_{\nu,bg} = (S_\nu - I_{\nu,bg})(1 - e^{-\tau_\nu})$$

where $I_{\nu,bg}$ is the intensity of the continuum against which the line is observed, τ_ν the optical depth of the line, and S_ν the source function. For positive optical depth (which is a usual situation for most of the lines), observation of an interstellar line in absorption against the cosmic background, obviously, implies the excitation temperature of the line, T_{ex}, to be less than 2.7 K.

Up to now, only two lines have been found in absorption against the cosmic 2.7 K background. The first one is the 1_{10} - 1_{11} transition of formaldehyde (H_2CO) at 4.831 GHz, which was found in absorption in several directions (Palmer and Zuckerman, 1969). However, in some cases, it has been seen in emission, and even as a maser line (Forster et al, 1980; Whiteoak, and Gardner, 1983). The second line found in absorption against the cosmic

J. Chela-Flores et al. (eds.), First Steps in the Origin of Life in the Universe, 135–138.

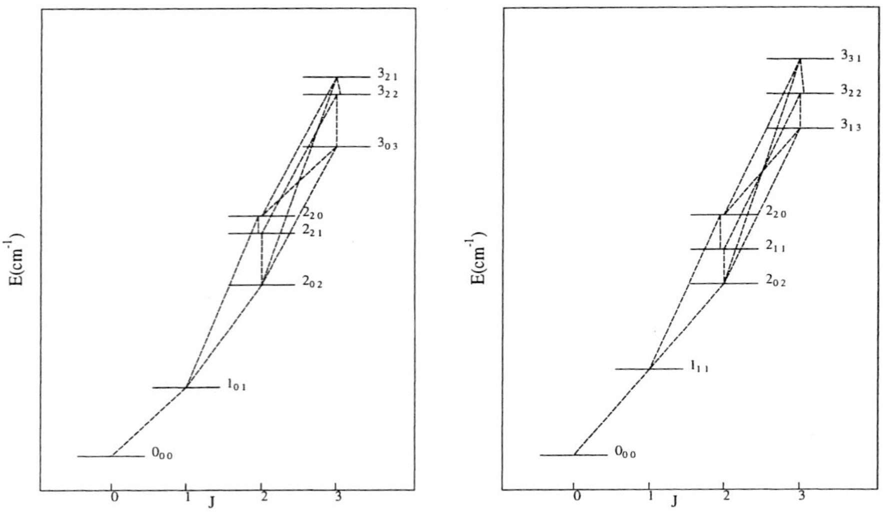

Figure 1. Rotational energy levels for rotational quantum numbers up to $J = 3$ (with arbitrary values of the energies E) in *a*-type and *b*-type asymmetric top molecules, along with the allowed radiative transitions.

2.7 K background, in a large number of cosmic objects, is the 2_{20} - 2_{11} transition of cyclopropenylidene (C_3H_2) at 21.587 GHz (Matthews et al., 1986; Madden et al. 1989) . Cox et al. (1987), however, reported the observation of this line in emission in the Planetary Nebula NGC 7027.

A good number of molecules, detected in cosmic objects, fall under the category of asymmetric top molecules, and have been detected through their emission spectra. In the cool cosmic clouds, kinetic temperature is not very high, and therefore, a number of molecules present in the clouds might not have been detected as they may not be capable to generate their emission spectra. Obviously, the observation of absorption line(s) from the molecule may be used as a technique for the detection of the molecules in cosmic objects. In order to deal with transfer of radiation through an interstellar cloud, one accounts for a multi-level system of energy levels of the molecule, and solves the statistical equilibrium equations coupled with the equations of radiative transfer. There are two simple limiting cases for the solution of the equations:

(i) when the collisional rates are negligible in comparison to the radiative ones ($n_{H_2} \rightarrow 0$), the T_{ex} becomes equal to the T_{bg}.

(ii) when the collisional rates are dominant over the radiative ones ($n_{H_2} \rightarrow \infty$), the T_{ex} becomes equal to the T_{kin}, assuming that $T_{kin} > T_{bg}$.

However, for intermediate values of n_{H_2}, T_{ex} not necessarily lie in be-

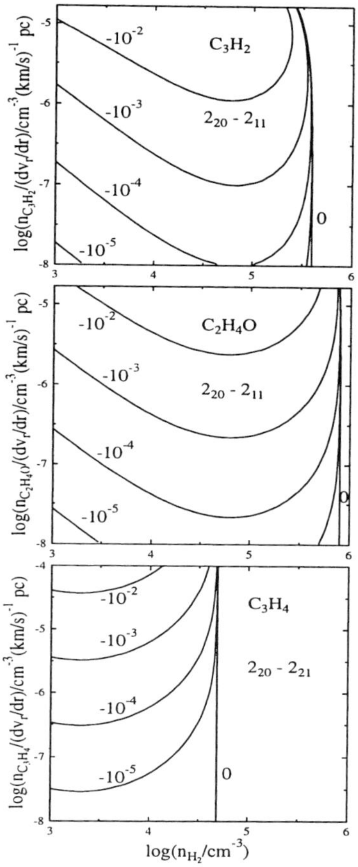

Figure 2. The iso-lines for intensity against the cosmic 2.7 K background, in the unit of Planck's function at the kinetic temperature of T (K), i.e., $(I_\nu - I_{\nu,bg})/B_\nu(T)$, for the 2_{20} - 2_{11} transition of C_3H_2 and C_2H_4O, and the 2_{20} - 2_{21} transition C_3H_4. Only negative values for the kinetic temperatures T = 30 K are plotted. For large molecular hydrogen densities (on the right-side of the iso-lines), the value becomes positive.

tween T_{bg} and T_{kin}. Obviously, the situation of anomalous absorption ($T_{ex} < T_{bg}$) is very unusual, where collisional rates should not play significant role. However, excitation to the higher levels would be due to collisions. As soon as collisional rates become significant, we have either $T_{ex} > T_{bg}$ or $T_{ex} < 0$, and the line may be observed in emission. On the other side, pure radiative transitions also do not allow the situation $T_{ex} < T_{bg}$. However, in asymmetric top molecules,when the levels with $J = 3$ are above the levels with $J = 2$, and the collisional deexcitations are negligible in comparison to the radiative ones, the two transitions $2_{20} - 2_{21}$ and $2_{20} - 2_{11}$ show the anomalous absorption (Figures 1 a and b).

In Figure 1 (a), the level 2_{21} can de-excite through the weak radiative transition $2_{21} \rightarrow 2_{02}$, whereas the level 2_{20} de-excites through the strong radiative transition $2_{20} \rightarrow 1_{01}$. Consequently, the molecules stay for a rather long time in the level 2_{21}, and wait for absorption of a frequency, corresponding to the transition $2_{20} - 2_{21}$, coming from the background source(s). In astronomical objects having low densities, the transition 2_{20} - 2_{21} may obviously show absorption, even against the cosmic 2.7 K background. Similar explanation is applicable for Figure 2 (b) for absorption of the transition 2_{20} - 2_{11}, even against the cosmic 2.7 K background.

As an example, we considered three molecules, C_3H_2, C_2H_4O and C_3H_4 at the kinetic temperature 30 K. The NLTE occupation numbers of the molecules are calculated in an on-the-spot approximation by using the escape probability method. The iso-lines of the intensity, I_ν, showing absorption against the cosmic background, $I_{\nu,bg}$, in the unit of Planck's function at the kinetic temperature T, for the transition 2_{20} - 2_{121} of C_3H_2 and C_2H_4O, and for the transition 2_{20} - 2_{21} of C_3H_4 are shown in Figure 2.

The investigations show that the transition 2_{20} - 2_{121} of C_3H_2 and C_2H_4O, and the transition 2_{20} - 2_{11} of C_3H_4 would show absorption, even against the cosmic 2.7 K background, in the cosmic objects having low density. The investigation for C_3H_2 is found in agreement with the observations of Matthews et al. (1986), Madden et al. (1989), and Cox et al. (1987).

Financial support from the I.C.T.P., Trieste, Italy, C.S.I.R., New Delhi, India, & S.R.T.M. Univ., Nanded, India is thankfully acknowledged.

References

Cox, P., Güsten, R. and Henkel, C. (1987) Detection of the hydrocarbon ring molecule C_3H_2 in the planetary nebula NGC 7027, *Astron. Astrophys.* **181**, L19 - L22.

Forster, J.R., Goss, W.M., Wilson, T.L., Downes, D. and Dickel, H.R. (1980) A formaldehyde maser in NGC 7538, *Astron. Astrophys.* **84**, L1 - L3.

Madden, S.C., Irvine, W.M., Matthews, H.E., Friberg, P. and Swade, D.A. (1989) A survey of cyclopropenylidene (C_3H_2) in galactic sources, *Astron. J.* **97**, 1403 - 1422.

Matthews, H.E., Madden, S.C., Avery, L.W. and Irvine, W.M. (1986) The C_3H_2 2_{20} - 2_{11} transition: Absorption in cold dark clouds, *Astrophys. J.* **307**, L69 - L73.

Palmer, P. and Zuckerman, B. (1969) Formaldehyde absorption in dark nebulae, *Astrophys. J.* **156**, L147 - L150.

Whiteoak, J.B. and Gardner, F.F. (1983) 4.8-GHz H_2CO maser emission in Sgr B2, *Mon. Not. Roy. Astron. Soc.* **205**, 27p - 32p.

CAN BIO-MOLECULES BE FORMED IN COLLAPSING INTERSTELLAR CLOUDS?

SANDIP K. CHAKRABARTI
S. N. Bose National Centre For Basic Sciences
JD Block, Salt Lake, Sector-III, Kolkata-700098 India
and Centre for Space Physics
114/v/1A Raja S.C. Mullick Rd., Kolkata-700047, India

AND

SONALI CHAKRABARTI
M. M. C. College
20 Ramkanto Bose Street, Kolkata-700003, India
and Centre for Space Physics
114/v/1A Raja S.C. Mullick Rd., Kolkata-700047, India

Abstract. Using parametric reaction rates of certain crucial reactions, we show that it may be possible to form significant biomolecules such as adenine, glycine etc. during the collapse process of the interstellar molecules. If correct, this would imply that the earth might have been contaminated by these bio-molecules. The formation of carbon based life may thus be more generic than what was previously thought.

1. Introduction

The cause of the origin of life on earth is not yet known. Certain study of isotopic ratios on carbon inclusions in ferric salt deposits in Greenland shows that life existed on earth before 3.85 billion years (Mojzsis et al. 1996; Holland 1997; Schidlowski 1993) or possibly even before 4 billion years (Mukhin & Gerasimov 1993). However, earth's age is supposed to be between 4.55 to 4.57 billion years (Faure 1986). Thus the time to develop life from scratch is perhaps very short (Crick & Orgel 1973). One of the ways, the formation of life on earth could be assisted if the building blocks of life, namely, amino acids and at least some of the bases of DNA could be

J. Chela-Flores et al. (eds.), First Steps in the Origin of Life in the Universe, 139–144.

produced well before the planetary formation, presumably during chemical evolution of interstellar clouds. A simple way to produce a base of DNA such as adenine ($H_5C_5N_5$) would be successive addition of HCN molecules in four steps (Volkenshtein, 1983). If the reaction rates were known, one could couple hydrodynamics and chemical evolution together to see how adenine abundance could evolve as the interstellar cloud collapses. Similarly, other bio-molecule formation could be studied. Details are presented in Chakrabarti & Chakrabarti (2000ab).

2. Reaction Network

We take the UMIST database (Millar, Farquhar & Willacy 1997; hereafter referred to as MFW97) as our basis of chemical reactants and reactions, but added several new reactions such as synthesis of amino acids (alanine and glycine), hydroxy-acids (glycolic and lactic acids), DNA base (adenine, see, Volkenshtein 1983), urea synthesis etc. These new reactions make the total number of species to be 421. In the UMIST database, the rate constant for a two body reaction is written as (MFW97),

$$k = \alpha(T/300)^{\beta}\exp(-\gamma/T) \ \ \mathrm{cm}^3 \ s^{-1} \tag{1}$$

where, α, β and γ are constants and T is the temperature. First we estimate amino acid synthesis rate from Fig. 8 of Schulte & Shock (1995). Urea synthesis rate is kept comparable to the rates given in UMIST table. The rate constants were taken to be $\alpha = 10^{-10}$, $\beta = \gamma = 0$ for each two-body reactions. In the first run we chose the rate constants for adenine synthesis to be similar to other two body reactions In §5, we revise this to include radiative association. Initial composition of the cloud is discussed in Chakrabarti & Chakrabarti (2000a).

3. Hydrodynamics During the Collapse

We start with a spherically symmetric interstellar cloud with generic initial conditions: initial size $r_0 = 3 \times 10^{18}$cm, average temperature $T = 10$K, and angular velocity $\Omega = 10^{-16}$ rad s^{-1}. The sound speed becomes $a_s = (kT/\mu m_H)^{1/2} \sim 19200$cm s^{-1} and corresponding initial density (Shu, Adams & Lizano 1987) is $\rho = a_s^2/2\pi G r^2 = 10^{-22}$g cm^{-3} and accretion rate is $\dot{M} = 1.06 \times 10^{20}$g s^{-1}. Here, $\mu = 2.3$ mean molecular weight, m_H is the mass of the hydrogen atom, and k is the Boltzmann constant. Initially the cloud is isothermal and density $\rho \propto r^{-2}$ (Chandrasekhar 1939). The velocity is constant. When opacity becomes high enough to trap radiation (say, at $r = r_{tr}$), collapse is adiabatic with $\rho \propto r^{-3/2}$. Because of rotation, centrifugal barrier forms at $r = r_c$, where centrifugal force balances gravity.

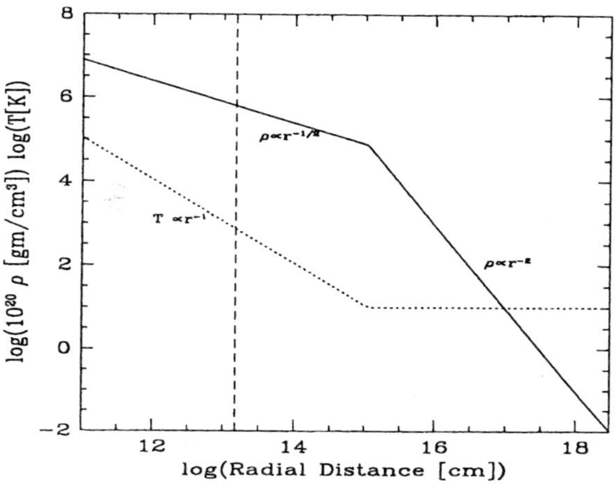

Fig. 1 : Log-log plot of temperature (dotted) and mass density (in units of 10^{20}) as functions of the radial distance. Vertical dashed line is drawn at 1AU.

Density falls off as $\rho \propto r^{-1/2}$ in this region (Hartmann, 1998). Details of the hydrodynamics work is in Chakrabarti & Chakrabarti (2000a). Figure 1 shows the temperature and mass density (in units of 10^{20}) distribution in a log-log scale.

4. Results

Figure 2a shows the log-log variation of mass fractions of some of the light and intermediate species of our simulation with radial distance (in cm). For clarity, we plot the curves alternately by solid and dotted type. In Fig. 2b, we plot the variation of more complex molecules. On the upper axis, we have put time (in seconds) elapsed since the beginning of collapse at radial distances of 10^{18}cm, 10^{17}cm, 10^{16}cm and 10^{15}cm respectively. Towards the end of the collapse, time spent is negligible. We note that around $r \sim$ 10AU, the mass fractions of adenine, urea and glycine are already significant. This computation, does not consider the destructions of adenine at a higher temperature region. It is likely that much of these contaminants are destroyed during collapse and formation of proto-earth. However, comets formed in the inner cloud could carry away these pre-biotic molecules and deposit them during future impacts on planets.

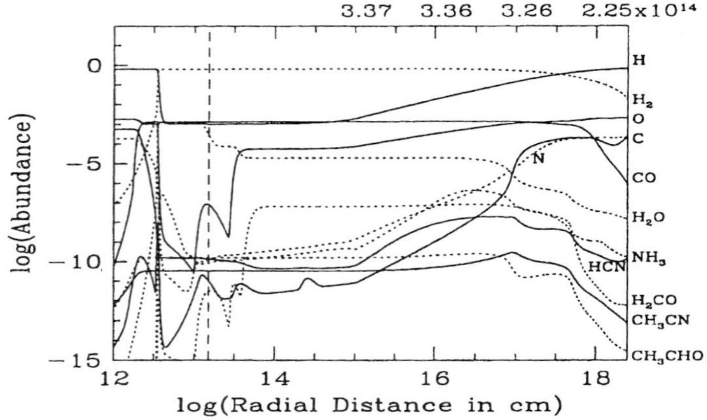

Fig. 2a : Log-log plot of the mass fraction of some of the lighter and intermediate mass species as functions of the radial distance. Alternate species have been plotted with dotted curves for clarity. Upper axis shows time elapsed in seconds since collapse began. Vertical dashed line is drawn at 1AU.

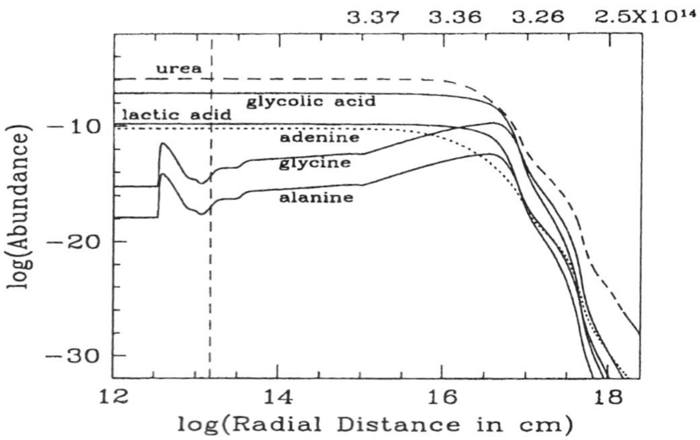

Fig. 2b : Log-log plot of the mass fractions of a few complex molecules (marked) as functions of the radial distance. Upper axis shows time elapsed in seconds since collapse began. Vertical dashed line is drawn at 1AU.

5. Effects of Radiative Association on Adenine Formation.

So far, we used an 'average' rate for HCNa addition. In the normal circumstances, in gas-phase reaction $HCN + HCN \rightarrow H_2C_2N_2$ rate would be small, since they must combine by radiative association, i.e., they must radiate a photon when combined together. This is a slow process and the probability of photon emission could be 1 in a few thousand to a few million

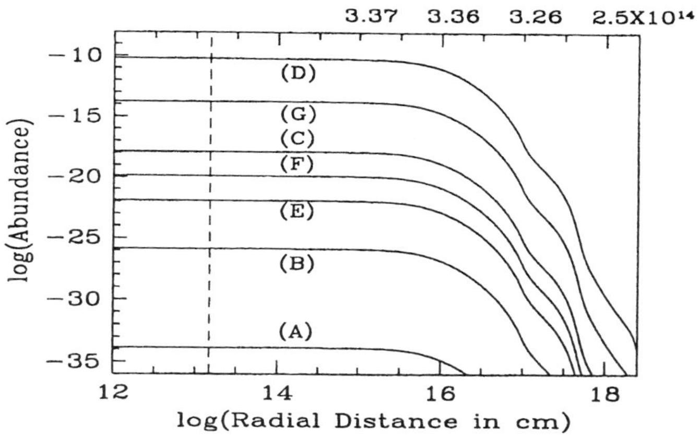

Fig. 3: Evolution of adenine abundance with radial distance and time (upper axis) in a $\alpha - f$ model. Various models are marked on the curve. See text for details.

(T. Millar, private communication). However as the size of the molecule gets bigger, the process becomes faster. Thus, it is likely that for a large enough molecule, the radiative association may take place at every collision and at this stage, the collisional rate may be used. One possibility is to assume that after every addition of HCN, the reaction rate goes up by a geometric factor of f (f is larger than 1). Hence we assume that $HCN + HCN \rightarrow H_2C_2N_2$ forms with a reaction rate of $\alpha_{Ad} \sim 10^{-16}$, but for $HCN + H_2C_2N_2$ the rate becomes $f \times 10^{-16}$, for $HCN + H_3C_3N_3$ the rate becomes $f^2 \times 10^{-16}$ and so on. f could also vary from one step to another — for instance, in the 1st step $f = f_1 \sim 4$; 2nd step $f = f_2 \sim 9/4$; 3rd step $f = f_3 \sim 16/9$ and 4th step $f = f_4 \sim 25/16$, if we assume that f varies with the cross-sectional area of the molecule. More favorable reactions could take place on ice, but the reaction rates of ice chemistry are not known. We believe that the best we could do is to study the formation of these important molecules as a function of two parameters, namely, α_{Ad} and f. We use the following models parameterized by $[log(\alpha_{Ad}), f]$: Model A $(-16, 1)$; Model B (−14, 1; Model C $(-12, 1)$; Model D $(-10, 1)$ (same as in Fig. 2b); Model E $(-16, 100)$; Model F $(-14, 10)$; Model G: $(-12, 5)$.

Fig. 3 shows the evolution of adenine abundance X_{Ad}. At $r = 10^{16}$cm, X_{Ad} is see to reach almost the saturated value. This is because, as the upper axis indicates, most of the time is spent in this region during the collapse. If the present detectability limit of abundance is around 10^{-11} (Combes et al, 1996), adenine processed in our method should not be detectable (except for Model D) even though it may be significant enough to contaminate and cause life to flourish in some planets. With a molecular weight of 135 for adenine, an abundance of 10^{-21} or less should really be considered as insignificant (less than atom/cc). In that case Models A, B and E must be rejected. Even when $\alpha_1 \sim< \alpha_{Ad} >$, f could be large enough to have eventual significant production (Model F).

There is controversy whether glycine is observed in interstellar matter. Miao et al. (1994) tentatively detected glycine in the massive star forming region Sgr B2(N) though this was later challenged by Combes et al. (1996). The latter suggested that when the sensitivity of the detector taken into account, the lines may be at the confusion limit and positive identification would require more sensitive instruments. Adenine has not been detected so far, however there may have been detection of adenine in several meteoritic samples (M. Bernstein, private communication).

6. Concluding remarks

If we assume that in the cool, tenuous interstellar cloud, adenine formation is energetically possible by successive addition of HCNs, then the reaction rates must be determined very accurately in order to estimate the final abundance. We show that this is sensitive to f and α_{Ad} parameters. Abundance of other bio-molecules could be significant. These may be brought to proto-earth by meteor impacts. Our suggestion also shows that life many be generic and most of the ingredients may be formed even before the planet formation.

References

Chakrabarti, S. & Chakrabarti S.K. 2000a, A&A 354, L6

Chakrabarti, S.K. & Chakrabarti S. 2000b, Ind. J. Phys. 74B, 97

Chandrasekhar S. 1939, An Introduction to Stellar Structure (Chicago: Univ. of Chicago Press)

Combes, F. Q-Rieu, N. and Wlodarczak, G. 1996, A&A 308, 618

Crick F.H.C. & Orgel L.E. 1973, Icarus 19, 341

Faure G. 1986, Principles of Isotope Geology, (John Wiley: New York)

Hartmann L. 1998, Accretion Processes in Star Formation (Cambridge Univ. Press: Cambridge)

Miao, Y., Snyder, L.E. Kuan, Y.-J. and Lovas, F. J., 1994, in *Proceedings of 184th Meeting of American Astronomical Society* No. 30.09

Millar T.J., Farquhar, P.R.A. & Willacy, K. 1997, A & A Suppl. Ser. 121, 139

Mojzsis et al. 1996, *Nat* **384**, 55

Mukhin L.M. & Gerasimov M.V. 1993 In: The Chemistry of Life's Origins, Greenberg J.M., Mendoza-Gómez C.X. & Pirronello V. (eds.) 185

Shidlowski M. 1993 In: The Chemistry of Life's Origins, Greenberg J.M., Mendoza-Gómez C.X. & Pirronello V. (eds.) 389

Shu, F.H., Adams F.C. & Lizano, S. 1987, ARA&A, 25, 23

Schulte M & Shock E. 1995, Origins of Life and Evolution of the Biosphere 25, 161

Volkenshtein M.V. 1983, Biophysics, Mir Publishers (Moscow)

INTERSTELLAR MOLECULES: AN EXOTIC CHEMISTRY

A. NEGRON-MENDOZA AND S. RAMOS-BERNAL

Instituto de Ciencias Nucleares, U.N.A.M.
Circuito Exterior, C. U. 04510, México D.F., México

Abstract

The processes that lead to the formation/destruction of molecules in the interstellar space are complicated and still poorly understood. First it is important to be acquainted with the extreme conditions in the interstellar space such as a) highly inhomogeneous environmental, b) very low temperature and c) extremely low density. These constrains allow only certain processes as plausible for the formation of these complex molecules. Three body collisions, endothermic reactions and reactions that have an activation energy are precluded. Chemical reactions may take place in gas phase or in the interstellar grains. In the latter catalysis plays an important role. The reactions are powered by the interstellar radiation field and by cosmic rays. In this work we present a brief summary of some chemical processes that may form complex molecules in the interstellar space.

1. Introduction

Molecules exist in the interstellar medium in great abundance and variety. The simplest and the most abundant interstellar molecule is H_2. The more complex molecules are located only in more dense region, but polar molecules such as CO and OH are widely distributed.

Interstellar chemistry studies encompass the physico-chemical processes involved in the synthesis / distribution / destruction of organic compounds found in the interstellar space. Theoretical calculations predicted that complex molecules could not survive in this medium. In 1968, a polyatomic molecule, NH_3, was detected. Since then many complex molecules have been identified in the interstellar space.

J. Chela-Flores et al. (eds.), First Steps in the Origin of Life in the Universe, 145–150.

1.1 WHY IS THE INTERSTELLAR CHEMISTRY AN EXOTIC CHEMISTRY?

To answer this question it is necessary to have a brief picture of the physical conditions of interstellar space. The experimental conditions are very special and different from those found on Earth. The main features are:
a) Low density. Interstellar gas is thought to have an average density of a few particles per cubic centimeter, b) Low temperature (10-100 K), c) highly inhomogeneous environment.

The interstellar molecules are observed in regions know as interstellar clouds. There are two types of these clouds: diffuse clouds and dense clouds (see Table I). Also, solid particles called interstellar dust or grains are found.

TABLE I. SOME CHARACTERISTICS OF DIFFUSE AND DENSE CLOUDS

TYPE OF CLOUD	DIFFUSE CLOUDS	DENSE CLOUDS
Amount of dust	Little dust	Abundant dust
Concentration of molecules and dominant species	Single hydrogen atoms. Low concentration of molecules	Molecular hydrogen gas
Temperature	10-100 K	10-50 K
Source of energy	Photo-dissociation	High energy flux of cosmi rays

Three different approaches have been used for the study of the interstellar molecules: a) observational analysis, b) theoretical models and c) laboratory simulations. In this work the basic question we addresses is the following: given that molecules are formed in the interstellar space, what are the chemical reactions that form and destroy those molecules?

It is generally believed that interstellar molecules are formed from simple constituents through a process of synthesis, rather than from a breakdown of more complicated systems. Both gas phase and surface reactions are important in interstellar chemistry. We assume that the chemical reactions begin with a gas containing various elements in their relative cosmic abundance. That means that hydrogen is by far the must abundant element. The other chemically important elements, such as carbon, nitrogen and oxygen are present to about 0.1% relative to the hydrogen abundance.

The interstellar medium represents a vast laboratory where reactions are occurring under a variety of conditions. The time scales are generally long compared with those used in the laboratories. These constrains have led to a model of the synthesis in dense clouds in which reactions are initiated by collisions of cosmic rays with hydrogen and helium. These encounters produce reactive, positively charged species that initiate chains of reactions leading to the formation of the observed molecules.

The probable mechanism for the formation of molecules involves a) gas phase processes, b) processes on surfaces of dust grains and c) photochemical processes in the grains themselves.

1.1.1 *Gas phase reactions*

Gas phase reactions are very important for the interstellar chemistry. However, there are some limitations of these types of reactions to occur: Reactions are restricted to binary processes. The low temperature found in interstellar space precludes endothermic reactions. Those reactions that have activation barriers are also forbidden. Some gas phase reactions are the followings:

Ion molecule Reactions. Ion-molecule collisions are very effective in forming new molecules. Reactions occur almost every time an ion and molecule met. They are drawn into interaction from relatively large separations by the induced polarization of the molecule. For initiating the reactions it is necessary to have both a molecule and an ion. In nearly all-astronomical conditions, the molecule H_2 is the primary step for chemical reactions. Once this molecule is available, then the effectiveness of ion-molecule chemistry is linked to the rate at which the ions can be created. Ionization can occur in various ways i.e. as ultraviolet or cosmic radiation.

Ultraviolet radiation with wavelengths of around 100 nm can ionize some atoms, e.g.,

$$C + \text{radiation} \rightarrow C^+ + e^-$$

Another possibility is that cosmic rays, which are mostly constituted by energetic protons, collide with atoms or molecules ejecting an electron, e.g.

$$H_2 + \text{cosmic rays} \rightarrow H_2^+ + e^-$$

Examples of this type of reactions are the following

$$A^+ + BC \rightarrow AB^+ + C$$

$$X^+ + BY \longrightarrow AB^+ + XY$$

$$H_2^+ + H_2 \longrightarrow H_3^+ + H$$

Neutral species that have a higher proton affinity than H_2 are protonated by a fast reaction:

$$H_3^+ + A \longrightarrow HA^+ + H_2$$

Additional hydrogen atoms are abstracted until saturated

$$He + \text{Cosmic rays} \longrightarrow He^+$$

He^+ can destroy most stable natural species via rapid dissociative charge transfer. A typical example is:

$$He^+ + CO \longrightarrow C^+ + O + He$$

This reaction is important in dense clouds as a source of C^+, because C^+ is not photoionized in dense clouds.

With ion-molecule reactions is possible to explain the formation and destruction of water molecule

$$H_2^+ + H_2 \longrightarrow H_3^+ + H$$

$$H_3^+ + O \longrightarrow OH^+ + H_2$$

$$OH^+ + H_2 \longrightarrow OH_2^+ + H$$

$$OH_2^+ + H_2 \longrightarrow OH_3^+ + H$$

$$OH_3^+ + \text{electron} \longrightarrow H_2O + H + OH + H_2$$

$$H_2O + \text{ions} \longrightarrow \text{products}$$

Table II summarizes other types of gas phase reactions relevant to interstellar chemistry.

TABLE II. SOME GAS-PHASE REACTIONS

NAME OF THE REACTION	EXAMPLE
Charge Transference	$A^+ + B \rightarrow A + B^+$ $H^+ + O \rightarrow H + O^+$
Radiative association	$A + B \rightarrow AB^* \rightarrow AB + h\nu$
Radiative recombination	$X^+ + e \rightarrow X + h\nu$
Negative ion reactions	$X + e \rightarrow X^- + h\nu$
Dissociative Recombination	$AB^+ + e^- \rightarrow A + B$ $HCO^+ + e \rightarrow CO + H$

1.1.2. *Grain catalyzed chemistry*

Another mechanism for the formation of polyatomic molecules is via heterogeneous catalysis on surfaces of interstellar dust. In the surface of the grains, atoms like A and B will be held long enough for a reaction to occur. Some of the energy liberated in the formation of the molecule is adsorbed by the grain. Such type of catalysis is responsible for the formation of molecular hydrogen; once this molecule is formed it can take part in a variety of gas phase reactions.

The steps in the formation of interstellar molecules in grains are the following: a) adsorption of gas phase atoms onto the grain surface, b) two adsorbed species must be in contact to form a chemical bond, c) the new molecule must be removed from the grain. Yet, there are some problems to explain the formation of heavier molecules in this kind of reactions because they are strongly bound.

1.1.3. Theoretical and Experimental model problems

The theoretical approach to study interstellar chemistry has the following problems to overcome: a) it is needed to get more data about the rate constants for potentially important ion-molecule reactions, b) theoretical models only consider the most abundant chemical elements, c) calculations of radiative decay rates of excited molecules are needed as an input for radiative association calculations, d) it is needed to make better models for processes on grain surfaces.

Some difficulties in the experimental approach can be outline as follow: a) gas phase laboratory spectroscopy is needed to expand the basis for identification of molecules in the interstellar space, b) it is necessary to propose studies of laboratory simulations on grains, c) it is needed the determination of rate constants for ion-molecule reactions in gas phase, d) the study of recombination, isotopic fractionation and radioactive association processes are scare. Since reactions in grain surfaces are due to gas-grain interactions and the storage of free radicals, then, a deeper knowledge of them is necessary.

2. Remarks

Interstellar chemistry produces molecules in great variety and abundance. Both gas phase and surface catalyzed reactions are very important in the synthesis of complex molecules. Organic chemistry occurs throughout the Cosmos, and the organic matter formed constitutes a molecular and isotopic record of the materials and processes involved in its formation. Time scales are generally long compared with those obtained in the laboratory.

In a similar way as biological evolution implies that all organisms on Earth have a common ancestor, also chemical evolution implies that all the matter in the solar system had a common origin. Interstellar molecules are a proof that organic chemistry is taken place in the universe and the formations of these molecules involve gas phase reactions and grains catalyze reactions.

Acknowledgement

The authors are very gratefully to the Organizing Committee for the financial support to attend the Conference.

3. References

Duley, W.W. and Williams, D.A. (1985) Interstellar Chemistry, Academic Press, New York. .

Green, S. (1981) Interstellar Chemistry: Exotic Molecules in Space, Ann. Rev. Phys. Chem. 32, 103-138.

Hartquist, T.W. and Williams, D.A. (1959) The Chemically Controlled Cosmos, Cambridge University Press.

Watson, W.D. (1976) Interstellar Molecule Reactions, Rew. Mod. Phys. 513-551.

PROTOPLANETARY DISKS, SOLAR SYSTEM MINOR BODIES AND ORIGIN OF LIFE

JOSEP Mª TRIGO-RODRIGUEZ[1,2]

1. Dept. Ciències Experimentals, Universitat Jaume I
2. Dept. Astronomia i Astrofísica, Universitat de València

Abstract: In the last decade, planetary science researchers have obtained significant advances on the origin and evolution of life on Earth. The increasing resolution and the development of new instruments (as space telescopes and adaptative optics) has provided first detailed images on protoplanetary disks and star formation regions. Moreover, in the last few years, doppler spectroscopic technique has confirmed the presence of extrasolar planets. These evidence shows planetary formation as a continuous and very common process in the Cosmos, indicating that our Solar System is not too special as was suggested in the past.

1. Introduction: The earth volatile depletion

Several research fields suggest the absence on volatiles in the primeval Earth. Such idea is not new, till now planetary formation models provide us with theoretical arguments to analyse the birth of our planet. Notwithstanding, the direct spectral analysis of protoplanetary disks confirms this scenario for last decade. According to the accepted accretion model of Cameron (1978), the planets are formed from a cloud of dust and gas that surrounded the Sun. Due to the rotation movement of the system, dust grains sedimented to the mid-plane of the primeval nebula, forming the protoplanetary disk. According to theoretical models in this disk dust accumulates forming large rings. These ideas have been confirmed directly, analysing images of other protoplanetary disks such Beta Pictoris. In these rings, the encounters between particles were most frequent and, consequently, they grew its size, forming major bodies. Probably these bodies formed planetesimals with tipical sizes between 50 to 500 km in only 100000

J. Chela-Flores et al. (eds.), First Steps in the Origin of Life in the Universe, 151–156.

years. Finally, from them the inner protoplanets were essentially completed in 50 million years according to Delsemme (2000).

In Cameron's model the maximum temperature reached by the grains of dust at a distance of 1 AU is on the order of 700 K (Cameron, 1978). Other models confirm similar or superior temperatures, generally in the interval from 700 to 1500 K (Saxena, 1981; Cameron, 1990, Delsemme, 1991). This would already be enough to degas the dust before it accretes into large chunks. Moreover the intense solar wind of the young Sun dispersed outside these volatile compounds. Consequently, primitive Earth was totally degassed and water and other volatiles must have been brought later by objects with more volatile material, such as comets and chondritic meteorites. These bodies were formed in the outer regions of Solar System where low temperatures preserved the ices.

2. Organics in the early Earth

On the basis of analysis of primeval terrestrial rocks, we know that life emerged very early on the primitive Earth and in very different conditions. At that period the atmosphere was very hot and dense, mainly composed by carbon dioxide (CO_2) and water (H_2O). Under prevailing high pressures, CO_2 is soluble in rainwater, becoming carbonic acid H_2CO_3. In the first atmosphere, long-lasting torrential rains transformed the CO_2 into solid carbonates, mostly limestone (carbonate of calcium) and dolomite (carbonate of calcium and magnesium). From old terrestrial rocks we have conclusive evidence that the amount of CO_2 in primeval atmosphere was higher than now. In fact, exist enormous sediments of limestone and dolomite 3.8 billions years old in Greenland f.e. If all these ancient sediments were heated enough, they would form an atmosphere of CO_2 comparable to that of Venus (Delsemme, 1998).

The chemical analysis of cometary comas made that Joan Oró proposed in 1961 the importance of these bodies in the development of terrestrial biosphere. At the same time, Solar System formation models proposed a bombardment stage as key to explain the enrichment of terrestrial biosphere. Now we found the first direct evidence of this period in other planetary systems, nowadays in broad formation. In these protoplanetary systems happen the enrichment of inner planets by cometary bombardment as reveals spectroscopy of zodiacal dust.

During last decades several authors have analysed the delivery of water and intact organics in the early Earth from comets and IDPs (Chyba et al., 1990). Using different assumptions, has been estimated f.e. that the magnitude order of this delivery was three or four higher than nowadays. For example, Chyba & Sagan (1992) calculated that comets and IDPs introduced between 10^7 and 10^9 Kg carbon yr^{-1} respectively to the early Earth. These rough estimations are hard dependent on the considered assumptions. We can assume the percent of organic composition of these bodies (aprox. a 15%) but we are especially ignorant of the flux number-mass distribution of matter and its variation through time. Lunar and planetary cratering record provide us with interesting information on the planetesimal spatial density during the planetary formation stages but this is only indicative. Other interesting research field for new techniques are quantifies the degradation of cometary and chondritic particles during its entry into the Earth atmosphere.

3. Empirical evidence of protoplanetary disks

The discovery and detailed analysis of main-sequence stars surrounded by circumstellar dust disks could be provide us with interesting data on the processes that dominate during planetary accretion. In this new view, the observed optical properties, spatial and size distribution, mineralogy and physics of the dust component of these protoplanetary disks could have a key role in our future ideas on exogenous enrichment of the early Earth. In fact, our understanding of the way the solar nebula evolved rests, partly on the study of the nature and evolution of young stellar objects (YSOs) surrounded by a circumstellar disk. Many of these YSOs exhibit infrared excess, indicating the presence of warm dust. This dust is not primordial, but must be replenished constantly as a consequence of the stellar wind intensity of these young stars. This necessary constant dust replenishment is only possible from the constant fragmentation and release from minor bodies, especially comets. Actually, we have technical difficulties to deduce if a cloud of comets similar to the Öort cloud surrounds these systems. Notwithstanding, considering the mean source of solar system zodiacal cloud, the presence of small particles around YSOs supports the existence of an important number of comets around these stars. The origin of these comets ought to be the outer protoplanetary disk.

The best studied of these circumstellar disks is β Pictoris that is more evolved and older than the primordial solar nebulae. The importance of this disk is remarkable because we found the first evidences of relatively young planets in its interior in the last decade. One of the most important processes observed in this disk is the collisional fragmentation of orbiting solid bodies, probably comets, leading to the eventual removal of micron-sized and smaller debris from the system by the radiation presure. The quantity of circumstellar material, including comets, is some orders of magnitude larger than those present in the Solar System. Recently Artymowicz (1997) proposed a total mass of solid planetesimals around β Pictoris and in our early solar system near to $120\pm40\ M_{Earth}$ while analysing the expected ice/dust ratio of solids.

The detailed analysis of the gaseous component of β Pictoris disk reveals that the likely source of gas is evaporating solid dust grains from macroscopic bodies, specially comets. From the spectroscopic observations we know that the albedo of β

Pictoris materials is higher than in typical solar system IDPs or astronomical silicates. Artymowicz (1997) proposes that it could be produced from a different condensation sequence, favouring the presence of high reflectivity silicates. In our opinion it could be also an evidence of interplanetary ices presence due to the constant cometary fragmentation. Again the presence of Falling Evaporating Bodies (FEBs) could explain one characteristic of this disk as the relative absence of gas and the high reflectivity of materials.

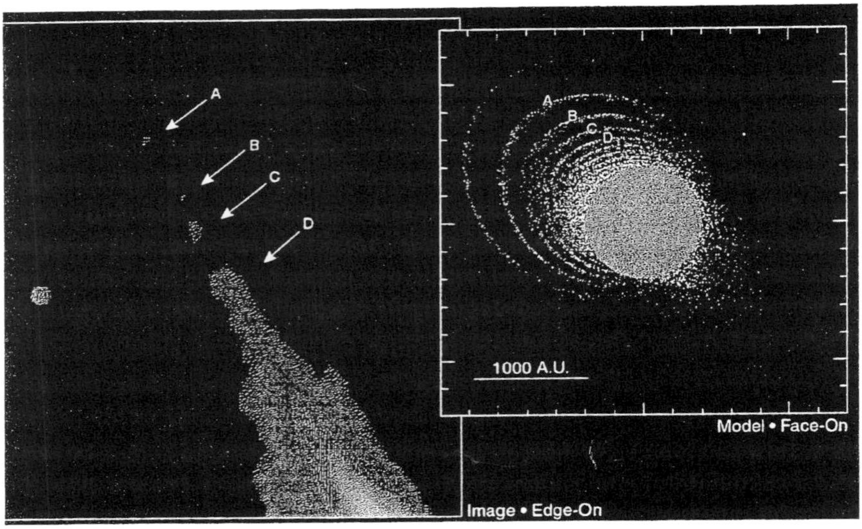

Figure 1. Recent observations of Beta Pictoris surroundings, which reveal the presence of an elliptical ring system, supporting planetary formation in this protoplanetary disk. (Image Paul Kalas-HST/NASA).

4. Conclusions: Carbon sources in the primeval nebula

The abundance and ability of carbon to form complex species provide a key role in the chemical evolution of the universe. The cosmic evolution of C from the interstellar medium into protoplanetary disks and planetesimals, and finally onto habitable planets is related directly to the study of the life's origin. Diverse fields bring that during early Earth stages, the amount of captured carbon compounds delivered by comets, chondritic meteorites or IDPs was extraordinary important. The ulterior evolution of these compounds onto the Earth surface generated the biochemical molecules that eventually led to the appearance of life on our planet. In situ analysis of cometary material during the next decade could contribute to increase our knowledge on its role in the development of life onto Earth.

On the other hand, the study of star and planetary formation in our galactic surroundings provides interesting clues on our origin. But due to interstellar distances, only Solar System minor bodies can be directly analysed in order to complete chemical and physical data on our planetary system. The direct analysis of comets and meteorites, as remnants bodies of this early period, has consequently a key role to complete this intricate puzzle. Especially remarkable would be knowing better the formation and degradation processes of comets and meteorites. For example, we don't know the origin in the Solar System of more of 99.9% of meteorites recovered because their entries in the Earth atmosphere were not registered. In this sense, photographic networks can provide interesting orbital and spectroscopic data on meteorite origin as explain our poster contribution (Trigo-Rodríguez et al, in this volume).

The analysis of primitive organic material in the solar system comes from the analysis of carbonaceous chondritic meteorites, interplanetary dust particles (IDPs) of presumable cometary origin collected in the upper atmosphere of Earth, and grains of comet P/Halley. In the future, especially important will be the NASA and ESA missions to collect pristine cometary material.

To conclude, the study of the origin and evolution of planetary systems is still in its infancy. As our theoretical models of the origin of planets and minor bodies become more complex, our sketches of plausible theories seem consistent with the observational data. Now the models are still far from the last word but they predict important aspects of the observed protoplanetary systems.

5. References

Beust H. et al. (1996); “The Beta Pictoris circumstellar disk: XXII. Investigations the model of multiple cometary infalls”, Astronomy & Astrophysics 310, pàgs. 181-198.

Boss A.P. (1998); “Temperatures in Protoplanetary Disks”, Annu. Rev. Earth Planetary Science 26, pàgs. 53-80.

Brown R.H. & D.P. Cruikshank (1997); “Determination of the composition and state of icy surfaces in the outher solar system”, Annu. Rev. Earth Planetary Science 25, pàgs. 243-277.

Delsemme A.H (2000); "Cometary origin of the biosphere", Icarus, in press.

Hodgson L.S. & A. Brandenburg (1998); “Turbulence effects in planetesimal formation”, Astronomy & Astrophysics 330, pàgs. 1169-1174.

Patterson C.C., (1956); “Age of meteorites and the Earth”, *Geochim. Cosmochim. Acta* 10, pp. 230-237.

Saxena S.K. (1981); "Planetary Phase Equilibria: Application to formation of Earth, Venus and Mercury" Geochimica et Cosmochimica Acta, vol. 45, pp. 813-820.

Wetherill, (1990); "Formation of the Earth", *Annu.Rev. Earth Planet. Sci.* 18, pp. 205-256.

Willacy K & T.J. Millar (1997); "Chemistry in oxygen-rich circumstellar envelopes", Astronomy & Astrophysics 324, pàgs. 237-248.

ELECTRIC DISCHARGE IN CARBONACEOUS METEORITES?

G. KLETETSCHKA

Code 691, NASA's Goddard Space Flight Center
Greenbelt, 20771, Maryland, USA

1. ABSTRACT

Discharge experiments produce amino acid distributions that are similar to distributions observed in carbonaceous meteorites (Oró, 1990). A recent work (Desch and Cuzzi, 2000) points out to the presence of lightning in the Solar Nebula. Magnetic measurements of chondrules reveal extraordinary magnetic acquisitions (Wasilewski and Dickinson, 2000) that can be explained simply by a presence of magnetic fields generated by lightning discharge (Wasilewski and Kletetschka, 1999). Thus we speculate that lightning discharge may be responsible for the origin of amino acid compounds in carbonaceous meteorites.

2. DISCHARGE EFFECTS

In 1953, Stanley Miller used electric discharge inside a mixture of hydrogen, methane, ammonia and water vapor. The presence of electric discharge converted the abiotic compounds in to certain proportions of amino acids. Similar proportions of amino acids were found in Murchison meteorite (e.g. Oró, 1990).

3. DISCHARGE MAGNETISM

Magnetic fields produced by an electric discharge are non-homogeneous with sharp gradients and of large intensity (Wasilewski and Kletetschka, 1999). A past presence of magnetic fields generated by the electric discharge can leave a magnetic signature in certain iron-bearing minerals. As a result these magnetic minerals may carry their own remanent magnetization in absence of magnetic field and in a sense these minerals behave like a small permanent weak magnets. The magnetic signature imposed in to the magnetic minerals by an exposure of unknown past magnetic field is called natural remanent magnetization (NRM). Each of the recording magnetic minerals has magnetic

J. Chela-Flores et al. (eds.), First Steps in the Origin of Life in the Universe, 157–159.

limit to which value they can be magnetized. When increasing the value of the exposing magnetic field beyond a certain quantity the remanent magnetization of magnetic minerals does not increase and stays at constant value. This constant value is called saturation isothermal remanent magnetization (SIRM). It is often the case that NRM of minerals is orders of magnitude lower than SIRM. However, NRMs of magnetic minerals exposed to magnetic fields generated by lightning discharge are approaching the SIRM. The ratio between NRM and SIRM is called REM and can be used as a magnetic indicator of an exposure to various magnetic fields. In case of the lightning discharge REM can be greater than 0.1 (Wasilewski and Kletetschka, 1999).

4. REMANENCE IN METEORITES

NRM/SIRM (REM) values for various types of meteorite range over 4 orders of magnitude (Sugiura and Strengway, 1988). The REM values of chondrules range also between 0.0001 and 1 (Wasilewski and Dickinson, 2000). Some of the chondrules that were extracted from meteorites were observed to contain very intense NRM. The magnetization of these chondrules was very close to SIRM. The values above 0.1 are associated with extraordinary conditions of magnetic remanence acquisition. A plausible explanation is an exposure to intense magnetic fields generated by lightning. Chondrules, however, are randomly oriented with respect to their magnetic direction. Thus a smaller samples (<1 mm) have intense magnetic remanence directions because the amount of chondrules in small samples is not large enough to randomize completely the remanence direction due to random orientation. When meteorite increases in size, magnetic moment is generally smaller.

5. REM INTERPRETATION

Magnetization of larger pieces of meteorites (Sugiura and Strengway, 1988) indicates mostly decreasing trend of REM when SIRM increases. Carbonaceous meteorites, however have SIRM fairly constant and with substantial variation of REM indicator. The decreasing trend of REM can be explained due to secondary contamination by chemical remanent magnetization (CRM) due to meteorite weathering in terrestrial conditions. Fine-grained secondary oxide minerals can precipitate on the surface or within the meteorite fractures and acquire terrestrial CRM. If the meteorite contained weak magnetization before the entry, than new weathering-related grains acquire a CRM, which is generally large due to presence of terrestrial magnetic field, which causes the REM value to increase. SIRM is indirectly related to the concentration of magnetic carriers in the sample. When the concentration is large (large SIRM) the secondary overprint due to CRM is not that significant and thus the REM value is lower because interplanetary fields are likely to be a small fraction of the terrestrial magnetic field. Non decreasing trend in carbonaceous meteorites indicates low level of the secondary contamination and stable NRM.

C. meteorites (Carbonaceous) have large spread of the REM indicator. Demagnetization experiments with the NRM indicate reasonably high degree of magnetic stability compared to other group of meteorites (Sugiura and Strengway, 1988). The NRM values are considered to provide the best evidence for the exposure to the original magnetic fields in the early solar system.

6. DISSCUSION AND CONCLUSIONS

Large variation of REM values that were found in carbonaceous meteorites (Sugiura and Strengway, 1988) suggests that this class of meteorite may have been subjected to strong magnetic fields that may be consistent with the presence of electric discharges in the early history of the carbonaceous meteorites. Magnetism of chondrules imbedded in various meteorites (Wasilewski and Dickinson, 2000) requires extraordinary magnetizing event consistent with lightning discharge (Wasilewski and Kletetschka, 1999). The presence of lightning in the solar nebula is supported by many features of chondrules (Desh and Cuzzi, 2000). Evidence of the amino acids occurring in a similar proportion as electric discharge experiments supports our speculation that some of the carbonaceous meteorites experienced electric discharges which have lead to the origin of the basic structural components required for evolution of life.

7. REFERENCES

Desch, S. J., and Cuzzi, J. N., (2000), The generation of lightning in the solar nebula. *Icarus*, **143**, 87-105.

Oró, J., (1990), The origin and early evolution of life on Earth. *Annu. Rev. Earth Planet. Sci.*, **18**, 317-356.

Sugiura, N., and Strangway, D. W., (1988), Magnetic Studies of Meteorites. in J. F. Kerridge, M. S. Matthews, *Meteorites and the early solar system*: Tuscon, US, Univ. Ariz. Press, 595-615.

Wasilewski, P., and Kletetschka, G., (1999), Lodestone - Natures Only Permanent Magnet, what it is and how it gets charged. *Geophysical research letters*, **26**, 2275-2278.

Wasilewski, P., and Dickinson, T., (2000), Aspects of the validation of magnetic remanence in meteorites. *Meteoritics and Planetary Science*, **35**, 537-544.

TERRENE METEORITES ON THE MOON

A SOURCE OF INFORMATION ABOUT THE ORIGIN OF LIFE IN THE EARTH?

JORDI L. GUTIÉRREZ
Dept. de Física Aplicada, Universitat Politècnica de Catalunya, Pla del Palau 18, 08003 Barcelona, Spain

Abstract.
The study of the origin of life on the Earth is impeded by the fact that very few ancient rocks from the era of the appearance of biological systems are preserved. The oldest sedimentary rock formations are about 3.9 billion years old, and even these show indications of the early emergence of life [1]. Impacts, tectonic activity and erosion have hopelessly destroyed all the information predating the aforementioned age. Nevertheless, here we show that the Moon could constitute an invaluable reservoir of very ancient terrestrial rocks, stripped off the surface of our planet during the great bombardment that took place shortly after the origin of the solar system. These rocks could store information about the origin of life on Earth that our own planet wiped aeons ago.

1. Introduction

The discovery on the Earth's surface of rocks coming from Mars (the SNC meteorites and ALH84001)and the Moon proved that there exist mechanisms to transport debris from the surface of a terrestrial body to another ([2]–[7]). The transport mechanism is such that the peak pressures (which gauge the stress experienced by the rock) recorded in several SNC and lunar meteorites are rather low, in the range 15–35 GPa [8]. Furthermore, studies of lunar meteorites show that their origin was located in the shallow surface of our satellite [5]. Taken together, these facts imply that the source of putative terrene meteorites was on, or near, the surface of the planet, the relevant zone for the origin of biological processes, and that biologically interesting macromolecules, and maybe even microorganisms, could survive the ejection event. While many works have addressed the pos-

J. Chela-Flores et al. (eds.), First Steps in the Origin of Life in the Universe, 161–166.

sibility of transfer of meteoroids (and even viable microorganisms) between Earth and Mars (see, for example, [9]), none has considered that, being much closer, it is far more likely that terrestrial rocks have fallen on our satellite at relatively moderate velocities.

2. The threats for the meteorites

The expulsion mechanisms suggested for other bodies of the solar system should also work for the Earth, even if scaled to our planet's deep gravitational well and thick atmosphere. It has been proposed [6] that the debris could escape through the atmospheric tunnel created by the impactor. During the great bombardment era, the huge impacts experienced by the Earth were powerful enough to create such a corridor, even through the much thicker primordial atmosphere [8], for the escaping debris. Since then, impacts of similar magnitude have become more and more scarce, as can be shown by the lack of fresh massive craters.

Numerical simulations [11] and many SNC meteorites indicate that part of the material that escapes a planet after an impact is very weakly shocked, with peak pressures of just a small percentage of the maximum impact pressure. This is a general trend expected of all the debris ejected of large bodies by means of this mechanism. On the Earth, debris coming from the Ries impact crater (Germany) have been found in Switzerland, more than 200 kilometres away [12]. To cover this distance, the debris had to be accelerated up to 1.4 km s^{-1}, but still show very light shock damage.

As it is more difficult to take away rocks from the surface of the Earth than from Mars or the Moon, the number of terrene meteoroids expelled should be correspondingly smaller than those related to the latter bodies. However, the closeness of the Moon converted our satellite into a trap for such terrestrial rocks: in the era of the great bombardment, the semimajor axis of the Moon's orbit was just about 2×10^4 kilometres [13]. Hence the Moon could have accreted a good deal of ancient terrestrial rocks, even if launched at significantly less than escape velocity. It is also important to note that some of these rocks did impact on the lunar surface at a rather moderate velocity, with a lower limit of about 1.5 km s^{-1} significantly smaller than the escape velocity of the Moon (2.4 km s^{-1}); this is because of the influence of the terrestrial gravitational field, which noticeably reduced the effective gravitational acceleration above the nearer lunar hemisphere. This reduction in the final impact velocity, resulted in a higher chance of survivability to the labile organic molecules. Moreover, Rubin [14] has found millimeter–sized unshocked chondritic fragments in lunar regolith, further supporting the likelihood of the terrene meteorites survival, especially when considering that being in heliocentric orbits, the impact speed of the original chondrite meteoroids would be notably higher than the lower impact velocity considered for terrene meteorites.

Nevertheless, the lack of a noticeable lunar atmosphere implies that the in–falling meteoroids would impact the surface at their full velocity. In the violent shock, the incident body would be heated since part (about a 20%, see [15]) of its kinetic energy would be transformed into heat. For the low velocities mentioned above, this temperature increase could be low enough to partially preserve the organic molecules and delicate structures such as microfossils. The importance of the increase of temperature can be further dampened if the meteoroid had cooled during its traverse through the circumlunar space; in this respect, it is interesting to note that in the young solar system, the luminosity of the Sun was significantly smaller than the current value, favouring a low temperature for the terrene meteoroid. On the other hand, it has been experimentally demonstrated [16] that aminoacids and nucleic acid bases can partially survive the exposure to temperatures of 500°C for short spans of time.

Cosmic rays impacting on the meteoroid when outside the magnetosphere of the Earth constitute a threat to biological molecules. Even so, the exposure time to cosmic rays (hereafter referred to as 4π CRE age) is limited because the rocks would spend a very short time freely flying in space, as its travel was restricted to the circumlunar space, where they should be either rapidly captured by the Moon or fall to the Earth (but some could enter into heliocentric orbits; if these debris were eventually captured by the Moon them should show a higher degradation by cosmic rays and a correspondingly higher 4π CRE age). In this respect, it is interesting to note that numerical simulations [3] have shown that most lunar debris fall onto Earth in few decades.

3. What information could be stored in the putative terrene meteorites?

Given the conditions prevalent in our satellite, the biogenically interesting molecules could be stored under the surface of the Moon for very long periods of time. An important factor that influences the lifespan of the molecules contained in terrene meteorites is the effect of cosmic rays (together with solar–particle events), but a few metres of rock or regolith are enough to shield the terrestrial rocks and molecules against them and the cascades of secondary particles produced by their interaction with matter. The same can be said about the extreme temperatures experienced on the surface of the Moon, with daily temperatures above 100°C and nightly below minus 150°C; the layer that protects the rocks against cosmic rays, plays an analogue role for temperature changes as the regolith has a small heat conductivity. DNA degradation is very sensitive to temperature [18], and its velocity of degradation is very small below 0^oC; Apollo experiments determined that the temperature of the regolith at a depth of 2 m is about -20^oC, implying that DNA molecules could survive there for geological ages. Finally, and this is an outstanding difference with the Earth, the Moon is devoid of free oxygen or water and is also completely sterile, and so no organisms can degrade the state either of the terrene meteorites or of the molecules contained inside them.

Some other problems associated with the origin of life could be addressed with terrene meteorites. In case of a low enough impact velocity, it is possible that terrestrial microorganisms could be deposited alive onto the Moon, where they would die after a short period of time. In any case, these microorganisms should become freeze–dried, and perhaps conserved for extended periods of time. On the other hand, during the final stages of the great bombardment, life could have originated on the Earth more than once, and been destroyed by the impacts of the last large bodies in Earth–crossing trajectories. If such is the case, terrene meteorites could provide indications in that sense, and perhaps allow us to have a look into the several biological experiments performed by nature, making possible to learn about the different possibilities (if any) for alternative biochemistries.

It is probable that some of the terrene meteorites contain microscopic bubbles of the ancient atmosphere of the Earth, and then them will be the only first–hand source to check the current theoretical models about the origin of the biosphere and the prebiological and protobiological molecules. Several authors ([19]–[22]) have proposed that part or most of the volatiles and organic material in the biosphere have been deposited by impacting comets. This hypothesis would solve at once the difficulties in forming biologically interesting compounds, something not evident with the current paradigm of the early Earth's atmospheric composition [10]. A direct confirmation of these theories is unlikely to derive only from data gathered on the Earth.

4. How to recognize terrene meteorites on the Moon?

Terrene meteorites must be found on the highlands (the oldest regions of the Moon, with rocks of ages up to 4 Gyr, showing that terrene meteorites of comparable ages could also exist) associated with small craters heavily eroded by 4 billion years of gardening of the regolith. However, it is likely that many terrene meteorites will have been covered by blankets of ejecta generated by nearby impacts; these blankets would protect the meteorite against further small impacts responsible of the gardening processes, allowing its survival up to our time.

It is possible that terrene meteorites are somewhat different of the typical lunar rocks; these differences could be used in order to identify them. On the first hand, the state of oxidation of iron should be higher (Fe^{+3}) than that of lunar rocks. Also, its appearance could be remarkable, specially if formed from sedimentary rocks with clearer color and linear patterns. Finally, the meteorites could be recognized by its possible fusion crust; this fusion crust would not be a consequence of the meteorite's passage through the incredibly subtle lunar atmosphere, but to its interaction with the gas still present in the atmospheric corridor created by the impactor. The fusion crust would be a telltale indication about its origin on a terrestrial planet with atmosphere, but on a such primitive epoch it can not be discarded an origin on Venus or Mars, as the atmospheres of those planets were in all likelihood very different than the current ones. However, due to its closeness, the Earth would be the most serious candidate as parent body of the meteorite.

Meteorite searching is not likely to be a high priority issue in the agenda of the first lunar settlers, but it is probable that meteorites will be a by–product of some lunar activities, such as ^{3}He mining or regolith excavation for other purposes. In any case, it seems obvious that a direct search for these objects is not a practical option due to its scarceness.

5. Conclusions

The lunar surface can be a reservoir of meteoritic material coming from the Earth, especially related to the era of the great bombardment, when the size and energy of the impactors were enough to inject debris, at least, into suborbital trajectories that intersected the ancient orbit of the Moon. These terrene meteorites could prove to be our best (and perhaps only) source of information about the primitive conditions prevailing on the Earth and about the origin of life on our planet.

Acknowledgements: The author gratefully acknowledges Antonio Lazcano Humberto Campins and Ignasi Casanova for fruitful comments about the subject and encouragement. This work has been supported in part by a grant of the European Union for young scientists.

References

1– Mojzsis, S. J. et al. (1996), Evidence for life on Earth before 3800 million years ago, *Nature* **384**, 55–59.

2– Gladman, B. (1997), Destination Earth: martian meteorite delivery, *Icarus* **130**, 228–246.

3– Gladman, B. et al. (1996), The exchange of impact ejecta between terrestrial planets, *Science* **271**, 1387–1392.

4– Gladman, B. & Burns, A. (1996), Mars meteorite transfer: simulation, *Science* **274**, 161–162.

5– Warren, P. H. (1994), Lunar and martian meteorites delivery systems, *Icarus* **111**, 338–363.

6– Melosh, H. J. (1993), Blasting rocks off the planets, *Nature* **363**, 498–499.

7– O'Keefe, J. A. & Ahrens, H. (1986), Oblique impact: a process for obtaining meteorite samples from other planets, *Science* **234**, 346–349.

8– Treiman, A. H. (1995), S?NC: Multiple source areas for martian meteorites, *JGR* **100**, 5329–5340.

9– Mileikowsky, C. et al. (2000), Natural transfer of viable microbes in space. 1– From Mars to the Earth and Earth to Mars , *Icarus* **145**, 391–427.

10– Kasting, J. F. (1993), Earth's early atmosphere, *Science* **259**, 920–926.

11– Gratz, A. J., Nellis W. J. & Hinsey, N. A. (1993), Observations of high-velocity, weakly shocked ejecta from experimental impacts, *Nature* **363**, 522–524.

12– McKinnon, W. B. (1993), *Nature* **363**, 211–212.

13– Sonnett, C. P. et al. (1996), Late proterozoic and paleozoic tides, retreat of the Moon and the rotation of Earth, *Science* **273**, 100–104.

14– Rubin, A. E. (1997), The Hadley Rille enstatite chondrite and its agglutinate–like rim: impact melt during accretion to the Moon, *Meteoritics Planet. Sci.* **32**, 135–141.

15– O'Keefe, J. A. & Ahrens, H. (1982), The interaction of Cretaceous/Terti extinction bolide with the atmosphere, ocean and solid Earth, *Geol. Soc. Am., Special Paper* **190**, 103–120.

16– Basiuk, V. A. & Navarro–González, R. (1998), Pyrolitic behaviour of amino acids and nucleic acid bases: Implications for their survival during extraterrestrial delivery, *Icarus* **134**, 269–278.

17– O'Keefe, J. A. (1976), *Tektites and their origin*, Elsevier, pp. 160–165.

18– Lindahl, T. (1993), Instability and decay of the primary structure of DNA, *Nature* **362**, 709–715.

19– Oró, J. (1961), Comets and the formation of biochemical compounds on the primitive Earth, *Nature* **190**, 389–390.

20– Delsemme, A. (1997), The origin of the atmosphere and the oceans, in Thomas, P. J., Chyba, C. F. & McKay, C. P. (eds.), *Comets and the Origin and Evolution of Life*, Springer–Verlag, New York, pp. 29–67.

21– Delsemme, A. (2000), Cometary origin of the biosphere, *Icarus* **146**, 313–325.

22– Chyba, C. & Sagan, C. (1992), Endogenous production, exogenous delivery and impact–shock synthesis of organic molecules: an inventory for the origins of life, *Nature* **355**, 125–132.

TWO POSSIBLE STEPS OF THE CHEMICAL EVOLUTION ON SURFACE OF SMALL BODIES IN THE SOLAR SYSTEM

M.B. SIMAKOV AND E.A. KUZICHEVA
Group of Exobiology, Institute of Cytology,
Russian Academy of Sciences, St.Petersburg 194064, RUSSIA

1. Introduction

The problem of life origins is a very intriguing question of the contemporary science. Geological data indicate that complex photosynthetic organisms possibly related to cyanobacteria existed for 3.5 billion years (Gyr) ago [1]. On the other hand, it is stated that impacts by meteorites and comets at early stage of the Earth history until 3.8-4.2 Gyr ago would have made our planets uninhabitable [2]. So we can see that life arose so rapidly on Earth once conditions became clement and this process could be very rapid, almost instantaneous on the geological time scale. Only 300 million years appear to be left for terrestrial chemical evolution to lead to primitive, yet fully developed organisms. There is a very short time interval for producing of the biochemical compounds in the atmosphere and on the surface of planet, but this kind of matter's evolution had to occur before the biological one. Besides, geological data and theoretical models point to the Earth's early atmosphere to be composed mainly of carbon dioxide, carbon monoxide, nitrogen and water vapor [3,4]. It is not easy to form biologically significant compounds such as components of peptides and nucleic acids from these gases by any energy sources.

We can suggest the chemical evolution which had proceeded at early stages of the Solar system origin, would lead to the appearance of a very complex organic compounds. These compounds may have been delivered to early Earth at the stage of the "late heavy bombardment" and later [5]. So, the highly complex organic molecules have been proposed as starting material for early biological activity on Earth, or on other planets in the Universe [6,7].

The main attention in experimental work devotes now to energetic chemistry in "simulated space ice conditions", such as icy mantles of the interstellar dust particles (IDP) [8,9] and cometary ices [10]. These laboratory and theoretical studies were stimulated by related astronomical observations showing the signature of complex carbonaceous material in circumstellar and interstellar environments [11]. The laboratory experiments yielded some interesting results [12]. However, the reactions of complex organic compounds could take place on the solid surface of the small bodies in the Solar system, such as asteroids, meteorites, IDP and satellites of the others planets (Mars, Jupiter, Saturn), too. These bodies have tremendous total surface and they are subjected to

J. Chela-Flores et al. (eds.), First Steps in the Origin of Life in the Universe, 167–170.

irradiation by ultraviolet photons, the solar wind, solar flare ions and galactic or extra-galactic cosmic rays.

In space the chemical evolution toward complex organic molecules takes place in the solid phase. The aim of our work is to understand better, using laboratory simulation and space crafts opportunities, the chemical evolution of biologically significant compounds on the surface of the small bodies in the Solar system. We present laboratory data relevant to ultraviolet irradiation and cosmic ray bombardment of the solid organic films in such environments. In our experiment in laboratory and on board of the space satellite "BION-11" we have planned to establish how far the chemical evolution could develop on the surfaces of small bodies under action of open space energy sources.

2. Experimental results

We investigated two types of reactions: (1) abiogenic synthesis of nucleotides from mixtures of nucleoside + inorganic phosphate; (2) abiogenic synthesis of dipeptides from mixtures of simple amino acids.

The reaction mixture in the form of a solid film contains nucleoside and dihydrogen phosphate or two different amino acids. Seven different nucleosides (thymidine, cytidine, uracil, adenosine or deoxyadenosine, guanosine or deoxyguanosine) and four mixtures of aromatic (tyrosine or triptophan) and aliphatic (glycin or alanine) amino acids were investigated. Mixtures were prepared by air drying of aqueous solutions and irradiated by different sources of energy: (1) VUV-light of 145 nm; (2) high energy protons (2-6 MeV); and (3) were installed on the surface of biosputnik in outstanding container (OC) when they were exposed to the action of all spectra of the open space energy sources during the entire time of flight - 327 hours (24.12.1996-07.01.1997).

As a result of VUV irradiation of the solid mixture of nucleoside and inorganic phosphate the natural monophosphates of corresponding nucleosides were found [13,14]. The main products were nucleoside-5'-monophosphates (5'NMP) and some amount of by-products (2'- and 3'-monophosphates, 2'3'- and 3'5'-cyclomonophosphates). The yields of products were small, within few per cents, however the effectiveness of the abiogenic synthesis on VUV irradiation is higher than on UV-radiation (λ=254 nm) and heat. When investigated films were irradiated with protons, the full mixture of nucleotides also was found [15]. The maximal summary yield was 9.43%. The main product was 5'AMP (3.19%) as under action of VUV. Preferable 5'-monophosphate formation is indicative of more advantageous spatial position of 5'-hydroxyl group in carbohydrate residue in comparison with 2'- and 3'-hydroxyl groups. The 2'3'-cyclomonophosphate has been formed in solid films as well. It is interesting to note the yield of 2'3'cNMP has values more than those of 2'- and 3'-monophosphates. In our previous papers we had proposed that interaction between all three nucleotides exist under VUV [14] and γ-radiation [16]. Kinetic analyses carried out by Reimann and Zubay [17] suggested that 2'3'-cyclomonophosphate formation arose from 3'- or 2'-nucleoside derivatives. In this work it was also shown that the 5'-OH site is inherently favored over the 3'- or 2'-OH in phosphorylation. We can see that 5'-nucleotides are predominantly formed under action of all investigated sources of energy as well as in open space. Other nucleotides (2' and 3') and cyclic nucleotides, such as 2'3'cNMP and 3'5'cNMP also have been found in the all investigated films.

The films containing a mixture of amino acids yielded various oligoptides after they were exposed to protons and VUV-radiation [18,19]. Such dipeptides as Trp-Trp, Trp-Gly, Gly-Trp, Ala-Trp, Trp-Ala, Tyr-Tyr, Tyr-Gly, and Gly-Tyr have been detected as products of irradiation with summary yields of ~2% in case of VUV-irradiation and ~2.5% for proton irradiation. Polymerization is an essential step in prebiological evolution and we have shown that this process probably could take place even at early stage of the Solar system formation, before planet accretion, on surface of small bodies.

It was important to test how far the process of chemical evolution could take place on the surface of space bodies under action of all energy sources of the open space. In space flight experiment on board of "BION-11" satellite the solid films from mixtures of different nucleosides and inorganic phosphate were exposed to space conditions. The abiogenic synthesis of the full set of the natural nucleotides is observed [20]. The total yields were 5.8% for adenosine, 4.7% for cytidine, 2.7% for deoxyadenosine and 2.0% for thymidine. The main products of the reactions in all experiments were 5'monophosphates of corresponding nucleosides as in the laboratory investigations.

3. Prebiological significance

The results of our laboratory experiments on action of UV radiation with different wavelengths (145-260 nm), high-energy particles, γ-radiation and heat show that many kinds of open space energy could be effective in the chemical evolution of nucleic acid precursors and oligopeptides. The abiogenic synthesis of nucleotides and dipeptides can proceed in open space conditions on early stages of the Solar system evolution. The results of experiment on board of "BION-11" enable us to conclude that 5'-nucleotides are predominantly formed by action of full spectra of space energy sources on dry thin films of nucleoside and inorganic phosphate. 5'NMP compounds are the most likely monophosphorylated intermediates in the prebiotic pathway to activated nucleotides [21]. Other derivatives were also have been found in the investigated films.

Of course, an irradiation of few hours in laboratory or a week spaceflight cannot reproduce billions of years of slow evolution on surface of the small bodies that will never be reproduced in the laboratory, however they might shed light on the basic tendency of chemical evolution. The complex bioorganic compounds are easily produced by ultraviolet radiation and cosmic ray components in solid state.

So, the organic compounds, which had been delivered on the primordial Earth would have had a very complex structure and reached the second stage of the chemical evolution — polymerization. The small bodies would have imported prebiotic, rather very complex, molecules such as nucleotides and olygopeptides, to early Earth, which have allowed the fast synthesis of all molecules thought to be necessary for the origin of life. As it was shown by Basiuk and Douda [22], while the safe delivery of these compounds depends on such factors as size of the space body, nature and altitude of the heating, their chemical composition, it may be very successful in some cases. The comets, carbonaceous chondrites and other primitive bodies of the Solar system including micrometeorites, interplanetary and interstellar dust particles are responsible for essentially all carbon compounds and water currently present in our planet [23].

4. References

1. Schopf, J.W. (1993) Microfossils of the early Archean Apex chert: New evidence of the antiquity of life. *Science,* **260**, 640-646.
2. Chyba, C.F. (1993) The violent environment of the origin of life: Progress and uncertainties. *Geochim. Cosmochim.Acta,* **57**, 3351-3358.
3. Holland, H.D. (1984) *The chemical evolution of the atmosphere and oceans*. Princeton University Press.
4. Kasting, J.F. (1993) Earth's early atmosphere. *Science,* **259**, 920-926.
5. Anders, E. (1989) Pre-biotic organic matter from comets and asteroids. *Nature,* **342**, 255-257.
6. Oro, J., Mills, T., and Lazcano, A. (1992) Comets and the formation of biochemical compounds on the primitive Earth - a review. *Orig.Life Evol.Biosphere,* **21**, 267.
7. Flynn, G.J. (1996) The delivery of organic matter from asteroids and comets to the early surface of Mars. *Earth, Moon and Planets,* **72**, 469-474.
8. Allamandola, L.J., Sandford, S.A., and Valero, G.J. (1988) Photochemical and thermal evolution of interstellar/precometary ice analogs. *Icarus,* **76,** 225–252.
9. Greenberg, J.M., Schutte, W.A., and Li, A. (1999) Space irradiation of interstellar dust analogs at cryogenic temperatures. *Adv.Space Res.,* **23**, 289-292.
10. Hudson, R.L. and Moore, M.H. (1999) Laboratory studies of the formation of methanol and other organic molecules by water + carbon monoxide radiolysis: Relevance to comets, icy satellites, and interstellar ices. *Icarus,* **140**, 451-461.
11. Irvine, W.M. (1998) Extraterrestrial organic matter: A review. *Orig.Life Evol.Biosphere,* **28**, 365-383.
12. Cottin, H., Gazeau, M.C., Raulin, F. (1999) Cometary organic chemistry: a review from observations, numerical and experimental simulations. *Planet.Space Sci.,* **47**, 1141-1162.
13. Kuzicheva, E.A., *et al.*, (1996) Role of vacuum ultraviolet (VUV) radiation in abiogenic synthesis of adenine nucleotides. *Adv.Space Res.,* **18**, 65-68.
14. Simakov, M.B., Kuzicheva, E.A., and Malko, I.L. (1997) Abiogenic synthesis of pyrimidine nucleotides in solid state by vacuum ultraviolet radiation. *Adv.Space Res.,* **19**, 1059-1062.
15. Simakov, M.B., Kuzicheva, E.A., Antropov, A.E., and Dodonova, N.Y. (2001) Abiogenic synthesis of nucleotides on the surface of small space bodies with high energy particles. *Adv.Space Res.,* in press.
16. Kuzicheva, E.A., Zhukova, V.F., and Simakov, M.B. (1993) Role of UV and g-radiation in the abiogenic synthesis of nucleotides in solid state. *Zhurn.Evol.Biokhimii i Fisiologii,* **29**, 339-344.
17. Reimann, R. and Zubay, G. (1999) Nucleoside phosphorylation: a feasible step in the prebiotic pathway to RNA. *Orig.Life Evol.Biosphere,* **29**, 229-247.
18. Simakov, M.B., Kuzicheva, E.A., Malko, I.L., and Dodonova, N.Y. (1996) Abiogenic synthesis of oligopeptides in solid state under action of vacuum ultraviolet light. *Adv.Space Res.,* **18**, 61-64.
19. Simakov, M.B., Kuzicheva, E.A., Dodonova, N.Y., and Antropov, A.E. (1997) Formation of oligopeptides on the surface of small bodies in solar system by cosmic radiation. *Adv.Space Res.,* **19**, 1063-1066.
20. Kuzicheva, E.A. and Simakov, M.B. (1999) Abiogenic synthesis of nucleotides in conditions of space flight of the biosputnik "BION-11". *Adv.Space Res.,* **23**, 387-391.
21. Ferris, J.P. and Ertem, G. (1992) Oligomerization of ribonucleotides on montmorillonite: reaction of the 5'-phosphoimidazolide of adenosine. *Science,* **257**, 1387-1389.
22. Basiuk, V.A. and Douda, J. (1999) Pyrolysis of simple amino acids and nucleobases: survivability limits and implications for extraterrestrial delivery. *Planet.Space Sci.,* **47**, 577-584.
23. Delsemme, A.H. (2000) Cometary origin of the biosphere. *Icarus,* **146**, 313-325.

Section 4:

Life without starlight

TETRA- AND NONAPEPTIDYL MOTIFS IN THE ORIGIN AND EVOLUTION OF PHOTOSYNTHETIC BIOENERGY CONVERSION

Possible implications for the molecular origin of phosphate metabolism

HERRICK BALTSCHEFFSKY[1], ANDERS SCHULTZ[1], BENGT PERSSON[2] AND MARGARETA BALTSCHEFFSKY[1]

[1]*Department of Biochemistry, Arrhenius Laboratories, Stockholm University, S-106 91 Stockholm , Sweden*

[2]*Stockholm Bioinformatics Center, Department of Biochemistry and Biophysics, Karolinska Institutet, S-171 77 Stockholm, Sweden*

"More than 50 % of known proteins recognize or bind phosphates as part of their function..." (la Cour et al, 1985)

1. Introduction

In this session on "Life without starlight" it may be useful to describe photosynthetic organisms with their light-induced transformation, conversion and conservation of energy as "life with starlight". To what extent could photophosphorylation, this fundamental process of energy transfer in photosynthesis, have originated and evolved in discrete molecular steps in the absence of light? An answer may be sought in the first and so far only known alternative biological photophosphorylation system, where the comparatively "primitive" energy donor inorganic pyrophosphate (PPi) is produced (H. Baltscheffsky *et al.*, 1966) instead of the more "complex" adenosine triphosphate (ATP), the well known central energy donor and "energy currency" of all now living cells. The gene for PPi synthase, which is the enzyme involved in the photosynthetic formation of PPi in chromatophores from the photosynthetic bacterium *Rhodospirillum rubrum*, has been cloned (M. Baltscheffsky *et al.*, 1998) and the primary structure and some other properties of this first bacterial representative of the membrane-bound family of proton-pumping inorganic pyrophosphatases (H^+-PPases) has been deduced (M. Baltscheffsky *et al.*, 1999). In the loop between the fifth and sixth putative transmembrane segments, in what appears to be a pyrophosphate binding region and the active site for the phosphorylation reaction, three different "primitive" tetrapeptidyl motifs have been located in two nonapeptidyl sequences. Their high content of the four "very early" proteinaceous amino acids glycine, alanine, valine and aspartic acid (G, A, V and D, respectively) and their very regular arrangement of amino acids with charged side groups, would seem to be of particular evolutionary significance. Our first tentative model of the active site has resulted in a structure, which may be useful for attempts to extrapolate back to the hitherto elusive molecular origin and evolution of the metabolism of phosphate compounds, specially those, which are directly involved in bioenergy conversion.

J. Chela-Flores et al. (eds.), First Steps in the Origin of Life in the Universe, 173–178.

2. General background

The primary structure of H^+-PPi synthase, the integrally membrane-bound, proton-pumping inorganic pyrophosphate synthase (M. Baltscheffsky *et al.*, 1998; 1999), has opened the possibility to explore in molecular detail various properties of this "primitive" alternative to the ubiquitusly occurring ATP synthases. In addition, certain apparently very early, unique and potentially important evolutionary characteristics have been observed (Schultz *et al.*, submitted). This could possibly lead to a clearer picture of the molecular origin and evolution of biological energy conversion with phosphate compounds as well as of the metabolism of phosphate compounds in general.

The PPi synthase from *Rhodospirillum rubrum*, as deduced from its cloned gene, has 702 amino acids. The loop between the fifth and sixth of the putative 15-16 transmembrane segments (loop 5-6) seems to contain the main site for the binding of phosphates and the active center for their reactions (M. Baltscheffsky *et al.*, 1999; Schultz *et al.*, submitted).

3. Methods

Figure 1 is based on topology prediction method TopPred II (Claros and von Heijne, 1994).

The model presented in Figure 2 was calculated with the program ICM (version 2.7, Molsoft LLC, La Jolla, CA) for the segment 185-222 of the PPi synthase. Distance constraints were given between Mg^{2+} ions and PPi and the putative liganding residues Asp 187, Asp191, Lys195, Arg206, Asp213, Asp217 and Asp221, whereafter the structure was subjected to energy minimisations.

4. Results and discussion

In Figure 1 we present some essentials of the amino acid sequence in the putative loop 5-6. Three sequence motifs and the charged residues of the loop are emphasized. In addition PPi, the product of the photophosphorylation reaction and the substrate for hydrolysis, is inserted, magnified for clarity, as are four magnesium ions. This number is in agreement with what is known from studies with crystals of the classical family of soluble PPases (Heikinheimo *et al.*, 1996). The great amount of charged residues would seem to suffice well for binding both the negatively charged PPi and the "surrounding" cations. The properties of the three sequence motifs, GGG, DVGADLVGK and DNVGDNVGD, which are shown as indicated by arrows 1, 2 and 3 in Figure 1, will be presented in some detail.

The motif GGG, which appears to be very close to or partially in transmembrane segment 5, depending upon the topology prediction method used (Claros and von Heijne, 1994; Sonnhammer *et al.*, 1998), has, other conditions permitting, the earlier known capability to act as a swinging entity. It may tentatively be assumed to function as an energy coupling structure between proton gradient and phosphorylation, involved in a conformational change mechanism. We suggest that such a mechanism in the photophosphorylation of Pi to PPi, catalyzed by the PPi synthase, may be more or less similar to the one operating in the photophosphorylation of ADP to ATP, catalyzed by ATP synthase.

Two remarkable properties of the two nonapeptide motifs (indicated by arrows 2 and 3 in Figure 1) are their high content of the four "very early" (Eigen and Schuster, 1978) proteinaceous amino acids G, A, D and V and the position of charged amino acid residues, which in both nonapeptides is 1, 5 and 9 and which in all cases are D except for amino acid 9 in the first nonapeptide, which is lysine, K (Schultz *et al.*, submitted). It is an open question whether the nonapeptides or the tetrapeptidyl parts of them, DVGA, DNVG and DLVG, have a common origin. However, the fact that both nonapeptides with their high content of "very early" amino acids in the H^+-PPase family are situated in the putative active site loop, in a way which appears to facilitate substrate, product and cation binding, has lead us to suggest that they indicate a very ancient origin (Baltscheffsky *et al.*, 1999; Schultz *et al.*, submitted). In addition, we have proposed that they reflect similar, early molecular arrangements in connection with the origin of life and its energy conversion with energy rich phosphates. As possible traces of a prebiotic selection of oligopeptides involved in reactions of inorganic phosphates they could be indicating a *sine qua non* in the origin and early evolution of that major part of phosphate metabolism, where the participation of divalent cations is required.

This reasoning leads to the question if clays or other classes of minerals were involved in some suitable prebiotic repertoire of conditions for the rise of early phosphate metabolism, constituting a preamble to a genetic takeover of early energy conversion with inorganic phosphate compounds. As we have pointed out recently (Baltscheffsky *et al.*, submitted), there seems to exist a plausible pathway from prebiotic homooligomers of the earliest proteinaceous amino acids, glycine and alanine, over a genetic takeover and two transitions, to the above mentioned tetrapeptide motif DVGA.

5. Epilogue and outlook

The H^+-PPi synthase may have had its origin as an early H^+-PPase, utilising environmentally available prebiotic PPi, which is known to be formed, for example, from volcanic P_4O_{10}, by partial hydrolysis of cooling magma (Yamagata et al., 1991). Subsequent lack of available PPi could have led to a change in kinetic constants, to facilitate the formation of the energetically required PPi. The demonstrated photosynthetic formation of PPi may have been an early anastrophic adaptation to a diminishing pool of environmental PPi and thus a strong need for biological formation of significant amounts of this energy donor.

In order to try to extrapolate backward from present life to early evolution it can be helpful to suggest possible models for the interactions between, as in this case, enzyme, metal ion and substrate. Figure 2 shows our attempt to obtain a plausible structure for the three-dimensional binding of four magnesium ions and PPi in loop 5-6 of PPi synthase. It is a very tentative picture of what is hoped to be a useful first step towards future attempts to grasp in more detail the essential early characteristics of molecular energy conversion with PPi in connection with the emergence of life on the earth.

6. Acknowledgements

Support from Carl Tryggers Stiftelse för Vetenskaplig Forskning, Magnus Bergvalls Stiftelse and Stiftelsen Wenner-Grenska Samfundet is gratefully acknowledged. B. P. gratefully acknowledges support from the Swedish Foundation for Strategic Research and the Swedish Research Council.

7. References

Baltscheffsky, H., von Stedingk, L.-V., Heldt, H.W., and Klingenberg, M. (1966) Inorganic pyrophosphate: Formation in bacterial photophosphorylation, *Science* **153,** 1120-1122.

Baltscheffsky, M., Nadanaciva, S., and Schultz, A. (1998) A pyrophosphate synthase gene: molecular cloning and sequencing of the cDNA encoding the inorganic pyrophosphate synthase from *Rhodospirillum rubrum*, *Biochim. Biophys. Acta* **1364**, 301-306.

Baltscheffsky, M., Schultz, A., and Baltscheffsky, H. (1999) H^+-PPases: a tightly membrane-bound family, *FEBS Lett.* ***457***, 527-533.

la Cour, T.F.M., Nyborg, J., Thirup, S., and Clark, B.F.C. (1985) Structural details of the binding of guanosine diphosphate to elongation factor Tu from *E. coli* as studied by X-ray crystallography, *EMBO J.* **4**, 2385-2388.

Eigen, M., and Schuster, P. (1978) The hypercycle, *Naturwiss*, **65**, 341-369.

Claros, M.G., and von Heijne, G, (1994) TopPred II: An improved software for membrane protein structure prediction, *CABIOS* **10**, 685-685.

Heikinheimo, P., Lehtonen, J., Baykov, A.A., Lahti, R., Cooperman, B.S., and Goldman, A. (1996) The structural basis for pyrophosphatase catalysis, *Structure* **4**, 1491-1508.

Schultz, A., Baltscheffsky, M., Pérez-Castiñeira, J.R., and Baltscheffsky, H. (submitted).

Sonnhammer, E., von Heijne, G., and Krogh, A. (1998) A hidden Markov model for predicting transmembrane helices in protein sequences, *Intell. Syst. Mol. Biol.* **6**, 175-182.

Yamagata, Y., Watanabe, H., Saitoh, M., and Namba, T. (1991) Volcanic production of polyphosphates and its relevance to prebiotic evolution, *Nature* **352**, 516-519.

Figure 1. Sketch of loop 5-6 in H^+-PPi synthase, describing some properties in broad terms. Three sequence motifs and charged residues are emphasized, around magnified PPi and four magnesium ions. The sequence is read from TM5 (transmembrane segment 5) to TM6, the first amino acid residue in the loop being L177. Numbers in parentheses in the loop indicate amount and location of amino acid residues, which are not of particular significance in the present context and thus not specified here.

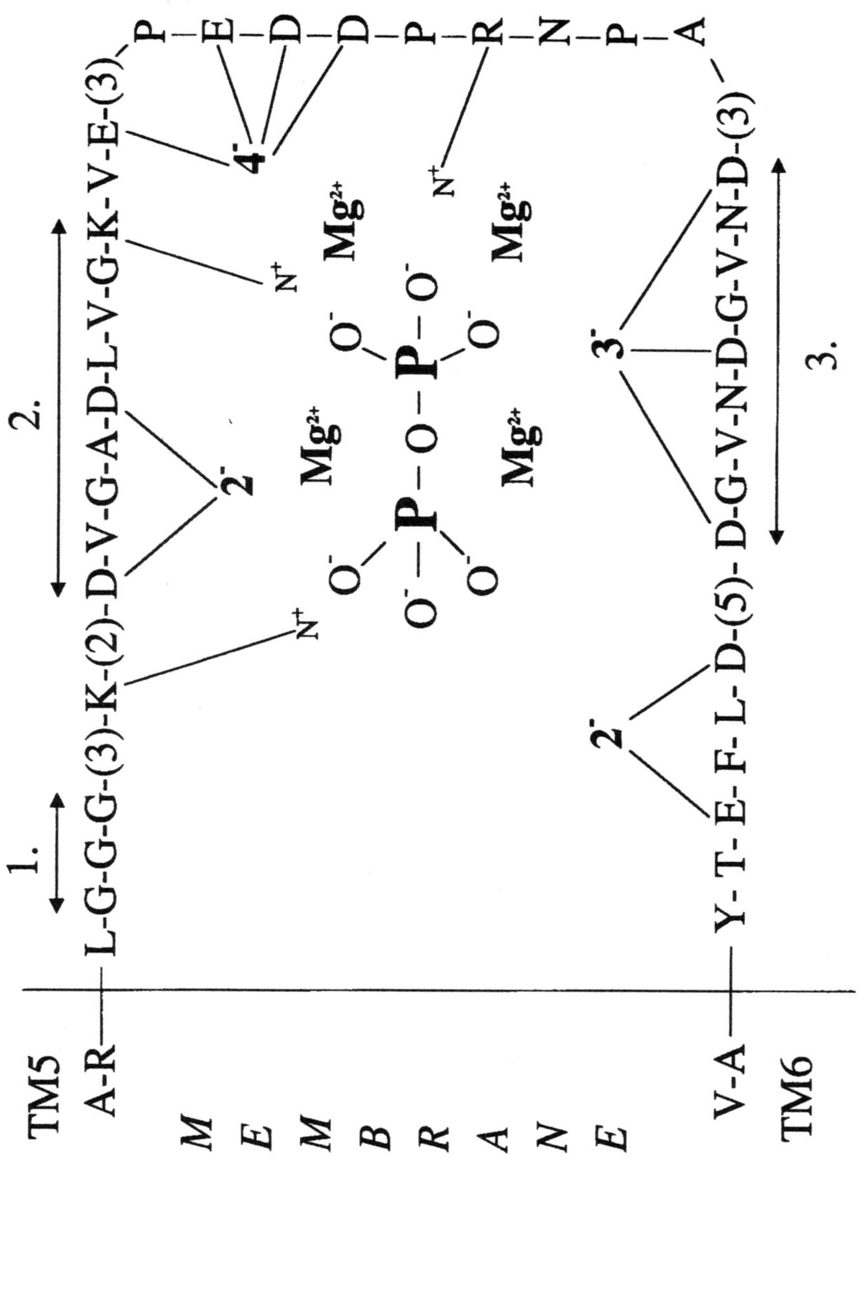

Figure 2. Possible structure of PPi binding in loop 5-6 of H^+-PPi synthase. The views in A and B are orthogonal and slightly tilted with respect to each other. Ribbon shows main chain of the segment 185-222, whereas side chains are represented by thin lines. The PPi is shown with bold lines and the Mg^{2+} ions as small spheres. Dashed lines show constraints between Mg^{2+} ions and PPi and the putative liganding residues Asp 187, Asp191, Lys195, Arg206, Asp213, Asp217 and Asp221. In the PPi, Lys195 binds to an oxygen atom at one of the phosphates, while Arg206 binds to oxygen atoms at the other phosphate. This model is partly based on the structure given in connection with a proposed mechanism for yeast soluble PPase, with four Mg^{2+} at the active site (cf. Fig. 7 in Heikinheimo *et al.,* 1996).

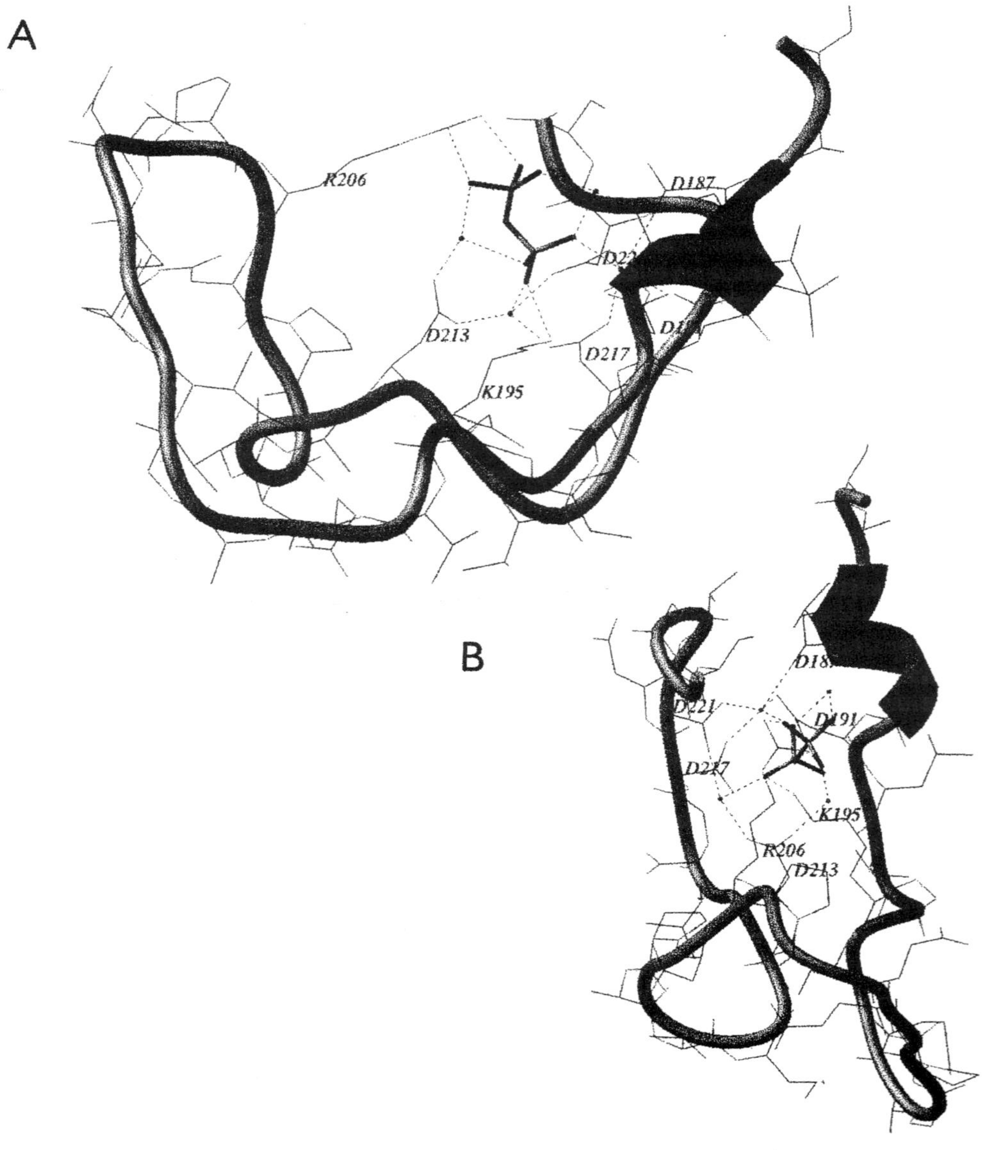

GEOCHEMICAL HABITATS IN HYDROTHERMAL SYSTEMS

EVERETT L. SHOCK

Group Exploring Organic Processes in Geochemistry
Department of Earth & Planetary Sciences
Washington University
St. Louis, MO 63130, USA

Deep Phylogeny and Geochemistry

There are many readings of the universal phylogenetic tree constructed from small subunits of RNA, such as that shown in Fig. 1. As a geochemist, two messages seem particularly clear to me from deep in the tree. First, the emphasis is on geochemical energy sources, which take the form of naturally occurring disequilibrium states engendered by geologic processes. Second, deep autotrophy does not require sunlight. These two messages suggest that early life was more likely to have lived in subsurface rather than surface environments. A logical deduction from the 16s rRNA phylogenetic tree (Fig. 1) is that the setting for the emergence of life had little if anything to do with the atmosphere or surface of the Earth. According to some estimates the subsurface biosphere *still* rivals by mass the surface biosphere (Whitman et al., 1998). We probably find this surprising because we ascribe great importance to the surface biosphere owing to our banishment there by our need for O_2. Nevertheless, deeply branching organisms in the universal phylogenetic tree show a proclivity for H_2, or at least for reduced conditions where even trace concentrations of O_2, sulfate, carbon dioxide and other oxidized compounds represent enormous energy sources.

Another message from deep in the tree involves high temperatures. All of the deepest branches are populated by hyperthermophiles and on branches that host both thermophiles and mesophiles the thermophiles diverge more deeply (Stetter, 1992). This message may be easily misunderstood, because similar chemical energy sources — rather than the thermal state of the system — may be the common link. In fact, high temperatures may be artifacts of the long-term poisoning of the atmosphere by the effluent (O_2) of photosynthesis. Perhaps the earliest organisms rapidly invaded all possible habitats, but their metabolic strategies became impossible once O_2 levels rose after oxygenic photosynthesis appeared. If so, it may be that hydrothermal systems are the only habitats that still host deeply branching organisms because those habitats remain largely unchanged since early in Earth history. This possibility can be tested by finding deeply branching organisms in low temperature environments that are otherwise chemically similar to hydrothermal systems.

Nevertheless, it is presently a fact that hydrothermal ecosystems host the extremophiles on the deepest branches of the 16s rRNA tree. Despite the possibility that this is an artifact of O_2 pollution, we can ask: what is it about hydrothermal ecosystems that make them such great places to live? I believe that answering this question will lead to new hypotheses about early life and the emergence of life that can be tested experimentally.

J. Chela-Flores et al. (eds.), First Steps in the Origin of Life in the Universe, 179–185.

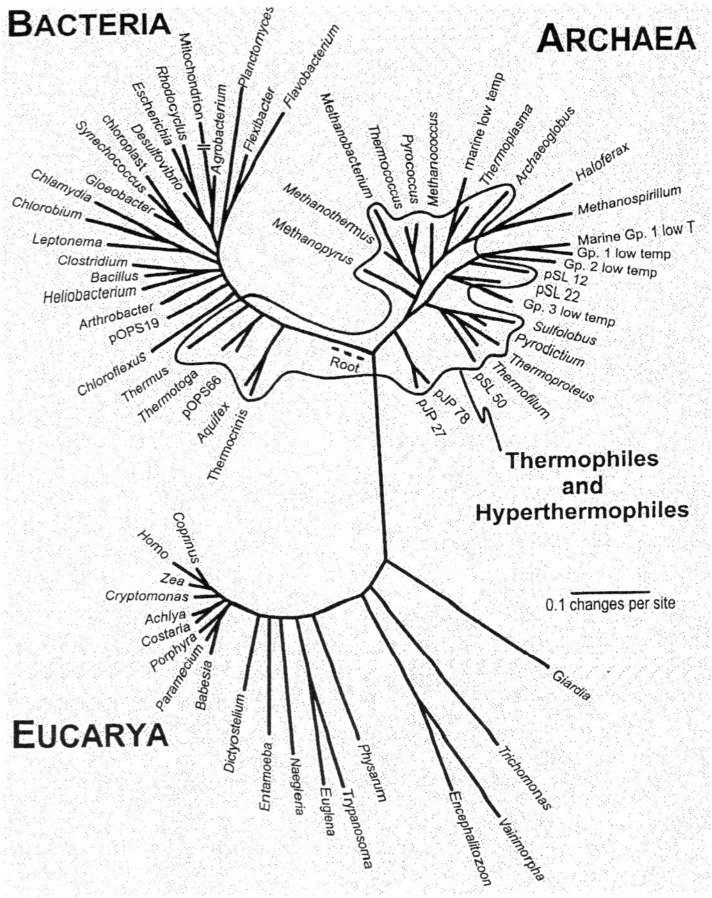

Fig. 1. (after Pace, 1997) Universal phylogenetic tree of life based on 16s rRNA data. Lengths of lines separating organisms represent the genetic difference between them. Deep, short branches of the tree are populated by thermophilic and hyperthermophilic organisms.

Metabolic Strategies in Hydrothermal Ecosystems

The metabolic strategies of deeply-branching hyperthermophiles provide some clues for understanding why hydrothermal ecosystems are such great places to live. I have picked three examples to illustrate the types of geochemical energy tapped by organisms near the base of the phylogenetic tree. A common theme uniting these deep metabolisms is the use of H_2.

The deepest branch in the Bacteria domain is populated by the *Aquificales*, which include *Aquifex pyrophilus*, *Aquifex aeolicus, Thermocrinis ruber*,

Calderobacterium hydrogenophilum, and *Hydrogenobacter thermophilus*, along with several unnamed organisms known only from environmental DNA samples (Kryukov et al., 1983; Kawasumi et al., 1984; Huber et al., 1992; 1998; Deckert et al., 1998; Hugenholtz et al., 1998; Yamamoto et al., 1998; Skirnisdottir et al., 2000; Reysenbach et al., 2000). As the name implies, members of the genus *Aquifex* gain energy by reacting oxygen and hydrogen to form water via the "knallgas" reaction

$$H_2(aq) + 1/2\ O_2(aq) = H_2O\ , \qquad (1)$$

where (aq) refers to the neutral solute in aqueous solution. One location where *Thermocrinis ruber* thrives is the runoff channel of Octopus Spring at Yellowstone National Park, USA, where it contributes to the pink filament community described by many investigators over the years (Setchell, 1903; Brock, 1978; Reysenbach et al., 1994; Huber et al., 1998). Our field measurements show that this community thrives at temperatures from about 80° to 90°C, at pH values near 8, and that there is from 0.7 to 2.3 mg/kg dissolved O_2 in the hot spring water running through the channel (Shock et al, 2001). Although there is not much H_2 present, we were able to detect about 0.02 volume % in two dry gas samples collected in 1999. If the dry gas represents 1% of the total (the rest being steam), which is probably a conservative estimate, then the equilibrium activity of dissolved H_2 would be about $10^{-8.8}$.

These analytical data, together with standard state thermodynamic data, allow a calculation of the amount of energy available per mole of H_2 consumed. Equations, data, and parameters for $H_2(aq)$ and $O_2(aq)$ (Shock et al., 1989) permit calculation of the standard state Gibbs free energy of reaction (1) at elevated temperatures and pressures ($\Delta G°_r$), which is equal to -259.60 kJ $(mol)^{-1}$ at 85°C and 1 bar (see also: Amend and Shock, 2001). Taking account of the relation

$$\Delta G_r = \Delta G°_r + RT\ \ln Q_r\ , \qquad (2)$$

where ΔG_r represents the overall Gibbs free energy of reaction and Q_r stands for the stoichiometric activity product for the reaction, and converting analytical data to activities, leads to a range of –183.1 to –184.9 kJ $(mol\ H_2)^{-1}$ available from reaction (1) at the conditions where *Thermocrinis ruber* lives at Octopus Spring.

Many hyperthermophilic Archaea are known to reduce native sulfur to H_2S in the laboratory, and it is likely that they conduct the reaction

$$H_2(aq) + S = H_2S(aq) \qquad (3)$$

in nature as well. *Pyrodictium occultum*, a chemolithoautotrophic member of the Euryarchaeota isolated from shallow marine hot springs in the Baia di Levante of Vulcano, Italy, was the first organism to grow above 100°C in pure culture (Stetter et al., 1983). Preliminary analytical data from our studies of these hot springs indicates that at temperatures near 100°C the mole fractions of H_2S are near 10^{-4}, and tend to exceed those of H_2 by about an order of magnitude (Amend et al., 2001; Amend, 2001). These data, converted to equilibrium activities of the aqueous neutral solutes and combined with standard state thermodynamic data, indicate that the overall Gibbs free energy for reaction (3) at the Baia di Levante is in the vicinity of –29. kJ (mol H_2)$^{-1}$ at locations likely to support *Pyrodictium occultum.* Unlike the case of the pink filament community at Octopus Spring described above, incomplete reporting of sample localities by researchers who isolated *Pyrodictium occultum* makes it impossible, at present, to connect analytical data with the actual habitat of this organism. Part of our ongoing field and laboratory work aims to resolve this inadequacy in hyperthermophile microbial ecology.

Autotrophic methanogens such as *Methanopyrus kandleri* and *Methanococcus jannaschii* are among the deepest-branching members of the Crenarchaeota. These chemoautotrophs gain energy by reduction of carbon dioxide to form methane via

$$4H_2(aq) + CO_2(aq) = CH_4(aq) + 2H_2O \quad . \tag{4}$$

Laboratory studies show that the optimum temperature for *Methanopyrus kandleri* is 100°C and that its maximum temperature is 110°C (Kurr et al., 1991). Corresponding values for *Methanococcus jannaschii* are somewhat lower: 85°C optimum in the lab and 94°C maximum (Jones et al., 1983). This organism was isolated from a submarine hydrothermal vent system on the East Pacific Rise, and no samples of 85°C water from these systems were collected. Nevertheless, attempts have been made to estimate the composition of fluids with intermediate temperatures given known compositions of seawater and high-temperature vent fluids (McCollom and Shock, 1997). This approach makes it possible to estimate the concentrations of $H_2(aq)$, $CO_2(aq)$ and $CH_4(aq)$ in mixed submarine hydrothermal fluids at 85°C (0.4 mmol, 3.1 mmol, and 0.02 mmol, respectively), and to calculate that the overall Gibbs free energy for methanogenesis at these conditions is equal to –26.6 kJ (mol H_2)$^{-1}$.

Limiting our focus to these metabolic strategies in these specific geochemical habitats, the knallgas reaction appears to yield 6 to 7 times more energy per mole of H_2 than sulfur reduction or methanogenesis. If the comparison is made per mole of electron (e-mol), the knallgas reaction yields about 92 kJ (e-mol)$^{-1}$, sulfur reduction yields about 14 kJ (e-mol)$^{-1}$ and methanogenesis yields about 3 kJ (e-mol)$^{-1}$. The thirty fold difference in energy supply between the knallgas reaction and methanogenesis may tell us something about why the knallgas reaction is imbedded so deeply in the tree. At first glance it seems peculiar that a deep metabolism would require O_2, which would have been a trace gas on the early Earth, and it is certainly the case that O_2 is much more abundant than H_2 in continental hot springs like those at Yellowstone. On the other hand, if the relative abundances of H_2 and O_2 were reversed, as they would be in cooling volcanic gases, the amount of energy available from the knallgas reaction would be nearly the same as the analysis for Octopus Spring (see: Shock et al., 2000). This suggests that deeply-branching microorganisms were initially dependent on volcanic gases as geochemical energy sources and have adapted to complementary habitats such as hot springs.

Geochemistry and Deep Metabolism

The reductive strategies employed by deeply-branching hyperthermophiles all yield abundant energy as the examples described above illustrate. Geochemical features of hydrothermal systems that make these strategies possible are the supply of H_2 from water/rock reactions, and rapid cooling and/or mixing of fluids. Measurable concentrations of H_2 are likely to be far from equilibrium with O_2, S and CO_2, with the result that there are enormous thermodynamic drives to form H_2O, H_2S and CH_4. Kinetic constraints cause these reactions to be sluggish without catalysts, and enzymatic processes play the catalytic role in each of these reductive metabolic strategies. Therefore, hyperthermophiles in hydrothermal ecosystems exploit disequilibria established by geochemical processes. It appears that they have done so since life first appeared on the Earth.

As plausible early metabolic strategies go, chemolithoautotrophy is a particularly attractive model. For instance, not only do chemolithoautotrophic methanogens gain energy by catalyzing the formation of methane from CO_2, they make all of their biomolecules by reducing CO_2 as well. Similarly, the *Aquificales* generate their biomolecules from CO_2. In many ways chemolithoautotrophs seem to be living examples of emergence of life experiments -- starting with inorganic carbon and making it into biomass, proteins, DNA, RNA, and so forth. It follows that it may prove profitable to examine the metabolism of chemolithoautotrophs to gain insight into how life emerged.

With this in mind, it is worth noting that the synthesis of many amino acids results in a net release of energy at the conditions where *Methanopyrus kandleri* and *Methanococcus jannaschii* live (Amend and Shock, 1998). Gibbs free energies of overall synthesis reactions of amino acids from CO_2, NH_4^+ and H_2 are negative for alanine, glutamate, isoleucine, leucine, lysine, methionine, phenylalanine, proline, tryptophan, tyrosine and valine at 100°C and 250 bars (Amend and Shock, 1998), and all are slightly more negative at 85°C and 250 bars where *Methanococcus jannaschii* thrives. The enzymes coded by the genome of *Methanococcus jannaschii* are somewhat enriched in these amino acids relative to the others, such that there is the potential for a net release of energy during synthesis of the amino acids required to form 99.4 % of its enzymes. Despite a published genome (Bult et al., 1996) much remains mysterious about the biosynthetic pathways used by *Methanococcus jannaschii*, including whether any of this energy can be conserved. Nevertheless, this analysis helps to illustrate how conditions in the habitats inhabited by hyperthermophiles differ fundamentally from those we find familiar, and how geochemical processes make these habitats particularly conducive to life.

Acknowledgements -- Thanks to the organizers for inviting me to speak, and especially to Julian Chela-Flores for his patience. Thanks to Gary Olsen and colleagues for help with the genome of *Methanococcus jannaschii.* Research supported by National Science Foundation grants OCE-9714288 and OCE-9817730. This is GEOPIG contribution #234.

References

Amend, J. P. (2001) Metabolic energy from sulfur redox for hyperthermophiles in the hydrothermal ecosystems on Vulcano, Aeolian Islands. *Extremophiles* (in prep.)

Amend, J.P. and Shock, E.L. (1998) Energetics of amino acid synthesis in hydrothermal ecosystems. *Science* **281**, 1659-1662.

Amend, J.P. and Shock, E.L. (2001) Energetics of overall metabolic reactions in thermophilic and hyperthermophilic Archaea and Bacteria. *FEMS Microbiology Reviews* **25**, 175-243.

Amend J. P., Shock E. L., Inguaggiato S., Italiano F., and Gurrieri S. (2001) Chemical analyses of hydrothermal seeps, wells, and springs in the Aeolian Islands, southern Italy. *Geochim. Cosmochim. Acta*, (in prep.).

Brock, T.D. (1978) *Thermophilic Microorganisms and Life at High Temperatures,* Springer-Verlag, New York.

Bult, C.J., White, O., Olsen, G.J., Zhou, L., Fleischmann, R.D., Sutton, G.G., Blake, J.A., FitzGerald, L.M., Clayton, R.A., Gocayne, J.D., Kerlavage, A.R., Dougherty, B.A., Tomb, J.F., Adams, M.D., Reich, C.I., Overbeek, R., Kirkness, E.F., Weinstock, K.G., Merrick, J.M., Glodek, A., Scott, J.L., Geoghagen, N.S., and Venter, J.C. (1996) Complete genome sequence of the methanogenic archaeon, *Methanococcus jannaschii*. *Science* **273**, 1058-1073.

Deckert, G., Warren, P.V., Gaasterland, T., Young, W.G., Lenox, A.L., Graham, D.E., Overbeek, R., Snead, M.A., Keller, M., Aujay, M., Huber, R., Feldman, R.A., Short, J.M., Olsen, G.J. and Swanson, R.V. (1998) The complete genome of the hyperthermophilic bacterium *Aquifex aeolicus*. *Nature* **392**, 353-358.

Huber, R., Wilharm, T., Huber, D., Trincone, A., Burggraf, S., Konig, H., Rachel, R., Rockinger, I., Fricke, H. and Stetter, K.O. (1992) *Aquifex pyrophilus* gen. nov. sp. nov., represents a novel group of marine hyperthermophilic hydrogen oxidizing bacteria. *Syst. Appl. Microbiol.* **15**, 340-351.

Huber R., Eder W., Heldwein S., Wanner G., Huber H., Rachel R., and Stetter K. O. (1998) *Thermocrinis ruber* gen. nov., sp. nov., a pink-filament-forming hyperthermophilic Bacterium isolated from Yellowstone National Park. *Appl. Environ. Microbio.* **64**, 3576-3583.

Hugenholtz, P., Pitulle, C., Hershberger, K.L., and Pace, N.R. (1998) Novel division level bacterial diversity in a Yellowstone hot spring. *J. Bacteriol.* **180**, 366-376.

Jones, W.J., Leigh, J.A., Mayer, F., Woese, C.R. and Wolfe, R.S. (1983) *Methanococcus jannaschii* sp. nov., an extremely thermophilic methanogen from a submarine hydrothermal vent. *Arch. Microbiol.* **136**, 254-261.

Kawasumi, T., Igarashi, Y., Kodama, T and Minoda, Y. (1984) *Hydrogenobacter thermophilus* gen. nov., sp. nov., an extremely thermophilic, aerobic, hydrogen-oxidizing bacterium. *Int. J. Syst. Bacteriol.* **34**, 5-10.

Kryukov, V.R., Savel'eva, N.D. and Pusheva, M.A. (1983) *Calderobacterium hydrogenophilum* gen. et sp. nov., an extremely thermophilic hydrogen bacterium and its hydrogenase activity. *Mikrobiologiya* **52**, 611-618.

Kurr, M., Huber, R., Konig, H., Jannasch, H.W., Fricke, H., Trincone, A., Kristjansson, J.K. and Stetter, K.O. (1991) *Methanopyrus kandleri*, gen. and sp. nov. represents a novel group of hyperthermophilic methanogen, growing at 110°C. *Arch. Microbiol.* **156**, 239-247.

McCollom, T.M. and Shock, E.L. (1997) Geochemical constraints on chemolithoautotrophic metabolism by microorganisms in seafloor hydrothermal systems. *Geochim. Cosmochim. Acta* **61**, 4375-4391.

Pace, N.R. (1997) A molecular view of microbial diversity and the biosphere. *Science* **276**, 734-740.

Reysenbach, A-L., Wickham, G.S. and Pace, N.R. (1994) Phylogenetic analysis of the hyperthermophilic pink filament community in Octopus Spring, Yellowstone National Park. *Appl. Environ. Microbiol.* **60**, 2113-2119.

Reysenbach, A-L., Ehringer, M. and Hershberger, K. (2000) Microbial diversity at 83°C in Calcite Springs, Yellowstone National Park: another environment where the *Aquificales* and "Korarchaeota" coexist. *Extremophiles* **4**, 61-67.

Setchell, W.A. (1903) The upper temperature limits of life. *Science* **17**, 934-937.

Shock, E.L., Helgeson, H.C., and Sverjensky, D.A. (1989) Calculation of the thermodynamic and transport properties of aqueous species at high pressures and temperatures: Standard partial molal properties of inorganic neutral species. *Geochim. Cosmochim. Acta*, **53**, 2157-2183.

Shock, E.L., Amend, J.P. and Zolotov, M.Yu. (2000) The early Earth vs. the origin of life. In: *The Origin of the Earth and Moon* (R. Canup and K. Righter, eds.) University of Arizona Press, p. 527-543.

Shock E.L., Amend, J.P, Meyer, D., Rogers, K., Osburn, G.R. Chan, G. and Fischer, T. (2001) Geochemistry of hydrothermal ecosystems at Yellowstone National Park. (in prep).

Skirnisdottir, S., Hreggvidsson, G. O., Hjörleifsdottir, S., Marteinsson, V. T., Petursdottir, S. K., Holst, O., and Kristjansson, J. K. (2000) Influence of sulfide and temperature on species composition and community structure of hot spring microbial mats *Appl. Envir. Microbiol*. **66**, 2835-2841.

Stetter, K.O. (1992) Life at the upper temperature border. In *Frontiers of Life* (Trân Thahn Vân et al., eds.) pp 195-219, Editions Frontières.

Stetter, K.O., Konig, H. and Stackebrandt, E. (1983) *Pyrodictium* gen. nov., a new genus of submarine disc-shaped sulphur reducing archaebacteria growing optimally at 105°C. *Syst. Appl. Microbiol.* **4**, 535-551.

Whitman, W.B., Coleman, D.C. and Wiebe, W.J. (1998) Prokaryotes: The unseen majority. *Proc. Natl. Acad. Sci. U. S. A.* 95: 6578-6583.

Yamamoto, H., Hiraishi, A., Kato, K, Chiura, H.X., Maki, Y and Shimizu, A. (1998) Phylogenetic evidence for the existence of novel thermophilic bacteria in hot spring sulfur-turf microbial mats in Japan. *Appl. Environ. Microbio.* **64**, 1680-1687.

DEEP-SEA HYDROTHERMAL VENTS: AN EXAMPLE OF EXTREME ENVIRONMENT ON EARTH

D. M. PRIEUR

Université de Bretagne Occidentale, UMR CNRS 6539
IUEM, Technopole Brest-Iroise, 29280 Plouzané, France

Introduction

In the solar system, Earth is the only object with liquid water at the surface. Seas and oceans cover 70% of Earth surface and have an average depth of 3800m. Although this huge amount of water seems homogeneous, the oceanic environment is extraordinary diversified.

The marine environment

All living organisms on Earth are exposed to many physio-chemical parameters, and particularly temperature, light, pH, salt concentration, oxygen, water activity or hydrostatic pressure. Any combination of values for these parameters defines a biotop, a geographic entity in which organisms adapted to this set of conditions carry out their life cycle. All these parameters may vary for oceanic waters. For instance temperature is in the range -2°C (polar oceans) to 35°C (Persian Gulf), and salinity (3,5% in average) may reach 4,1% in the Red Sea and salt saturation in seawater evaporating ponds. Also oxygen concentration may reach saturation in surface waters while some deep layers and sediments are totally anoxic. Hydrostatic pressure that increases of 1 Mpa for 100 m is about 110 Mpa for the deepest trenches (11km). Finally, sunlight, besides variations due to latitude or day/night alternance does not penetrate beyond several hundred meeters. There, it is complete darkness and no photosynthesis may occur. Organisms living in cold (2°C) deep oceans, are exposed to elevated hydrostatic pressure and depend on more or less bacteria-degraded food particles sinking from the euphotic zone. For all these reasons, until rather recently, only low density animal communities were known on the deep sea-floor.

The definition of an extreme environment is biased by our own perception of the environment. However, any biotope in which a very low or high value has been measured for a major physic-chemical parameter may be called extreme. The deep ocean, dark, cold, oligotroph and exposed to elevated hydrostatic pressure is one of theses terrestrial extreme environments where nevertheless thrive communities of bacteria, invertebrates and fishes.

J. Chela-Flores et al. (eds.), First Steps in the Origin of Life in the Universe, 187–193.

Deep-sea hydrothermal vents

The discovery in 1977 of deep-sea hydrothermal vents has totally modified our knowledge about the deep ocean. Hydrothermal processes are a consequence of plate tectonics, and principally exist on submarine ridges which cover 60 000kms all around the planet. Analysis of seawater above certain ridge segments revealed several anomalies and particularly a slight increase of temperature and manganese concentrations. These features were explained by an hydrothermal circulation: seawater penetrates through cracks of the sea floor, is heated when it approaches the magma chamber, leaches deep rocks, is enriched with minerals and rises up because of its low density.

When the US manned submersible Alvin dove at a depth of 2000m near the Galapagos Islands in 1977, pilot and scientists were fascinated by an unexpected landscape. Surrounding vents of warm waters, novel luxuriant animal communities were observed for the first time. Two years later, in 1979, an other Alvin dive on the East Pacific Rise revealed the existence of the black smokers, spectacular mineral constructions from which particle rich hot fluids vent out at temperature up to 350°C (water remains liquid because of elevated hydrostatic pressure) and speeds sometimes above 2 ms-1. Probably the most extreme environment on Earth was discovered.

Deep-sea hydrothermal vents were then discovered in tectonically active zones of the sea floor (ridges and back arc basins), in both Pacific and Atlantic oceans, at depths from 800 to 3500m. Hydrothermal processes are roughly similar in all cases. After a complex sub-seafloor circulation, acidic, reduced hydrothermal fluids enriched with heavy metals, methane, hydrogen sulphide and other inorganics, vent out. If they mix before venting out with cold seawater, their temperatures are about 10-40°C. If no mixing occurs, hot fluids are emitted in cold oxygenated seawater, and minerals precipitate and build the so-called black smokers. Such ecosystems could not be ignored by microbiologists.

Efficient symbiosis

The most surprising feature linked to the discovery of deep-sea hydrothermal vents is the occurrence of spectacular animal communities living around emissions of moderate temperature fluids: vestimentifera tube worms, bivalves, annelids, gastropods, according to the vent location. How to explain these high densities of animals, in an environment where no photosynthesis can exist? Biochemical analysis and TEM examination of vent invertebrate tissues rapidly gave conclusive answers: these invertebrates live in close associations with intracellular chemo-litho-autotrophic bacteria which use energy obtained from sulphide oxidation to transform carbon dioxide into organic carbon through the Calvin cycle.

This symbiosis is particularly efficient for the Vestimentifera. These tube worms (the giant tube worm *Riftia pachyptila* may reach 2 meters high) live in proteo-chitineous tube from which escape a plume of red gills. The animal has no digestive tract, but a special organ, the trophosome, full of autotrophic sulfo-oxidizing bacteria. The tube worm takes from the environments all nutriments required for bacterial life: carbon dioxide as a carbon source, hydrogen sulphide and molecular oxygen as electron donor and acceptor respectively. These two last molecules are transported on two different sites of the blood pigment.

Bivalve mollusks also harbor bacterial symbionts, but they live in the gill filaments of the mollusks. These symbionts also are sulphide-oxidizers, but some mussels from Atlantic deep vents harbor also a second type of symbiont which oxidize methane. Up to now, all symbionts escaped to laboratory cultures and are only known through their enzymes involved in sulphide oxidation and carbon dioxide fixation or their rRNA (5S and 16S) sequences which indicated they belong to the gamma subdivision of Proteobacteria.

One of the main questions about these symbiosis is the transmission of symbionts. Molecular techniques based on PCR amplification of 16S rRNA genes and *in situ* hybridizations showed a vertical transmission trough oocytes in the case of the bivalve *Calyptogena magnifica*, but rather a post metamorphosis environmental infestation in the case of the tube worm *R. pachyptila*.

While Vestimentifera and bivalve live around moderate temperature vents, other invertebrates live close to, or on hydrothermal chimneys (black smokers) where they are exposed to highest temperatures. Particularly , on the East Pacific Rise, the Polychaetous annelid *Alvinella pompejana* builds its tube on the outer wall of active smokers where temperature is probably in the range 40-60°C, although one individual was observed twisted around a temperature probe, operated by a submersible and measuring 100°C. These worms show a spectacular bacterial epibiosis located on the dorsal integuments. Several morphotypes have been cultured and revealed a metabolic diversity. However, the dominant filamentous types although not cultured, have been assigned to the epsilon subdivision of Proteobacteria.

In the Atlantic, deep-sea vents are characterized by dense clouds of shrimps, swimming around active chimneys. They belong to several species but particularly the species *Rimicaris exoculata* has been studied. These shrimps have hypertrophied mouthparts on which are filamentous and rod-shaped epibiotic bacteria, again belonging to the epsilon subdivision of Proteobacteria. For the two cases of epibiosis (worms and shrimps) the trophic role of these epibiosis has not been elucidated.

Life at high temperature

The existence of hydrothermal fluids, still liquid at temperatures up to 350°C particularly motivated microbiologists interested by life at high temperature, and they focus on search for novel hyperthermophiles.

NOVEL BACTERIA SPECIES

However, incubation of smoker debris suspended in culture media at temperatures below 80°C allowed to isolate several strains belonging to the Bacteria domain (in 1977, Woese established that living organism are distributed into three domains, the Eukarya, the Bacteria and the Archaea). Some are strictly anaerobic, ferment organic matter, and belong to the genera *Thermotoga* and *Thermosipho*. More recently, a novel species has been described: *Desulfurobacterium thermolithotrophicum*, that utilizes hydrogen as electron donor and elemental sulphur (S°) as electron acceptor. Aerobic thermophiles also have been reported, and assigned to the genera *Thermus* and *Bacillus*. Within hydrothermal systems, temperature gradients are very steep, and a few centimeters only may exist between the hot fluid circulating inside a black smoker, and the cold deep water, while the temperature at the border between aerobiosis and anaerobiosis is about 30°C. These features were not in favor of aerobic thermophiles. However, some mineral structures called "beehives" are rather porous and allow some cold water to enter the structure, producing a warm (70°C) slightly oxygenated water circulation, and providing a suitable habitat for aerobic thermophiles.

NOVEL SPECIES OF ARCHAEA

The Archaea consist of two main lineages called Euryarchaeota and Crenarchaeota. The first lineage consist of organisms with various metabolic types. The most frequently isolated belong to the Thermocoocales with two main genera: *Pyrococcus* and *Thermococcus*, and many species. All are hyperthermophilic and grow optimally above 80°C by fermentation of organic compounds, especially peptides. Under laboratory conditions many strains require elemental sulphur or a thio-amino acid, and produce hydrogen sulphide. The prevalence of Thermococcales among the collection of Archaea from hydrothermal origin is rather surprising and rises the question of their real dominance in the ecosystem. The relatively high number of species described certainly corresponds to the diversity of this taxon, but also to the relatively easiness of their culture in the laboratory. Although they are strictly anaerobic, they can resist to an exposure to oxygen, if it occurs at low temperature, and to carbon starvation, all these features making them more easy to handle in the laboratory. Recently, a novel genus of Thermocoocales has been described, which uses ferric iron as electron acceptor, when elemental sulphur is absent: *Paleococcus ferrophilus*.

Still among Euryarchaeota, two species of *Archaeoglobus* (sulphate reducers) and several methanogens have been isolated. The methanogens belong to the genera *Methanococcus* or *Methanopyrus*, are strict autotrophs and produce methane from molecular hydrogen and carbon dioxide. *Methanopyrus kandleri* that grows optimally at 106°C, with a maximum temperature for growth of 110°C is the most thermophilic methanogen.

The second lineage, the Crenarchaeota, only encompasses hyperthermophiles which metabolize sulphur or sulphur compounds. Several species have been isolated from hydrothermal vents, within the genera *Desulfurococcus*, *Staphylothermus*, *Pyrodictium* or *Pyrolobus*. This last genus with the species *P. fumarii* isolated from the mid-Atlantic ridge is the most thermophilic organism on Earth with a maximum temperature for growth of 113°C: it also resist 2 hours in an autoclave!

If one compare the list of thermophiles and hyperthermophiles isolated from deep-sea hydrothermal vents and other hot environments (terrestrial or coastal), very few differences are noted. Some species such as *Staphylothermus marinus* have been isolated from both coastal and deep areas, and probably have an ubiquitous distribution. For most of the genera, novelty was found at the species level as testified by species names like *abyssi*, *profundus* or *hydrothermalis*. Some genera were only isolated from deep-sea hydrothermal vents (*Pyrolobus*, *Paleococcus* or *Desulfurobacterium*), but they may also exist in coastal areas. Although hydrothermal fluids have a low pH, no thermoacidophilic organism has been isolated yet. As for temperature, the pH gradient is very steep, and the suitable niche for acidophiles, if it exists, is probably rather narrow.

The existence of hydrothermal fluids with temperatures above the boiling point for water under atmospheric pressure was one of the motivation for high temperature microbiologists. However, if the most thermophilic organism, *P. fumarii* comes from the mid Atlantic ridge, its upper temperature limit for growth (113°C) is only three degrees above that of *Methanopyrus kandleri* (also isolated from vent at 100m depth off the coast of Iceland) or *Pyrodictium occultum* isolated from a shallow vent in Vulcano (Sicilia).

Responses to hydrostatic pressure

Among all parameters discussed above, hydrostatic pressure does not vary within a given vent site. The effects of hydrostatic pressure on deep-sea animals and bacteria have been intensively studied for deep-sea habitats. Barophilic bacteria and even strictly barophlic bacteria have been described, studied for their physiology, and adaptative mechanisms have been studied at the molecular level. On the contrary, deep vent thermophiles have been less intensively studied. It is not so easy to collect a smoker sample using a pressure-retaining container (any small piece of material on the moving parts of the collector may cause a leak). So, all samples have been collected and depressurized when lift to the surface, and subsequent microbiological enrichment cultures performed under atmospheric pressure conditions (in some cases, a slight gas pressure avoids liquid boiling).

Nevertheless, responses of deep vent thermophiles to hydrostatic pressure have been investigated. When hyperthermophilic organisms previously isolated under atmospheric pressure are exposed to hydrostatic pressure, some appear to be baro-sensitive (they grow more slowly), and others appear to be baro-tolerant (growth is not affected by pressure). But most of the strains studied are barophilic: hydrostatic pressure increases optimum and maximum temperature for growth (4 to 5°C) and the growth rate as well. For instance, for *Pyrococcus abyssi*, while maximum temperature for growth under atmospheric pressure is 102°C, hydrostatic pressure allows the organism to growth up to 106°C, and the optimum shifts from 96 to 100°C.
Similar observations have been reported by several authors, and the published data were analyzed by Baross and Deming who noted that for all the baro-hyperthermophilic strains, the pressure allowing the maximum growth rate was always above the pressure existing in situ. For instance, the optimum pressure for an organism isolated from a sample collected at 2600m was not 260 bars, but 400 bars. This feature is remarkable since for deep-sea psychrophiles, it had been noted that optimum pressure for growth was always lower than in situ pressure.

Another experimental approach consisted in carrying out all the enrichment culture process under hydrostatic pressure. Using this method, a novel Archaeon, *Thermococcus barophilus* has been isolated. This organism does not show any taxonomic novelty but interesting physiological properties. Although it was isolated under hydrostatic pressure, it can grow under atmospheric pressure, but at a lower growth rate. Again maximum temperature for growth is increased of a few degrees, but optimal temperature does not change. Further information was obtained from a protein study. A 35 kD protein with an unknown function is expressed when *Th. barophilus* is grown under hydrostatic pressure. But under atmospheric pressure, a 60 kD protein is expressed which corresponds to a stress protein already known for other hyperthermophilic Archaea. All these data may confirm that *Th. barophilus* is a true baro-hyperthermophile.

Arguments for a deep biosphere

When they noted that all baro-hyperthermophiles had an optimum pressure higher than their *in situ* pressure, Baross and Deming asked the question: " are black smokers windows to a deep biosphere?". This novel concept of deep or subterranean biosphere has been progressively confirmed by the discovery of living bacteria within deep aquifers, deep rock formations, deep on-shore and off-shore oil reservoirs, deep marine sediments, etc.

These new data have many important results. About Earth, it is now clear that the volume of the biosphere is considerably larger than previously expected, and all the novel biotopes mentioned above are exposed to elevated hydrostatic and lithostatic pressures. But in addition, for an exobiology perspective, these terrestrial discoveries may have more spectacular consequences.

Deep biotopes on other solar bodies?

If the surface of Mars is too much oxidized for allowing life, what about the deep layers, if some water still exists? About Europa, the icy Jupiter satellite, if it really harbors a deep ocean of liquid water, it may have allowed some life to rise up. For centuries, human beings tried to get answers to these questions. Nobody can tell if answers will be obtained within this century, but technology for such fascinating explorations are now available. On Earth, research carried out by this novel kind of biological explorers push more and more the geographical and physico-chemical limits for life. It is less and less unreasonable, but rather more and more necessary to explore these remote solar bodies and to search for past or present life.

References

1. Review articles

Prieur, D. 1992. In Molecular biology and biotechnology of extremophiles (Herbert, R.A. and Sharp, R.J., eds) pp 163-197. Blackie.

Prieur, D., Erauso, G.& Jeanthon, C. 1995. Hyperthermophilic life at deep-sea hydrothermal vents. Planet. Space Sci., 43: 115-122.

Prieur, D. 1997. Microbiology at dep-sea hydrothermal vents. Trends in Biotechnology, 15: 242-245.

Prieur, D. & Parkes, J. 1999. La biomasse cachée des fonds océaniques. La Recherche, 317: 58-61.

Prieur, D. & Marteinsson, V.T. 1998. Prokaryotes living under elevated hydrostatic pressure. Advances in Biochemical Engineering/Biotechnology, 61: 23-35.

Jeanthon, C. 1999. Molecular ecology of hydrothermal vent microbial communities. Antonie van Leuwenhoek, 77: 117-133.

2. Recent publications not quoted in the review articles:

Marteinsson, V.T., Birrien, J.L., Reysenbach, A.L., Vernet, M., Marie, D., Gambacorta, A., Messner, P., Sleytr, U.W. & Prieur, D. 1999. Thermococcus barophilus sp.nov., a new barophilic and hyperthermophilic archaeon isolated under high hydrostatic pressure from a deep-sea hydrothermal vent. Int. J. Syst. Bacteriol., 49: 351-359.

Marteinsson, V.T., Reysenbach, A.L., Birrien, J.L. & Prieur, D. 2000. A stress protein is induced in the deep-sea hyperthermophile Thermococcus barophilus when grown under atmospheric pressure. Extremophiles, 3: 277-282.

Takai, K., Sugai, A., Itoh, T. & Horikoshi, K. 2000. Palaeococcus ferrophilus gen.nov., sp.nov., a barophilic, hyperthermophilic archaeon from a deep-sea hydrothermal chimney. Int. J. Syst. Evol. Microbiol., 50: 489-500.

CATALYTIC CAPABILITY OF OLIGOPEPTIDES IN THE INTERFACE CHEMISTRY IN HYDROTHERMAL ENVIRONMENTS

KOICHIRO MATSUNO
Department of BioEngineering
Nagaoka University of Technology
Nagaoka 940-2188, Japan

Abstract. Oligopeptides synthesized from glycine and alanine in the flow reactor simulating hydrothermal circulation of seawater through hot vents in the ocean on the primitive earth demonstrated that the activity of dissecting the peptide bond of glycylalanine coincided with the buildup of those oligomers of alanylalanine, alanylgylcine, alanylalanylalanine, alanylalanylalanylalanine. Hydrothermal environments can provide oligopeptides with both the capacities of integration and disintegration. Such a coexistence of both the capacities comes to enhance the specificity of oligomers to be synthesized in hydrothermal environments.

1. Introduction

Hydrothermal vents on the sea floor of the primitive earth have been proposed as likely locales among others for prebiotic synthesis (Corliss et al., 1979). What makes hydrothermal environments significant prebiotically is constant circulation of seawater from hot vents to surrounding cold seawater and back again repeatedly. In particular, the interface between hot and cold seawater in hydrothermal environments can serve as a selective sieve for saving chemical products synthesized in hot vents if the products are immediately transferred into the cold region before they are disintegrated while remaining in the hot region. The selective enhancement of the synthesis could become multiplicative if the products can revisit hot vents repeatedly.

We already attempted an experimental model simulating geological conditions for hydrothermal circulation of seawater through hot vents (Matsuno, 1997; Imai et al., 1999), and observed an exponential growth of oligopeptides with the elapse of time at least initially when the circulating reaction solution was initially glycine dissolved in water (Imai et al., 1999). One specific implication of the significance of the interface chemistry in the simulated hydrothermal environment was that the yields of triglycine synthesized from gylcine were more than 1000 times greater than the yields expected from equilibrium reactions at the high temperature available from the simulated hot vents (Ross, 2000).

A principal feature of the flow reactor we constructed for simulating hydrothermal circulation of reaction solution was that a high-temperature, high-pressure fluid is injected into a low-temperature chamber maintaining the same high pressure while the

J. Chela-Flores et al. (eds.), First Steps in the Origin of Life in the Universe, 195–199.

whole fluid is circulated in a closed manner in the system with a fixed turnover rate. The high-temperature, high-pressure fluid was prepared in a pressurized and heated section of the closed circuit of the fluid, whereas the low-temperature chamber was maintained by immersing it in an external cooling apparatus. The high-temperature, high-pressure fluid jet from a nozzle into the low-temperature chamber was intended to simulate a submarine hydrothermal vent. At the same time, the fluid flowing out of the low-temperature chamber was first depressurized to normal atmospheric pressure to sample a very small fluid volume for the purpose of measurement at a given time interval and then followed by conversion back to the high-temperature, high-pressure fluid.

2. Results

We first prepared 500 *mℓ* solution of 100*mM* glycine without and with 10*mM* diketopiperazine (DKP) at pH 2.5 adjusted by HCl at room temperature. The addition of DKP was intended to see how intermediary reaction products would contribute to the on-going synthetic reactions. The pressure of the high-temperature, high-pressure chamber with a volume 15*mℓ* was set at 24.0 *MPa* which is only slightly above the pressure of the critical point of water (22.1*MPa*). The flow rate was about 10 *mℓ/min*. This gave the cycle time 34 seconds of reactants rounding the closed flow pathway in stirred conditions. The temperature of the high-temperature chamber was set at 250°*C*. Figure 1 demonstrates the time courses of the yields of oligoglycines without (a) and with (b) DKP. There is observed a significant enhancement of the synthesis of the longer oligoglycines due to the addition of DKP as the intermediary reaction products.

In order to see how the intermediary reaction products would contribute to further elongation of oligopeptides thus synthesized, we tried 500 *mℓ* solution of 40*mM* *L*-alanine and 10*mM* glycine at pH 2.5 adjusted by HCl at room temperature, while other conditions were the similar to the previous one. Figure 2 demonstrates the time courses of the yields of oligopeptides identified. We have identified at least six different oligpeptides; Ala-Gly, Gly-Ala, Ala-Ala, Gly-Ala-Ala, Ala-Ala-Ala, Ala-Ala-Ala-Ala, where Ala for *L*-alanine and Gly for glycine. We then observed that the rises of Ala-Gly, Ala-Ala-Ala, Gly-Ala-Ala, and Ala-Ala-Ala-Ala coincided with the decrease of Gly-Ala. It is thus suggested that the activity of dissecting the Gly-Ala bond could enhance the production of the four oligopeptides. The driving factor for selective elongation of oligopeptides supplemented by the activity of dissecting the intermediary products is the interface chemistry between the hot and cold regions through constant circulation of the reaction solution from the hot to the cold repeatedly.

3. Concluding Remarks

Activity of dissecting intermediary oligopeptides for further elongation, that is similar to a protease activity, can in turn become highly selective in the oligopeptide elongation when it is further supplemented by an exponential growth of the products. Constant circulation of seawater through hot vents into surrounding cold seawater can install such a multiplicative amplification of the synthesized products if the interface chemistry

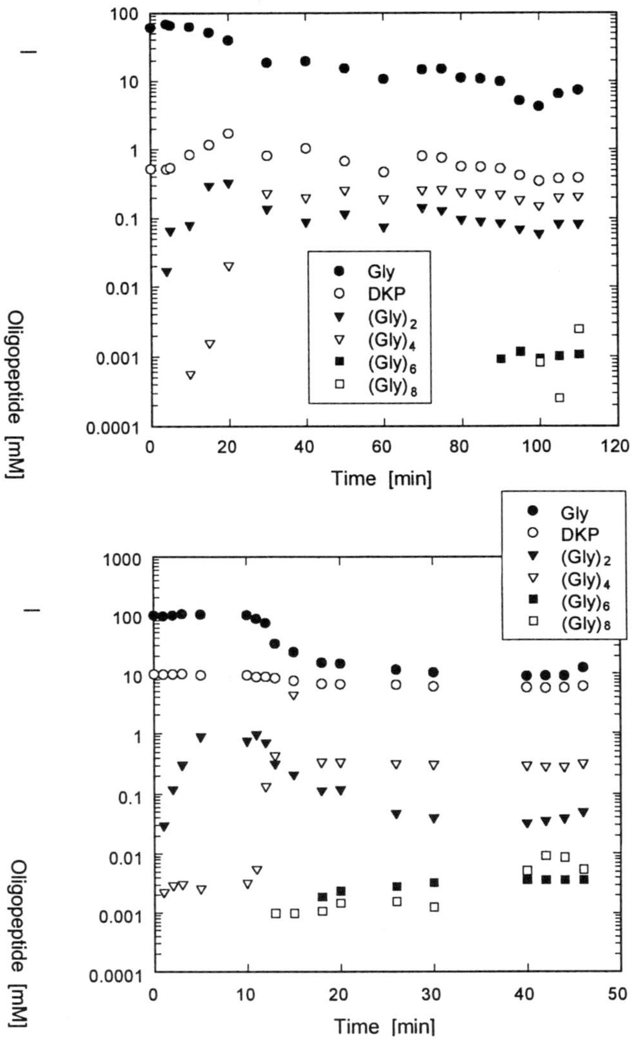

Figure 1. Time courses of the yields of the oligomers from the reaction solution of 100 mM glycine and 10 mM $CuCl_2$ at pH 2.5 adjusted by HCl at room temperature for (a), and further addition of 10 mM diketopiperazine to the initial solution for (b). The origin of the time coordinate was taken to be the time point when the temperature of the high-temperature, high-pressure chamber reached the designated 250 _. Gly for glycine, DKP for diketopiperazine, $(Gly)_2$ for diglycine, $(Gly)_4$ for tetraglycine $(Gly)_6$ for hexaglycine, and $(Gly)_8$ for octaglycine. Identification of the oligomers in the products was accomplished by two independent methods. One was comparison with the HPLC elution profiles of standards up to octaglycine, and the other was due to LC mass spectroscopy up to triglycine (Ogata et al, 2000).

capable of rapidly transferring the synthesized products into the cold water is certainly operative. As a matter of fact, cycle time for those reaction products to revisit any one of the hot vents in the primitive ocean could certainly have remained finite, even though it may have been over tens of thousands of years. In particular, those oligopeptides carrying proto-protease activities may have been synthesized even prebiotically in hydrothermal environments on the primitive earth or on the similar planets elsewhere.

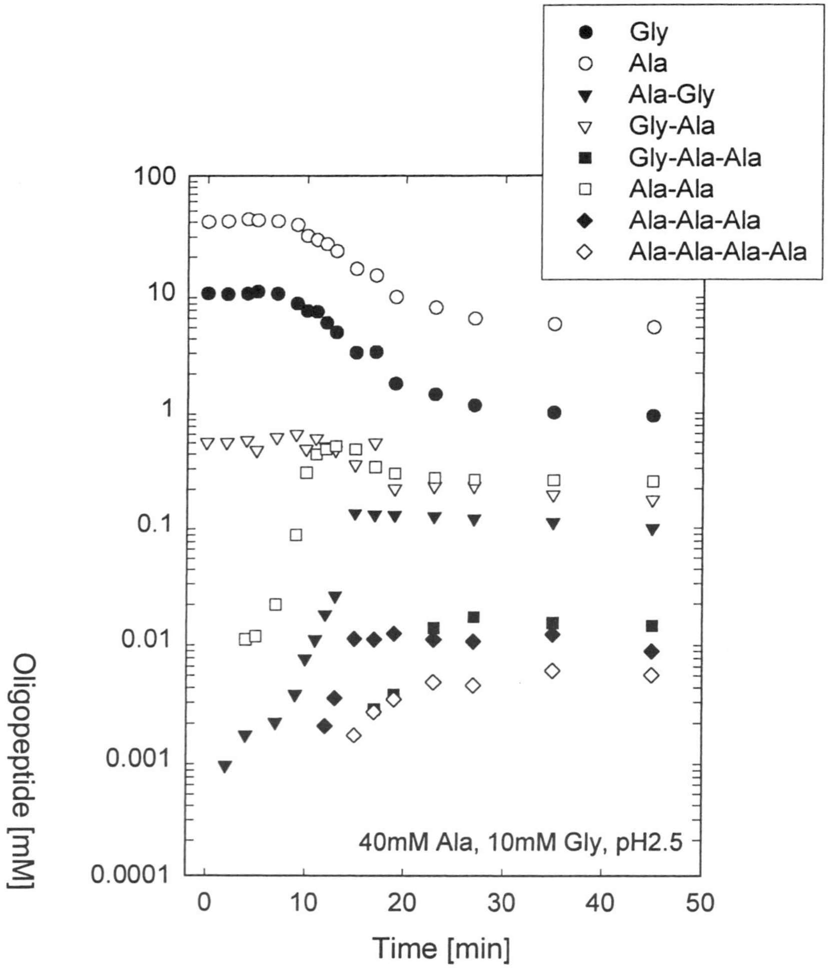

*Figure2*_Time courses of the yields of the oligomers from the reaction mixture of 40 mM L-alanine and 10 mM glycine at pH 2.5 adjusted by HCl at room temperature. The origin of the time coordinate was taken to be the time point, when the temperature of the high-temperature, high-pressure chamber reached the designated 250_. The abbreviation is Ala for L-alanine, and Gly for glycine (Ogata et al, 2000).

Corliss, J. B., Dymond, J., Gordon, L. I., Edmond, J. M., von Herzen, R. P., Ballard, R. D., Green, K. K., Williams, D., Bainbridge, A., Crane, K. & van Andel, T. H. Submarine thermal springs on the Galapagos rift. *Science* **203**, 1073-1083 (1979).

Imai, E., Honda, H., Hatori, K., Brack, A. & Matsuno, K. Elongation of oligopeptides in a simulated hydrothermal system. *Science* **283**, 831-833.

Matsuno, K. A design principle of a flow reactor simulating prebiotic evolution. *Viva Origino* **25**, 191-204 (1997).

Ogata, Y., Imai, E., Honda, H., Hatori, K. & Matsuno, K. Hydrothermal circulation of seawater through hot vents and contribution of interface chemistry to prebiotic synthesis. *Origins Life Evol. Biosphere* (2000), in press.

Ross, D. Private communication (2000).

VOLCANIC LIGHTNING AND THE AVAILABILITY OF REACTIVE NITROGEN AND PHOSPHORUS FOR CHEMICAL EVOLUTION

RAFAEL NAVARRO-GONZÁLEZ AND ANTÍGONA SEGURA
Laboratorio de Química de Plasmas y Estudios Planetarios, Instituto de Ciencias Nucleares, Universidad Nacional Autónoma de México. Circuito Exterior, Ciudad Universitaria. Apartado Postal 70-543, México, D.F. 04510 MEXICO.

Abstract. Nitrogen and phosphorous are two essential elements for life that belong to the same chemical group in the periodic table. Nitrogen is believed to have been present mainly in the early atmosphere as molecular nitrogen, an inert gas under most atmospheric conditions. Phosphorous was present mainly in the lithosphere as the relatively insoluble apatite. If these two elements played a role in the origins of life, some mechanisms for making them available for chemical evolution must have operated in the early history of the Earth and Mars. We review here a novel mechanism to fix atmospheric nitrogen and to reduce phosphate into phosphite (a soluble compound) by lightning discharges occurring inside explosive volcanic clouds.

1. Introduction

Nitrogen and phosphorus are fundamental elements for life. Nitrogen is present in structural (*e.g.*, proteins), catalytic (*e.g.*, enzymes and ribozymes), energy transfer (*e.g.*, ATP) and information storage (RNA and DNA) biomolecules. Atmospheric and planetary models suggest that nitrogen was abundant in the early atmospheres of Earth and Mars as dinitrogen (N_2), an inert gas under normal atmospheric conditions. To be available for prebiotic synthesis it must be converted into hydrogen cyanide, ammonia and/or nitrate, in a process referred as nitrogen fixation. Due to the strength of the triple bond in N_2, nitrogen fixation, while thermodynamically favored is kinetically restricted (Navarro-González *et al.*, 2001). In a reducing atmosphere dominated by CH_4-N_2, thunderstorm lightning efficiently produces HCN and NH_3 (Toupance *et al.*,1975; Stribling and Miller, 1987; Chameides and Walker, 1981). Nevertheless, photochemical and geochemical constraints strongly suggest that the early atmosphere was weakly reducing, dominated by CO_2 and N_2 with traces of CH_4, CO, and H_2 (Kasting, 1993). Under these conditions, HCN is no longer synthesized in the lightning channel and instead NO is formed (Navarro-González, *et al.*, 2001 and references therein). NO has not yet been implicated in the syntheses of amino acids, purines or pyrimidines under prebiotic conditions. The lack of formation of HCN by thunderstorm lightning introduces serious barriers to the process of chemical evolution in the early Earth.

Phosphorous, on the other hand, is present in ribozymes, ATP, RNA and DNA molecules. This element occurs naturally as apatite ($Ca_5(PO_4)_3(F, Cl)$), an insoluble mineral (Glindemann *et al.*, 1999). In order to be available for prebiotic process, phosphorus must be reduced to hydrophosphites or phosphites. Recent experimental work conducted in this laboratory and elsewhere show that the lightning formed in volcanic clouds during highly explosive eruptions may be an important source of reactive nitrogen and phosphorus. This paper reviews the available knowledge on the role of volcanic lightning in chemical evolution

J. Chela-Flores et al. (eds.), First Steps in the Origin of Life in the Universe, 201–210.

2. Explosive volcanism

Volcanic plumes were proposed as a viable environment for prebiotic synthesis because volcanism was a very common phenomenon in the Archean (Fox and Harada, 1961; Harada and Fox, 1964). The range of styles of volcanic activity on a planet is a complex function of the physical properties of magma (for example, volatile content and viscosity) and the environmental conditions, mainly atmospheric pressure and acceleration due to gravity (Wilson and Head, 1983). Explosive volcanism is caused by (1) degassing of juvenile volcanic gases dissolved at high pressure, (2) the interaction with magma with external water (ground, lake or sea) or (3) some combination of both (Sparks *et al.*, 1997). During these eruptive episodes, the magma and part of the volcanic system are fragmented in particles known as pyroclasts. A volcanic plume is formed by the discharge of a mixture of pyroclasts and magmatic gases into the atmosphere. Such plumes typically exhibit strong electric fields and generate copious lightning discharges of hundreds of meters in length at rates of $10\text{-}10^2$ flash min^{-1} in and around the plume near the eruption site (Navarro-González *et al.,* 1996). The compounds formed in the volcanic lightning channel can escape the high temperature zone of the eruption site at sonic or supersonic speeds and be injected into the troposphere and stratosphere where they would undergo photochemical processing. These characteristics allow volcanic plumes to serve as natural chemical reactors. The chemical compounds formed by volcanic lightning depend on the composition of magmatic gases. Composition of the volcanic gases is controlled by the oxidation state of the upper mantle that is characterized by the ratio of iron oxides to metallic iron (Whittet, 1997 and references therein). The oxidation state of a planet is a result of its accretionary history and later evolution. There are two accepted models for the accretion of terrestrial planets: heterogeneous and homogeneous. Roughly, the type of accretion describes the variation in time of the accreting material. In a homogenous accreted planet, with metallic iron present in its upper mantle, the volcanic gases would be reduced. A planet heterogenously accreted would lead to volcanic emissions composed mainly by H_2O and CO_2. The latter is the case of Earth (Ahrens *et al.*, 1989) and the homogenous one corresponds to Mars (Dreibus and Wänke, 1989)

2.1. VOLCANISM ON EARLY EARTH

Volcanism was a widespread phenomenon in the early Earth as indicated through the geologic record (BVSP, 1981). More than 50% of the rocks deposited during the Archean are volcanic, particularly of mafic and ultramafic origin. Basaltic volcanism is not explosive unless magma has a significant content of dissolved volatiles or interacts with underground or ground water causing fragmentation of the magma (Sparks *et al.,* 1997). The explosiveness depends also on the rate of magma eruption. Accretion planetary models and geologic evidence indicate that early Earth had a mantle richer in volatiles, abundant surface waters and had very high eruption rates that may have favored magma fragmentation and therefore the formation of volcanic plumes (Navarro-González *et al.*, 2000).

2.2. VOLCANISM ON EARLY MARS

Volcanic activity on Mars was extended for a long period and volcanic surfaces cover more than half of Mars (Mouginis-Mark *et al.*, 1992). Numerical simulations have shown that explosive activity would occur if the magma volatile content exceeds 0.01 weight percent on Mars (Wilson and Head, 1983). Because of the lower martian gravity, nucleation of magma would occur at systematically greater depths than on Earth since volatile solubility is pressure-dependent, and consequently volatile release would likely be more efficient (Mouginis-Mark, *et al.*, 1992). More vigorous gas release and higher eruption rates in martian magmas compared with those on Earth imply that it may have been common for basaltic magmas to produce plinian eruptions (Mouginis-Mark, *et al.*, 1992). The oldest volcanic units identified on Mars are plateau plains that were formed by fissure eruptions (Greeley and Spudis, 1981). It has been shown that basaltic fissure eruptions can produce buoyant plumes of several kilometers when interaction with water is present (Woods, 1993). Then, because groundwater is an important part of Mars's upper crust (Wilson and Head, 1994) volcanic plumes may have been common on this planet.

3. Volcanic Lightning: Characteristics and Simulation

3.1. MEASURED PROPERTIES OF VOLCANIC LIGHTNING

Despite several reports of volcanic lightning in the past (Navarro-González, *et al.*, 1996), few researchers have been interested in studying volcanic lightning. Based on reports of electric field potentials in volcanic plumes it has been deduced that the injection of volcanic steam and ash into the atmosphere results in an abnormal positive electrostatic field that differs from the regular fine-weather field of +100-130 V m^{-1} by several orders of magnitude. Eruptions where field measurements of lightning activity and/or electric fields have been carried out are summarized in Table 1. The electric field data collected during the eruptions of Aso volcano (Japan, 1959) and Sakurijama volcano (Japan, 1992) led to the interpretation that plumes have a dipole structure, being positive at their upper part and negative in their lower region (Sparks et al., 1997 and references therein). The lightning discharges are hundreds of meters in length. On Earth the typical length is ~500 m (Navarro-González *et al.,* 1996; Basiuk and Navarro-González, 1996). Due to its lower gravity, for Mars it has been calculated that this length may be about 3 km (Segura and Navarro-González, 2001).

TABLE 1. Recorded characteristics of electric activity in explosive eruptions (data from Navarro-González *et al.*, 1996; Basiuk and Navarro-González, 1996 and references therein).

Volcano	Location	Flash Rate	Electric Potential
Surtsey	Surtsey, Iceland	20 flashes hr^{-1}	30 kV m^{-1}
Heimaey	Heimaey, Iceland		>7 kV m^{-1}
Mt. Usu	Hokkaido, Japan		15 kVm^{-1} about 5 km from the volcano
Redoubt	Alaska, U.S.A.	16 flashes min^{-1}	
Mt. St. Elena	Washington, U.S.A.	11 flashes min^{-1}	20 kVm^{-1} at the ground and 10 kVm^{-1} in an aircraft at 175 km from the volcano

3.2. EXPERIMENTAL SIMULATION AND THEORETICAL MODELING OF VOLCANIC LIGHTNING

The sudden deposition of electric energy along the lightning channel produces a plasma thermally equilibrated at temperatures that exceed 10,000 K. The rapid heating of the surroundings produces a cylindrical shock wave that propagates outward. As the electric discharge column expands, it is cooled by expansion, radiative heat loss and entrainment of the surrounding gas (Borucki *et al.*, 1984). During the expanding and cooling processes, the chemical species in the hot channel maintain a state of chemical equilibrium as long as their relaxation times are shorter than the cooling rate of the hot channel. Eventually a temperature is reached at which the rates of reactions become too slow to keep chemical equilibrium and the mixing ratios of the chemical species freeze-out at this point. This freeze-out temperature (T_f) is likely to be characteristic for each chemical species (Navarro-González *et al.,* 2000). The cooling of the lightning channel generates most of the chemical products; however, the intense ultraviolet light emitted by the lightning flash can generate additional chemical species by photolysis outside the channel. For the case of nitric oxide, T_f has been found to be about 2400°C for lightning in the contemporaneous Earth's atmosphere (Borucki and Chameides, 1984). In order to model the products of lightning and in the absence of further experimental data, Navarro-González *et al.* (1998) assumed as a first order approximation that T_f was similar to all the species formed by volcanic lightning. This assumption may result in as much as an order of magnitude error in the freeze-out mixing rations of the chemical species other than NO. Typically lightning is commonly simulated in the laboratory by arcs or sparks. These experiments have the disadvantage of sublimating the electrodes used and contaminating the plasma (Borucki *et al.*, 1988 and references therein). More sophisticated and expensive experiments now generate a plasma free of metal contamination that can simulate the hot lightning channel using lasers or microwave cavities. Laser induced plasmas (LIP) accurately simulate lightning discharges without introducing wall-effects and contamination (Borucki *et al.*, 1988).

4. Nitrogen fixation by volcanic lightning

Navarro-González *et al.* (1998) examined volcanic lightning as a mechanism to fix nitrogen on early Earth. For their experiment, it was assumed that chemical composition of Archean volcanic gases was similar to that of Hawaiian volcanoes because they arise from a primordial, undegassed reservoir deep in the Earth's mantle. The following composition was used as Archean magmatic gases: 50% H_2O, 30% CO_2, 11% N_2, 4.5% CO, and 4.5% H_2. Volcanic gases generally undergo rapid mixing with underground or

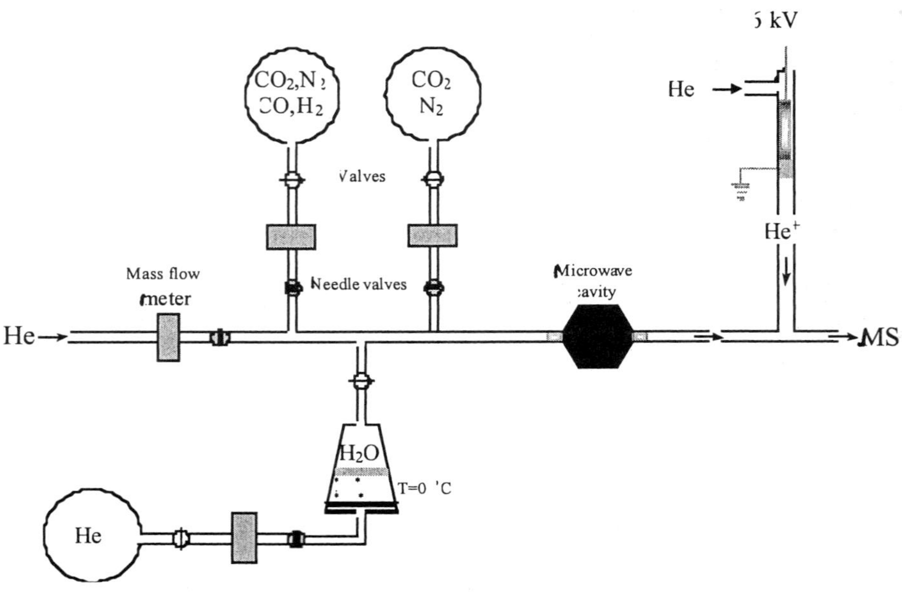

Figure 1. Experimental design for simulating volcanic lightning in a flow tube in which volcanic gases were excited by a microwave discharge and the products were analyzed downstream by positive chemical ionization mass spectrometry.

ground water or with the surrounding atmosphere present outside the volcanic vent environment. Consequently, the effects of dilution of volcanic gases with water and the atmosphere were investigated, assuming that early Earth's atmosphere was composed by 80% carbon dioxide and 20% molecular nitrogen (Kasting, 1990).

The experimental designed for the irradiation of magmatic gases is shown in Figure 1. The gas mixture flows into a microwave discharge cavity where the gases were excited.

Analysis of the products was performed by chemical ionization mass spectroscopy and infrared spectroscopy. A quantitative study of the experimental production efficiency of NO is given in Figure 2 as a function of dilution of the volcanic gas mixture with water and atmospheric gases. In addition, this figure also includes the predicted trends for NO and other minor species expected to form in the cooling channel of volcanic lightning.

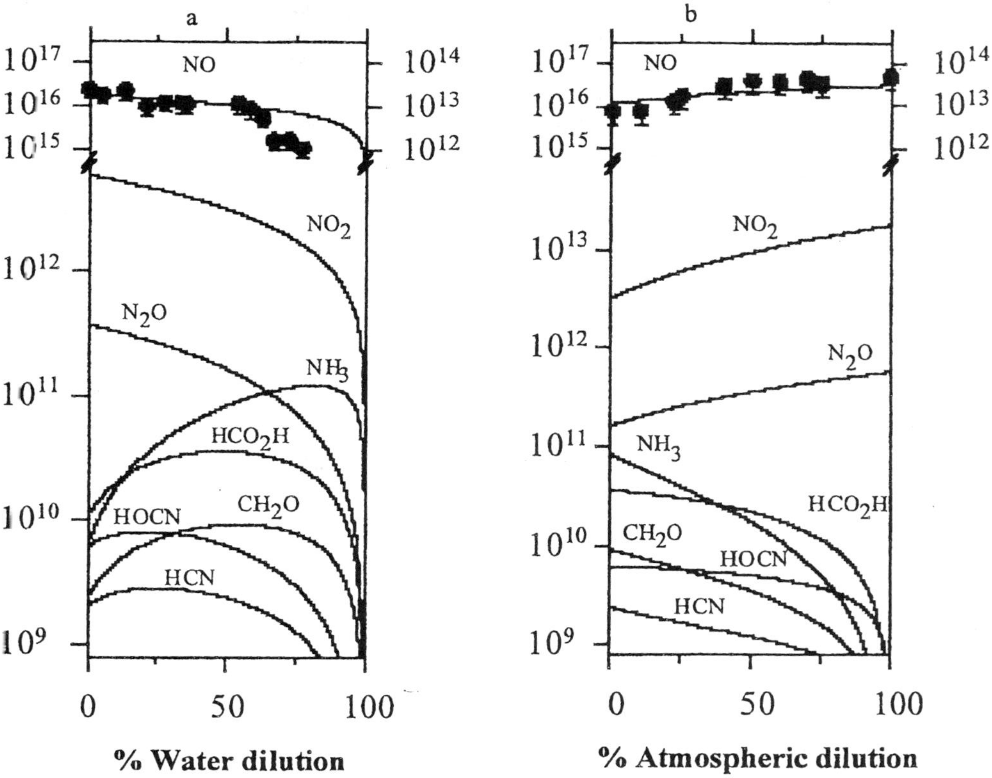

Figure 2. Production yields for several volcanic lightning products as a function of dilution of volcanic gases with (a) water and (b) the surrounding atmosphere. Lines are predicted trends from the thermochemical-hydrodynamic model whereas symbols are the experimental values derived in a microwave discharge. The axes are on the left Energy yield (molecule J)$^{-1}$ and on the right NO production (molecule J) $^{-1}$

The experimental and predicted trends for NO agree quite well under most conditions except at high dilutions of water vapor ($\geq$ 60%). This discrepancy is attributed to the injection of excess water in the form of aerosols when a high helium flow passes through the water bubbler. The predicted NO energy yield is estimated to be ~ 1 $\times$ 10^{16} molecule J^{-1} for volcanic lightning produced in a Hawaiian style volcanic mixture. This value can increase to ~ 2 × 10^{16} molecule J^{-1} in an anhydrous volcanic mixture or to ~ 3 × 10^{16} molecule J^{-1} at 97.5% dilution of the magmatic gases with the surrounding atmosphere. If the magmatic gases are significantly diluted with water, this value can decrease down to ~ 2 × 10^{15} molecule J^{-1} at 97.5% dilution. Other minor products predicted (see Fig. 2) were not detected experimentally on account of their extremely low product energy yields. A detailed analysis of volcanic lightning compared with other endogenous and exogenous sources showed that this source was the most important to fix nitrogen in early Earth (Navarro-González *et al.*, 2000).

Segura and Navarro-González (2000) examined the effect of volcanic lightning in Martian volcanic plumes. The chemical composition of the gas mixture used in these experiments was derived from the accretion model developed by Kuramoto and Matsui (1996) that was applied to Mars by Kuramoto (1997). The composition of volcanic gases was 64% CH_4, 24% H_2, 10% H_2O, 2% N_2. This was irradiated by focusing a high-energy infrared laser beam to simulate lightning. Figure 3 shows the experimental arrangement to

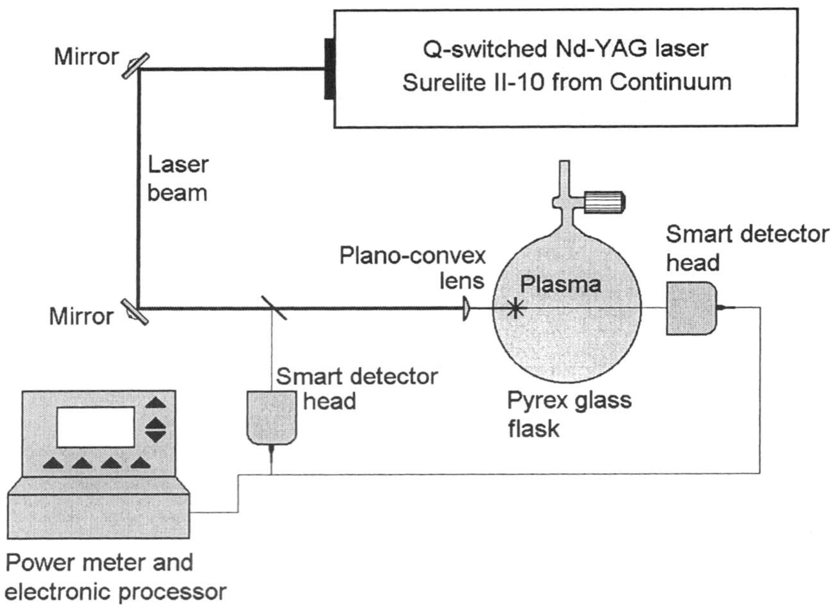

Figure 3. Experimental design for simulating volcanic lightning using a high power pulsed laser. The laser delivers a beam of 1.06 μm photons with an energy of 480 mJ per pulse in 5-7 ns at 10 Hz. The beam was focused inside a Pyrex reactor with a plano-convex optical glass lens coated with an anti-reflecting film and obtaining a focal spot size of ~9.7 μm. The energy deposited into the system was determined by the difference between the input laser energy and that transmitted by the LIP, using an energy measurement system

irradiate volcanic gases. The products of LIP irradiation were analyzed by a gas chromatograph interfaced in parallel with an infrared detector and a quadrupole mass spectrometer equipped with electron impact and chemical ionization modes. Figure 4 is the SIM chromatogram of the irradiation products. The most relevant compound for nitrogen fixation and prebiotic chemistry was HCN with an energy yield of 6×10^{14} molecules J^{-1}.

5. Phosphorus reduction by volcanic lightning

Glindemann *et al.* (1999) and de Graaf and Schwartz (2000) studied the phosphorus reduction by volcanic lightning using the experimental assembly showed in Figure 5. In a first set of experiments, Glindemann *et al.* (1999) used samples as solution of (Na_2HPO_4) or as pastes (mixtures of fluorapatite with montmorillonite) in water-saturated with a nitrogen atmosphere containing 1-10% CH_4. A spark discharge was generated in the device using a domestic microwave oven. The products were analyzed by gas chromatography-mass spectrometry and the results are summarized in Table 2. In 10%

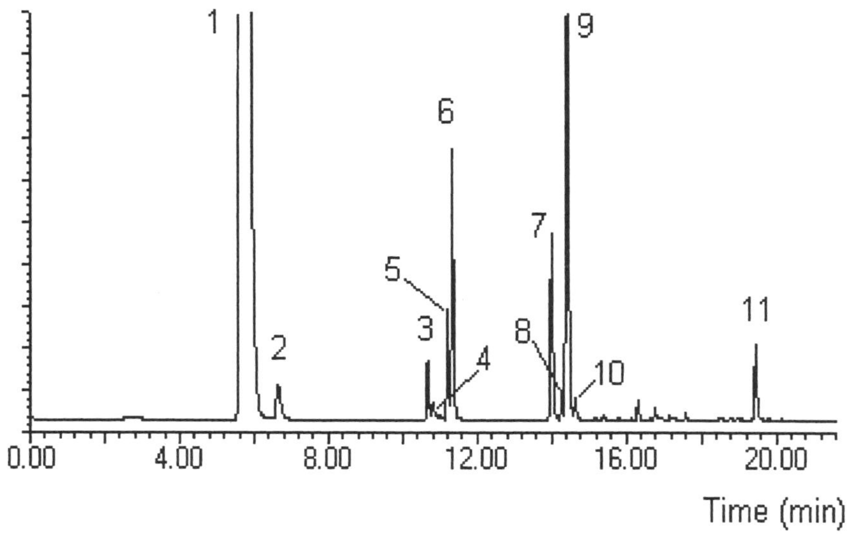

Figure 4. SIM gas chromatogram using 26, 27, 39, 40, 41, 50, 52, 54, 66, 78 m/z, of compounds produced by 25 minutes LIP discharge irradiation of a gas mixture composed by CH_4 (64%), H_2 (24%), H_2O (10%) and N_2 (2%).Peak identification: 1. Acetylene + ethylene; 2. Ethane; 3. Propene; 4. Hydrogen cyanide; 5. 1,2-propadiene; 6. Propyne; 7. 1-buten-3-yne; 8. 1-butyne; 9. 1,3-butadiyne; 10. 2-butyne; 11. Benzene.

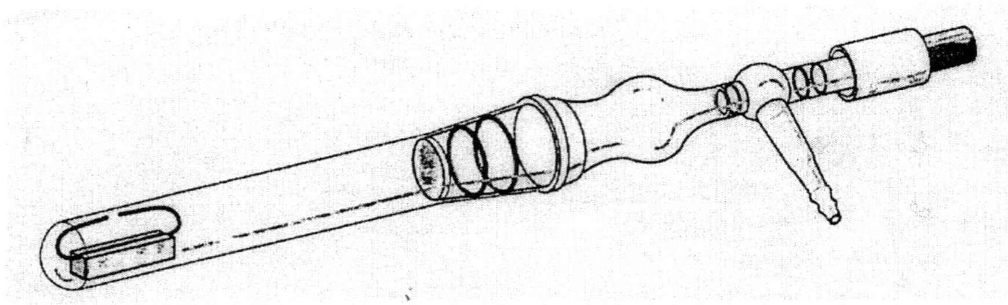

Figure 5. Experimental assembly to simulate the effects of ash in volcanic lightning. Samples were applied in quartz caps (not showed) covering the ends of the open tungsten loop and filling the system with the gas mixture. The assembly was placed into a 1 kW commercial (domestic) microwave oven. A spark discharge between the ends of the loop occurs several seconds after applying a microwave field. Reproduced from Glindemann *et al.* (1999).

methane an average yield of 11% phosphite was obtained using Na_2HPO_4 as the source of phosphorous. As the content of methane in the atmospheric mixture decreased, the production efficiency of phosphite dropped to 4% in 1% CH_4 and to 0.1% in pure N_2. In order to extend these results to a geophysical more plausible model, Glindemann *et al.* (1999) examined the reduction of fluorapatite, the most important phosphate mineral in igneous rocks. In these experiments, a yield of 7% phosphite was obtained in the presence of 10% methane. As with the experiments of Na_2HPO_4, the yield of phosphite decreased but was still substantial. Small amounts of methylphosphonic acid could also be detected in small quantities in these experiments. As was proposed by Segura and Navarro-González (2000, 2001), these reduced compositions may be geologically relevant for Mars, were phosphorus is an abundant element. Accretion models predicted concentrations exceeding those on Earth for volatile and moderatlely volatile elements (Dreibus and Wänke, 1985). Martian meteorites analyses (Banin *et al.*, 1992) and Mars Pathfinder mission results (Dreibus *et al.*, 1999) have verified the high phosphorus content on Mars. Apatite grains have been detected in the Martian meteorites (Leshin, *et al.*, 1996, Watson, *et al.*, 1994); this was an available source of phosphates on Mars.

TABLE 2. Formation of phosphite by reduction of orthophosphate in $CH_4 + N_2$ mixtures[a]. (from Glindemann *et al*, 1999)

Phosphate source	Matrix[b] (wgt % phosphate)	CH_4 (%in N_2)	Phosphite produced (μmol/mg sample x 10^2)	Yield (%)
Na_2HPO_4	-	10	30.0±9	11
Na_2HPO_4	-	1	11.0±3	4
Na_2HPO_4	-	0	0.3±0.4	0.1
Hydroxyapatite	Mont. (20)	10	7.0±1	6
Fluorapatite	Mont. (20)	10	9.0±0.7	7
Fluorapatite	Mont. (20)	1	4.0±2	4
Fluorapatite	Mont. (20)	0	0.6±0.6	0.5
-	Mont.	10	<0.1	-

[a] The yield of phosphite is based in the amount of phosphate deposited on the loop. For hydroxyapatite and fluorapatite the theoretical formula weights have been used.
[b] Mixtures were prepared with montmorillonite #26 containing the indicated wgt5 of the phosphate mineral shown.

De Graaf and Schwartz (2000) extended the research using gas mixtures contained 60% CO_2, 22-40% N_2 and variable concentrations of H_2 and CO. These mixtures are based on a volcanic outgassing model studied by Navarro-González *et al.* (1998, 2000). For these experiments, fluorapatite was mixed in a clay mineral matrix in order to have a more representative composition of igneous rocks. Samples undergo the same experimental and analysis processes that in the experiment developed by Glindemann *et al.* (1999). Their results demonstrated that several percent reduction of apatite occurs even in the presence of as little as 1% H_2 and CO_2 in the gas mixture. In addition, they demonstrated the formation of polyphosphate production (see Table 3) as a result of heating the mineral apatite in the presence of other minerals by the lightning discharge under various atmospheric chemical compositions.

TABLE 3. Analysis for orthophosphate (Pi), pyrophosphate (PPi) and tripolyphosphate (PPPi) in water extracts after exposure to electric discharge.

Matrix[a]	Conditions	Pi	PPi	PPPi	Total recovery %
Mont. 26	60% CO_2 + 40% N_2 + water vapor	36±8	10±3	1.7±0.8	48
Mont. 27	as above	39±1	13±2	1.7±0.5	54
Bentonite	as above	30±2	10±1	1.7±0.8	42
Mont. 31	as above	12±3	4±1	-	16
Mont. 26	as above but - water vapor	5±1	5±1	-	10
Mont. 26	100% N_2 + water vapor	35±5	12±3	-	47

[a]Samples of fluorapatite were mixed with the matrix indicated (1:4).Recoveries are based on the total amount of the apatite initially present in the sample.

6. Concluding remarks

The experiments presented above demonstrate that nitrogen fixation is an efficient process in volcanic lightning. Nitric oxide and hydrogen cyanide are the major volcanic lightning products expected for early Earth and Mars, respectively. In the case of phosphite, it is produced in high yields even in neutral atmospheres. Therefore electrical discharges associated with volcanic eruptions supplied a pathway by which nitrogen and phosphorous atoms were incorporated into prebiotic molecules needed for the emergence and sustainability of life. We hope that the work described in this review may motivate additional research on volcanic lightning.

References

Ahrens, T.J., O'Keefe, J.D, and Lange, M.A. (1989) Formation of atmospheres during accretion of the terrestrial planets in S.K. Atreya, J B. Pollack and M.S. Matthews (eds.), *Origin and Evolution of Planetary and Satellite Atmospheres*,University Arizona Press, Tucson, pp. 328-385.

Banin, A., Clark, B.C., and Wänke H. (1992) Surface Chemistry and Mineralogy in H.H. Kieffer, B.M. Yakosky, C.W. Snyder, M.S. Matthews (eds.), *Mars*, The University Arizona Press, pp. 594-625.

Basaltic Volcanism Study Project (1981) *Basaltic Volcanism on the Terrestrial Planets*, Pergamon Press, Inc, New York.

Basiuk, V.A., and Navarro-González, R. (1996) Possible role of volcanic ash-gas clouds in the Earth's prebiotic chemistry, *Origins Life Evol. Biosph.* **26**, 173-194.

Borucki, W.J., and Chameides, W.L. (1984) Lightning: Estimates of the rates of energy dissipation and nitrogen fixation, *Rev. Geophys. Space Phys.* **22**, 363-372.

Borucki, W.J., McKay, C.P., and Whitten, R.C. (1984) Possible production by lightning of aerosols and trace gases in Titan's atmosphere, *Icarus* **60**, 260-273.

Borucki, W.J., McKay, C.P., Scattergood, T., and Parris, J.E. (1988) Lightning production of hydrocarbons and HCN on Titan: Laboratory measurements, *Icarus* **76**, 125-134.

Chameides, W.L., and. Walker, J.C.G (1981), Rates of fixation by lightning of carbon and nitrogen in possible primitive atmospheres, *Origins Life Evol Biosph.* **11**, 291-302.

De Graaf, R.M., and Schwartz, A.W. (2000) Reduction and activation of phosphate on the primitive Earth, *Origin Life Evol. Biosph.* **30**, 405-410.

Dreibus, G., and Wänke, H. (1985) Mars, a volatile-rich planet, *Meteoritics* **20**, 367-381.

Dreibus, G., and Wänke, H. (1989) Supply and loss of volatile constituents during the accretion of terrestrial planets in S.K. Atreya, J B. Pollack and M.S. Matthews (eds.), *Origin and Evolution of Planetary and Satellite Atmospheres*, University Arizona Press, Tucson, pp. 268-288.

Dreibus, G., Brückner, J., and Wänke, H. (2000) Phosphorus in Martian rocks and soils and the global surfaces chemistry of Mars derived from APXS on Pathfinder. 31 Lunar and Planetary Science Conference, abstract 1127.

Glindemann, D., De Gaaf, R.M., and Schwartz, A. W. (1999) Chemical reduction of phosphate on the primitive Earth, *Origins Life Evol. Biosph.* **29**, 555-561.

Kasting J. F. (1993) Earth's early atmosphere, *Science* **259**, 920-926.

Kasting, J.F (1990) Bolide impacts and the oxidation state of carbon in the Earth's early atmosphere, *Origins Life Evol. Biosph.* **20**, 199-231.

Kuramoto K. and Matsui T. (1996) Partitioning of H and C between the mantle and core during the core formation in the Earth: its implications for the atmospheric evolution and redox state of early mantle, *J. Geophys. Res.* **101** (E6), 14, 909-14,932.

Kuramoto, K. (1997) Accretion, core formation, H and C evolution of the Earth and Mars, *Phys. Earth Planet. Int.* **100**, 3-20.

Leshin, L.A., Harvey, R.P., McCoy, T.J., and McKeegan K.D. (1996) *Meteoritics Planet. Sci.* **31**, A79-A80.

Mouginis-Mark, P., Wilson, L., and Zuber, M. (1992) The physical volcanology of Mars in H.H. Kieffer, B.M. Yakosky, C.W. Snyder, M.S. Matthews (eds.), *Mars*, The University Arizona Press, pp. 425-452.

Navarro-González, R., Basiuk, V.A., and Rosembaum, M. (1996) Lightning associated to Archean volcanic ash clouds in J. Chela-Flores and F. Raulin (eds.*), Chemical Evolution: Physics of the Origin and Evolution of Life*, Kluwer Academic Publishers, Netherlands, pp 123-142.

Navarro-González, R., M. J. Molina, and L. T. Molina, (1998) Nitrogen fixation by volcanic lightning in the early Earth, *Geophys. Res. Lett.* **25**, 3123-3126.

Navarro-González, R., Molina, M.J., and Molina, L.T. (2000) The chemistry of Archean volcanic lightning in M Akobayashi, N. Fujii and R. Navarro-González (eds.), *The Role of Radiation in the Origins of Life and Evolution of Life*. Kyoto University Press, pp. 121-141.

Segura, A., and Navarro-González, R. (2000) Experimental simulation of volcanic lightning in early Mars in J. Chela-Flores, G.A. Lemarchand and J. Oró (eds.), *Astrobiology: Origins from the Big bang to Civilisation*, Kluwer Academic Publishers, Netherlands, pp 293-296.

Segura, A., and Navarro-González, R. (2001) Experimental simulation of early Martian volcanic lightning, *Adv. Space. Res.* (in press).

Sparks, R.S.J., Bursik, M.I., Carey, S.N., Gilbert, J.S., Glaze, L.S., Sigurdsson, H., and Woods, A.W. (1997) *Volcanic Plumes*, Wiley, New York, USA.

Stribling, R., and Miller, S.L. (1987) Energy yields for hydrogen cyanide of formaldehyde synthesis: The HCN and amino acid concentrations in the primitive ocean, *Origins Life* **17**, 261-273.

Toupance, G., F. Raulin, and Buvet, R. (1975) Formation of prebiochemical compounds in models of the primitive Earth's early atmosphere. I: CH_4-NH_3 and CH_4-N_2 atmospheres, *Origins Life Evol. Biosph.* **6**, 83-90.

Walker (1977) *Evolution of the Atmosphere*, Macmillan, New York.

Watson, L.L., Hutcheon, I.D., Epstein, S., and Stolper, E. (1994) Water on Mars: Clues from deuterium/hydrogen and water contentes of hydrous phases in SNC meteorites, *Science* **265**, 86-90.

Whittet, D.C.B. (1997) Is extraterrestrial organic matter relevant to the origin of life on Earth? *Origins Life Evol. Biosph.* **27**, 249-262.

Wilson, L. and Head J.W. (1994) Mars: Review and analysis of volcanic eruption theory and relationships to observed landforms, *Rev. Geophys.* **32**(3), 221-263.

Wilson, L., and Head, J.W (1983) A comparison of volcanic eruption processes on Earth, Moon, Mars, Io and Venus, *Nature* **301**, 663-669.

Woods, A.W. (1993) A model of the plumes above basaltic fissure eruptions, *Geophys. Res. Lett.* **20(**12), 1115-1118.

PLANETARY HABITABLE ZONES: THE SPATIAL DISTRIBUTION OF LIFE ON PLANETARY BODIES

ABEL MÉNDEZ
University of Puerto Rico at Arecibo
Department of Physics and Chemistry, Arecibo, PR 00613, USA
email: a_mendez@cuta.upr.clu.edu

Abstract: In this work, the biophysical requirements of microbial life and the mean global physical environment of Earth and Mars are compared to suggest potential habitable zones on both planets. On average, life on Earth is restricted to the troposphere up to 8 km above mean sea level. In the crust, life is limited down to 4 km below the surface while in the oceans is limited to 10 km below mean sea level. On Mars, a biophysical appropriate environment for extant life is the crust between 2 and 10 km below its surface. In general, a planetary environment with a near-surface thermal evolution far from 310 K implies a more restricted spatial distribution of life, a lower biomass and less biodiversity on a planetary wide scale.

1. Introduction

Biodiversity is distributed heterogeneously across the Earth. There is a high proportion of species in tropical forests while other areas, like deserts and polars regions, are almost devoid of life. The causes of these global patterns have long been a central problem of ecologist and biogeographers. In the past decade, there have been an increase of studies documenting the broad-scale spatial patterns in biodiversity, its causes, and implications (Gaston, 2000). The average species richness as function of latitude has been documented for a wide variety of taxonomic groups including protists, plants, birds, insects and primates (Stevens, 1989, and Gaston, 1996). In a global sense, most biodiversity is concentrated toward the tropical and near-surface environments. Explaining such spatial distributions is not easy matter. This is because the environments of organisms depend of various components that interact in complex ways.

The causes of the broad-scale spatial patterns in biodiversity are current topics of high interest and research. Understanding these causes concerns the past, present, and future of life on Earth. More than 25 different mechanisms have been suggested to explain why the tropics are highly diverse (Brown and Lomolino, 1998). Relationships between species richness and environmental energy have mostly been found to be associated with latitudinal, elevational and depth gradients (Rohde, 1992). However, no single mechanism fully explains the global distribution of life on Earth.

J. Chela-Flores et al. (eds.), First Steps in the Origin of Life in the Universe, 211–214.

temperature. About 85% of the microbial diversity has an optimum growth temperature between 295 and 315 K, 14% has an optimum growth temperature over 315 K and only 1% below 295 K.

The database includes data about the microbial diversity that has been cultured in the laboratory. It shows a strong tendency of microbial diversity toward mesophilic conditions above Earth's mean surface temperature of 288 K. Those conditions are more consistent with tropical and subsurface environments. Assuming that the database is a representative sample of the microbial population of the biosphere, it means that although Earth provided a large habitat for prokaryotes, they did not evolve to optimally growth in all the available environments. This suggests that the biophysical limitation of life have been more important in the biodistribution than the evolutionary adaptations.

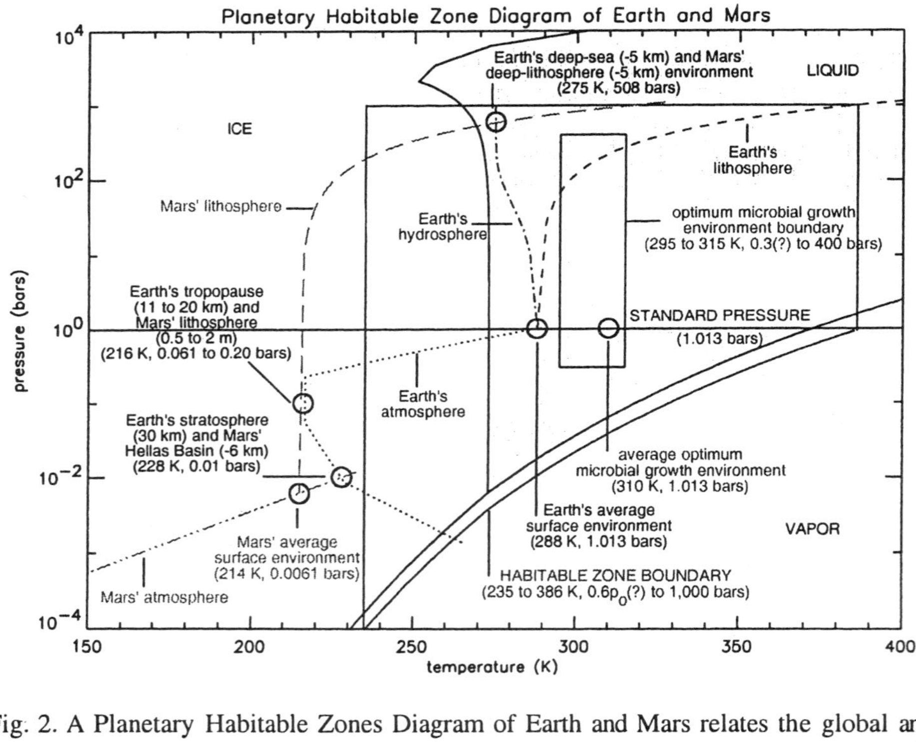

Fig. 2. A Planetary Habitable Zones Diagram of Earth and Mars relates the global and average near-surface environment of both planets with the water state and microbial life. This type of diagram gives a general panorama of the spatial limits of habitats as a function of the physical environment state described in terms of pressure and temperature (Méndez, 2000).

Availability of Liquid Water. Liquid water is the major component of living systems. It constitutes 70 to 90% of a cell mass and provides a solvent and transport medium for biochemical reactions. An environment where liquid water could be stable does not mean that water or the other requirements of life are available. Life is

This work discusses a new model to explain the spatial distribution of life on Earth and its potential distribution on other planetary bodies. The model is based on the relationship between the physical environment and the growth rate of prokaryotes. Only those physical environment variables related to the stability and availability of liquid water are considered because they are usually better known at planetary scales. The model implications on the past, and current state of life on Earth and Mars are discussed.

2. Model Description

The stability of liquid water on planetary bodies has been considered a criterion of habitability (Rassol, 1970, Hart, 1978 & 1979, Kasting, Whitmire, and Reynolds, 1993, Frank, *et. al.* 2000). The habitability criterion is here expanded to include the stability and availability of liquid water for prokaryote growth. Many prokaryotes are primary producers or sustain mutualistic relations with primary producers like plants (i.e. nitrogen fixation and mychorrhiza). Thus, in general, an environment that supports a high growth rate for prokaryotes will sustain a larger food chain and biodiversity.

Physical Environment State. There are many physical, chemical and biological variables that affect the growth rate of prokaryotes. In particular, all are affected by temperature. They thrive within a thermal range and perform less well as threshold temperatures are approached. However, there is considerable variation between species in the relative importance of pressure in their physiology. The environment total pressure limits the state and availability of liquid water. However, steady hydrostatic pressure changes have little or no effect on the growth and metabolism of most microorganisms. Growth of most terrestrial microorganisms is retarded between 300 and 400 bars at 303 K (ZoBell and Johnson, 1949). Similar lower pressure limits have not been established, but in any case, the total environment pressure should be larger than the vapor pressure of water to sustain life processes. Therefore, both temperature and pressure have effects on the growth rate of prokaryotes.

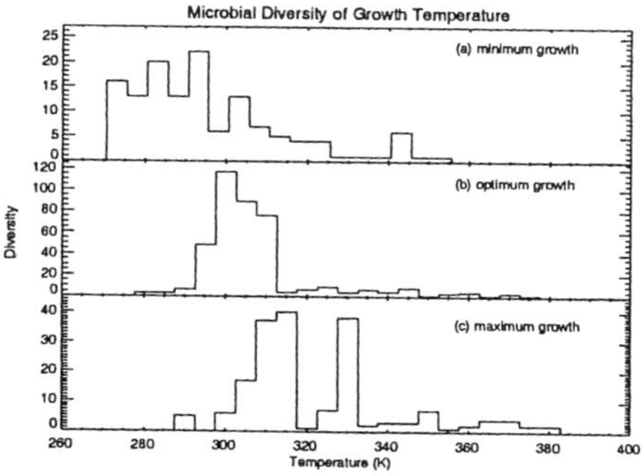

Fig. 1. Growth temperature diversity of prokaryotes. The plot shows the minimum growth temperature diversity for 134 prokaryote genera (a), the optimum growth temperature diversity for 400 prokaryote genera (b), and the maximum growth temperature diversity for 181 prokaryote genera (c).

In this work a microbial database of various prokaryotes physiological properties was extended and used to interpret the role of temperature on the biodiversity (Méndez, 2000). The database includes temperature growth conditions for about 600 well-known prokaryote genera. The mean minimum, optimum, and maximum growth temperatures of prokaryotes are 296, 310, and 325 K respectively. In general, a gaussian distribution (Fig. 1) can describe the distribution of prokaryotes as function of also limited by the temporal availability of liquid water. Water activity (i.e. relative humidity) is a good descriptor of the available liquid water in the environment for biological purposes (Griffin, 1981). Most bacteria require a water activity over 0.96 but xerotolerant fungi can growth at 0.60 using water vapor from the atmosphere. In a practical sense, this gives a biophysical limit of 0.6 times the vapor pressure of water as the minimum environment pressure to support life. A Planetary Habitable Zones diagram shows these biophysical limits (Fig. 2).

3. Conclusion

Based on the physical limits imposed by supercooled water, growth of life on Earth is restricted in the atmosphere up to 8 km above mean sea level. Life may also be possible between 35 to 45 km in the stratosphere in a gaseous phase under vapor-saturated conditions and UV protection. A xerophilic prokaryote may have the physiology to survive such environment. In the crust, life is limited down to 4 km below the surface by temperature and pressure. Life in the hydrosphere is limited down to 10 km below mean sea level by pressure, but not by temperature. On Mars, life will be limited in the crust from 2 to 10 km below its surface by temperature and pressure, respectively. This might include vapor-saturated caves, hydrothermal systems, and large water reservoirs. The physiology at such depths may be more similar to the communities of Earth's deep-sea habitats.

About 85% of the microbial species have optimum growth temperatures between 295 to 315 K, just out-side Earth's average surface environment of 288 K. Therefore, the current surface state of our planet is not in an optimum condition for prokaryotic life, which are or sustain primary producers. Thus, a planet with an average surface temperature closer to 310 K might provide a better habitat and sustain a larger biodiversity than Earth's current state. Such conditions might be representative of some extrasolar planets or Earth's past environments.

4. References

Brown, J. H., and Lomolino, M. V. (1998) *Biogeography*, 2nd edn. (Sinauer, Sunderland, MA).

Franck, S., von Bloh, W., Bounama, C., Steffen, M., Schönberner, D. & Schellnhuber, H. J. (2000) Determination of Habitable Zones in Extrasolar Planetary Systems: Where are Gaia's Sisters?, *J. of Geo. Res.* **105**, E1 1651-1658.

Gaston, K. J. (2000) Biodiversity—Latitudinal Gradients. *Prog. Phys. Geogr.*, **20**, 466-476.

Gaston, K. J. (2000) Global Patterns in Biodiversity, *Nature*, **405**, 220-227.

Griffin, D. M. (1981) Water and Microbial Strees. *Adv. in Microbiol. Ecol.* **5**, 91-136.

Hart, M. H. (1978) The Evolution of the Atmosphere of the Earth, *Icarus* **33**, 23-39.

Hart, M. H. (1979) Habitable Zones about Main Sequence Stars, *Icarus* **37**, 351-357.

Kasting, J. F., Whitmire, D. P. & Reynolds, R. T. (1993) Habitable Zones around Main Sequence Stars, *Icarus* **101**, 108-128.

Méndez, A. (2000) Planetary Habitable Zones on Earth and Mars in *Astrobiology: Origins from the Big-Bang to Civilisation*, eds. J. Chela-Flores, G. A. Lemarchand & J. Oró (Kluwer, The Netherlands), pp. 303-306.

Rassol, S. I. & DeBerg, C. (1970) *Nature (London)* **226**, 1037-1039.

Rohde, K. (1992) Latitudinal Gradients in Species Diversity: The Search for the Primary Cause. *Oikos* **56**, 514-527.

Stevens, G. C. (1989) The Latitudinal Gradient in Geographical Range: How so Many Species Co-exist in the Tropics, *Am. Nat.*, **133**, 240-256.

ZoBell, E., and Johnson, F. H. (1949) The Influence of Hydrostatic Pressure on the Growth and Viability of Terrestrial and Marine Bacteria, *J. of Bacteriol.* **57**, 179-189.

Section 5:

Evolution of life in the universe

GEOLOGICAL AND GEOCHRONOLOGICAL CONSTRAINTS FOR THE AGE OF THE OLDEST PUTATIVE BIOMARKERS IN THE EARLY ARCHAEAN ROCKS OF WEST GREENLAND

STEPHEN MOORBATH
Department of Earth Sciences, Oxford University,
Parks Road, Oxford, OX1 3PR, U.K.

Abstract. There has been much debate on the age of metamorphosed sedimentary rocks from West Greenland which contain the oldest putative terrestrial biomarkers. Recent evidence is summarised here showing that these rocks were deposited within the age range ca. 3.75-3.65 Gyr, and are at least 50-150 million years younger than the termination of the period of lunar and, by implication, terrestrial bombardment at ca. 3.8 Gyr, widely regarded as capable of frustrating the origin and survival of earliest life.

1. Introduction

Recent popular and professional scientific writings commonly claim that life on earth started at 4 Gyr ago or, more precisely, at least 3.85 Gyr ago. Unfortunately, the older of these two ages is at worst apocryphal and at best an unjustified "rounding off", whilst the younger age is based on data which do not bear close scientific scrutiny. Whilst life *might* already have existed on earth at this time, the purpose of this paper is to show that there is simply no convincing evidence for it.

The above ages are significant in context of the intense impact bombardment of the moon which peaked during the period 4.0-3.8 Gyr, [1, 2] and which is widely regarded as having affected the earth as well, in direct proportion to its much greater gravitational sink. Whilst such impacts probably frustrated the origin and survival of life [3, 4], it has also been argued [5] that such conditions might have permitted the survival of thermophilic bacteria in environmental extremes of temperature, pressure and pH. Indeed, it has even been specially pleaded from astronomical calculations [6] that because of the supposed overlap in age of earliest life on earth with impacts on the moon, together with probable impact frustration of life, the earth avoided lunar-type impacts altogether at 4.0-3.8 Gyr ago.

Here we summarise recent work (mainly since 1997) bearing on the age of two localities in West Greenland where ancient metamorphosed sediments (metasediments) contain ^{13}C-depleted graphite microparticles arguably regarded as genuine biomarkers [7, 8]. It will become evident that the rocks are significantly younger than claimed, and post-date lunar (and, by implication, terrestrial) impacts by at least 50-150 million

J. Chela-Flores et al. (eds.), First Steps in the Origin of Life in the Universe, 217–222.

years. This paper is essentially a continuation of the previous one in this series [9] which forms an essential background to it.

2. The age of the Akilia association metasediments and their included graphite microparticles, Akilia island

The claim that life is older than 3.85-3.87 Gyr [7, 10] was entirely based on ion-probe U-Pb dates on zircon grains from a single 60-80 cm-wide granitoid (quartz-diorite) sheet regarded, from supposedly discordant (cross-cutting) field relationships, as intrusive into Akilia association metasediments in which an iron-rich unit contains the calcium-phosphate mineral apatite with ^{13}C-depleted graphite microparticles. Because of the postulated discordant relationship of the quartz-diorite sheet, its zircon U-Pb date of 3.85-3.87 Gyr must be a minimum age for the Akilia metasediments. It should be noted that, although individual zircon grains from the quartz-diorite gave a range of dates from 3.87 to 3.53 Gyr, it was proposed [10, table 2] that the oldest dates represent the true age of magmatic crystallisation and emplacement of the quartz-diorite sheet.

In order to test these proposals, Whitehouse et al. [11] not only carried out further ion-probe U-Pb analyses on zircons from the quartz-diorite sheet, but also made detailed cathodoluminescence-imaging studies (a well-established technique) on every analysed zircon grain. In this way, it was shown that some grains contained i) cores with ages of up to ca. 3.85 Gyr, interpreted [11] as inherited from older rocks which are not exposed or have not yet been discovered, ii) a central sector surrounding the core, with oscillatory zoning of magmatic origin and an age of 3.65 Gyr, interpreted [11] as the true age of the host rock, iii) a patchy outer zone with an age of 2.8-2.7 Gyr, corresponding to the last major, well-documented metamorphic event suffered by the host rock.

In fairness, it must be pointed out that the interpretation of Whitehouse et al. [11] is strongly opposed by Nutman et al. [12, 13], who continue to maintain that the zircon cores at ca. 3.85 Gyr represent the age of magmatic emplacement of the host rock, whilst the two outer zones represent metamorphic episodes.

Detailed Pb/Pb isotope studies on whole rocks and separated feldspars from the regional (Amîtsoq) gneisses from the Godthaabsfjord region of West Greenland have also yielded an age of ca. 3.65 Gyr, interpreted as the age of mantle differentiation and emplacement of the magmatic precursors of the gneisses [14]. The rapid change in ^{207}Pb/^{206}Pb ratio during early earth history (because of the vastly different radioactive decay rates of ^{235}U and ^{238}U) provides a powerful constraint for temporal resolution of early Archaean crustal evolution. No Amîtsoq gneiss ever analysed, not even the critical quartz-diorite sheet on Akilia island, commenced its Pb-isotope evolution in the crust prior to 3.65 Gyr, and certainly nowhere near 3.85 Gyr [14]. The difference between the initial ^{207}Pb/^{206}Pb ratios of the two dates is ca. 13%, which far exceeds any possible analytical error. Despite the claims of Nutman et al. [12] that the Amîtsoq gneisses were emplaced over a lengthy time interval of ca. 300 million years (ca. 3.9-3.6 Gyr), we [e.g. 11, 14] maintain that the Amîtsoq gneisses of the Godthaabsfjord region were emplaced close to 3.65 Gyr, and that those in and around Akilia island which yield significantly older U-Pb dates [10, 11] are based on inherited zircons which do not reflect the age of magmatic emplacement of the host rocks. The regional magmatic age of ca. 3.65 Gyr also agrees much better with the voluminous Rb-Sr, Sm-Nd, U-Pb and Pb/Pb age work on the Amîtsoq gneisses published during the past thirty years, which cannot be referenced here.

Having ascertained that the emplacement age of the supposedly discordant quartz-diorite sheet is only 3.65 Gyr and not 3.85 Gyr, it follows that the minimum age for the Akilia metasediments containing the ^{13}C-depleted graphite microparticles is reduced accordingly.

The few direct age measurements on the Akilia rocks themselves were reviewed earlier [9]. An ion-probe zircon U-Pb date on an Akilia association metasediment gave an age of 3685±8 Myr [15], whilst a Sm-Nd isochron age (MSWD<1) on Akilia association mafic rocks, based on data by Bennett et al. [16], and plotted by Moorbath and Kamber [9], gave 3677±37 Myr. This evidence shows that the crucial Akilia metasediments may be as young as ca. 3680 Myr, nearly two hundred million years younger than the age claimed by Nutman et al. [10]. Zircons extracted directly from the iron-rich rocks which host the ^{13}C-depleted graphite particles yielded ion-probe zircon U-Pb dates of only ca. 2.6-2.7 Gyr, interpreted as a strictly metamorphic age [10].

Up to now, it had been assumed that the crucial quartz-diorite sheet is actually cross-cutting and discordant. Whilst the evidence for this has never been compelling [e.g. 10], recent small-scale field mapping by Myers and Crowley [17] has now demonstrated beyond any doubt that previous mapping [10] was inadequate and erroneous, and that the quartz-diorite sheet is structurally and tectonically completely concordant with the immediately surrounding rocks. It <u>cannot</u> therefore be used to set a lower age limit for them. In other words, the relative age relationship between the quartz-diorite sheet and the surrounding rocks is indeterminate. Myers and Crowley [17, page 112] state that “the rocks on Akilia indicate a complex history of repeated intense ductile deformation in which most geological features were attenuated, rotated and transposed into a new tectonic fabric.” Myers and Crowley [17] also emphasise the high metamorphic grade and intense metamorphic recrystallisation of all the Akilia association rocks on Akilia island and further state that “the possibility of any trace of early Archaean life that may have existed in southwestern Akilia surviving these multiple episodes of intense ductile deformation and high-grade metamorphism, and being recognisable as early Archaean, appears miraculous.” In this connection, we note that Sano et al. [18] determined U-Pb and Pb/Pb isotopic ages of the graphite-bearing apatites from Akilia island and obtained an age of ca. 1500 Myr. They concluded that “either the apatites grew about 1500 Myr ago, or they grew earlier than that but were substantially affected by recrystallisation and/or diffusive exchange with the environment, which reset the U-Pb system of the samples.” Sano et al. [18] estimate that this metamorphic event reached ca. 600°C, and formed part of a previously documented regional thermal event. In view of all this, there are now considerable doubts whether ^{13}C-depletion in the graphite microparticles in apatite from the Akilia rocks is biologically or metamorphically (thermally) mediated.

Buick [19] has questioned the biogenic interpretation [7] of the large range of reported ^{13}C-isotope depletions in the Akilia graphite particles on a variety of technical grounds and concludes that "until similar data have been obtained by similar methods from similar Phanerozoic and Proterozoic rocks, perhaps judgement should be suspended on whether the carbonaceous inclusions in Greenland apatite are biogenic". In fact, Buick [19] regards the oldest terrestrial rocks with compelling carbon-isotope evidence for life as the ca. 3.52 Gyr-old Coonterunah Group of Australia.

3. The age of the Isua greenstone belt (IGB)

Rosing [8] has reported the occurrence of graphite microparticles with $\delta^{13}C$‰ of ca. -19 in "turbiditic and pelagic sedimentary rocks" from the IGB. Rosing states that "these data and the mode of occurrence indicate that the reduced carbon represents biogenic detritus, perhaps derived from planktonic organisms". The rocks are described as having suffered variable degrees of deformation as well as intense metamorphism (amphibolite-facies) with production of garnet. The degrees of deformation and metamorphism are probably not as high as on Akilia island but, in our view, the rocks lack sufficiently detailed field and petrological work to establish their tectonic structure and the true nature of their protolith. All that can be said at present is that claims for biogenicity of the Isua graphite microparticles are only marginally more compelling than those from Akilia island.

The age of the IGB was briefly reviewed by Moorbath and Kamber [9], and not much additional work is available. Since the early 1970's, conventional dating by Rb-Sr, U-Pb, Pb/Pb and Sm-Nd methods [reviewed in 20] has given an age range of ca. 3.70-3.75 Gyr for IGB lithologies. Further age data are as follows: garnet-mica schist whole-rock Sm-Nd regression age, 3742±49 Myr [21]; banded iron-formation Pb/Pb regression age 3691±22 Myr [22]; schists, felsic rocks, banded iron-formation U-Pb zircon dates, close to 3.71 Gyr [23]; mixed rock suite Sm-Nd regression age 3712±26 Myr [24]; mixed rock suite Lu-Hf regression age 3729±16 Myr [24], recalculated here with the recently re-determined ^{176}Lu decay constant [25]. In some contrast, Rosing et al. [8] reported a Sm-Nd regression age on a mixed suite of samples of 3779±81 Myr. Furthermore, a small possibility remains that the IGB contains some >3.79 Gyr lithologies [23], but these are not strictly relevant for discussion of the rocks with putative biogenic components.

4. Conclusions

Much additional, multidisciplinary evidence is required to establish whether the ^{13}C-depleted graphite microparticles in the Akilia and Isua metasedimentary rocks are truly biogenic or not, or whether the carbon isotope abundance variations result from inorganic processes. It is emphasised again that this question is by no means settled.

The most compelling measured ages for the Akilia and Isua rock suites are in the ranges of ca. 3.65-3.70 Gyr and 3.70-3.75 Gyr respectively, although the individual age uncertainties would just about permit an overlap. However, there is no fundamental reason why the Akilia and Isua rock suites should be identical in age. At any rate, these ages are some 50-150 Myr younger than the termination of major impact on the moon and, by implication, on the earth at $\geq$3.8 Gyr [1, 2]. This post-impact interval is almost surely long enough to enable enhanced mantle and crust mobility in the early Archaean to erase surface impact features. It is also a more plausible time period for the origin and survival of life. By the time of deposition of the Akilia and Isua marine sediments and volcanic rocks [26] at ca. 3.75-3.65 Gyr, surface conditions in this region had become quiescent, and the rocks show no obvious diagnostic structural or geochemical evidence for major bolide impact.

Recent work by Cohen et al. [27], published after completion of the above text, provides strong evidence from ^{40}Ar-^{39}Ar age studies of lunar impact melts supporting the concept of a short, intense period of bombardment in the earth-moon system at ca. 3.9 Gyr. This would further increase the probable time interval between impact and deposition of the Akilia and Isua sediments to between 150 and 250 Myr.

Acknowledgements

This is a contribution to the Isua Multidisciplinary Research Project, supported by the Danish National Research Council, the Geological Survey of Denmark and Greenland, the Commission for Scientific Research in Greenland, and the Mineral Office of the Greenland Government.

References

[1] Wilhelms, D.E.: The geologic history of the Moon. *U.S. Geol. Surv. Prof. Paper*, **1348** (1987), 127p.

[2] Ryder, G.: Lunar samples, lunar accretion and the early bombardment of the Moon. *EOS (Trans. Amer. Geophys. Union)*, **71** (1990), 322-323.

[3] Maher, K.A. and Stevenson, D.J.: Impact frustration and the origin of life. *Nature*, **331** (1988), 612-614.

[4] Sleep, N.H., Zahnle, K.J., Kasting, J.F. and Morowitz, H.J.: Annihilation of ecosystems by large asteroid impacts on the early Earth. *Nature*, **342** (1989), 139-142.

[5] Mojzsis, S.J. and Harrison, T.M.: Vestiges of a beginning: clues to the emergent biosphere recorded in the oldest known sedimentary rocks. *GSA Today*, **10** (2000), 2-6.

[6] Arrhenius, G. and Lepland, A.: Accretion of Moon and Earth and the emergence of life. *Chem. Geol.*, **169** (2000), 69-82.

[7] Mojzsis, S.J., Arrhenius, G., McKeegan, K.D., Harrison, T.M., Nutman, A.P. and Friend, C.R.L.: Evidence for life on Earth before 3800 million years ago. *Nature*, **384** (1996), 55-59.

[8] Rosing, M.T.: ^{13}C-depleted carbon microparticles in >3700 Ma seafloor sedimentary rocks from West Greenland. *Science*, **283** (1999), 674-676.

[9] Moorbath, S. and Kamber, B.S.: Re-appraisal of the age of the oldest water-lain sediments, West Greenland. In: J. Chela-Flores and F. Raulin (eds), *Exobiology: Matter, Energy, and Information in the Origin and Evolution of Life in the Universe.* Kluwer Academic Publishers, Dordrecht, (1996), 81-86.

[10] Nutman, A.P., Mojzsis, S.J. and Friend, C.R.L.: Recognition of •3850 Ma water-lain sediments in West Greenland and their significance for the early Archaean Earth. *Geochim. Cosmochim. Acta*, **61** (1997), 2475-2484.

[11] Whitehouse, M.J., Kamber, B.S. and Moorbath, S.: Age significance of U-Th-Pb zircon data from early Archaean rocks of West Greenland – a reassessment based on combined ion-microprobe and imaging studies. *Chem. Geol.*, **160** (1999), 201-224.

[12] Nutman, A.P., McGregor, V.R., Friend, C.R.L., Bennett, V.C. and Kinny, P.D.: The Itsaq Gneiss Complex of southern West Greenland: The world's most extensive record of early crustal evolution (3,900-3,600 Ma). *Precamb. Res.*, **78** (1996), 1-39.

[13] Nutman, A.P., Bennett, V.C., Friend, C.R.L. and McGregor, V.R.: The early Archaean Itsaq Gneiss Complex of southern West Greenland: The importance of field observations in interpreting age and isotope constraints for early terrestrial evolution. *Geochim. Cosmochim. Acta*, **64** (2000), 3035-3060.

[14] Kamber, B.S. and Moorbath, S.: Initial Pb of the Amîtsoq gneiss revisited: implication for the timing of early Archaean crustal evolution in West Greenland. *Chem. Geol.*, **150** (1998), 19-41.

[15] Schiøtte L. and Compston, W.: U-Pb age pattern for single zircon from early Archaean Akilia association south of Ameralik fjord, southern West Greenland. *Chem. Geol.*, **80** (1990), 147-157.

[16] Bennett, V.C., Nutman, A.P. and McCulloch, M.T.: Nd isotopic evidence for transient, highly depleted mantle reservoirs in the early history of the Earth. *Earth Planet. Sci. Lett.*, **119** (1993), 299-317.

[17] Myers, J.S. and Crowley, J.L.: Vestiges of life in the oldest Greenland rocks? A review of early Archaean geology in the Godthaabsfjord region, and reappraisal of field evidence for >3850 Ma life on Akilia. *Precamb. Res.*, 103 (2000) 101-124.

[18] Sano, Y., Terada, K., Takahashi, Y. and Nutman, A.P.: Origin of life from apatite dating? *Nature*, **400** (1999), 127.

[19] Buick, R.: Life in the Archaean. In: D.E.G. Briggs and P.R. Crowther (eds), *Palaeobiology II.* Blackwells Science U.S., (2001), 13-21.

[20] Moorbath, S., Taylor, P.N. and Jones, N.W.: Dating the oldest terrestrial rocks – fact and fiction. *Chem. Geol.*, **57** (1986), 63-86.

[21] Kamber, B.S., Moorbath, S. and Whitehouse, M.J.: Extreme Nd-isotope heterogeneity in the early Archaean – fact or fiction? Case histories from northern Canada and West Greenland – Reply. Chem. Geol., **148** (1998), 219-224.

[22] Frei, R., Bridgwater, D., Rosing, M. and Stecher, O.: Controversial Pb-Pb and Sm-Nd isotope results in the early Archaean Isua (West Greenland) oxide iron formation: Preservation of primary signatures versus secondary disturbances. *Geochim. Cosmochim. Acta*, **63** (1999), 473-488.

[23] Nutman, A.P., Bennett, V.C., Friend, C.R.L. and Rosing, M.T.: ~3710 and •3790 Ma volcanic sequences in the Isua (Greenland) supracrustal belt; structural and Nd isotope implications. *Chem. Geol.*, **141** (1997), 271-287.

[24] Blichert-Toft, J., Albarède, F., Rosing, M., Frei, F. and Bridgwater, D.: The Nd and Hf isotopic evolution of the mantle through the Archaean. Results from the Isua supracrustals, West Greenland, and from the Birimian terranes of West Africa. *Geochim. et Cosmochim. Acta*, **63** (1999), 3901-3914.

[25] Nir-El, Y. and Lavi, N. Measurement of half-life of ^{176}Lu. *Appl. Radiat. Isot.*, **49** (1998), 1653-1655.

[26] Appel, P.W.U., Fedo, C.M., Moorbath, S. and Myers, J.S.: Recognisable primary volcanic and sedimentary features in a low-strain domain of the highly deformed, oldest known (~3.7-3.8 Gyr) Greenstone Belt, Isua, West Greenland. *Terra Nova*, **10** (1998), 57-62.

[27] Cohen, B.A., Swindle, T.D. and Kring, D.A.: Support for the lunar cataclysm hypothesis from lunar meteorite impact melt ages. *Science*, **290** (2000), 1754-1756.

THE CENANCESTOR AND ITS CONTEMPORARY BIOLOGICAL RELICS: THE CASE OF NUCLEIC ACID POLYMERASES

Luis Delaye, Héctor Vázquez, and Antonio Lazcano
Facultad de Ciencias, UNAM
Apdo. Postal 70-407
Cd. Universitaria, 04510 México D.F.
MÉXICO
E-mail: alar@hp.fciencias.unam.mx

1. Introduction

The recognition that different macromolecules may be uniquely suited as molecular chronometers in the construction of nearly universal phylogenies has widened the range of phylogenetic studies to previously unsuspected heights. In particular, the use of small subunit ribosomal RNAs (16/18S rRNA) as molecular markers led to the construction of a trifurcated, unrooted tree in which all known organisms can be grouped in one of three major monophyletic cell lineages: the eubacteria, the archaebacteria, and the eukaryotic nucleocytoplasm, now referred to as the domains *Bacteria*, *Archaea*, and *Eucarya*, respectively (Woese et al., 1990). The construction of the rRNA tree showed that no single major branch predates the other two, and all three derive from a common ancestor. It was thus concluded that the latter was a progenote, which was defined as a hypothetical entity in which phenotype and genotype still had an imprecise, rudimentary linkage relationship (Woese and Fox, 1977). According to this view, the differences found among the transcriptional and translational machineries of eubacteria, archaebacteria, and eukaryotes, were the result of evolutionary refinements that took place separately in each of these primary branches of descent after they have diverged from their universal ancestor (Woese, 1987).

From an evolutionary point of view it is reasonable to assume that at some point in time the ancestors of all forms of life must have been less complex than even the simpler extant cells, but our current knowledge of the characteristics shared between the three lines suggests that the conclusion that the last common ancestor was a progenote may have been premature. Pending the issue of horizontal gene transport (Figure 1), a partial description of the last common ancestor (LCA) of eubacteria, archaebacteria, and eukaryotes may be inferred from the distribution of homologous traits among its descendants. Ten years ago, the set of such genes that had been sequenced and compared was still small, but the sketchy picture that had emerged suggested that the most recent common ancestor of all extant organisms, or *cenancestor*, as defined by Fitch and Upper (1987), was a rather sophisticated cell (Lazcano, Fox and Oró, 1992) with at least (a) DNA polymerases endowed with proof-reading activity; (b) ribosome-

J. Chela-Flores et al. (eds.), First Steps in the Origin of Life in the Universe, 223–230.

mediated translation apparatus with an oligomeric RNA polymerase; (c) membrane-associated ATP production; (d) signalling molecules such as cAMP and insulin-like peptides; (e) RNA processing enzymes; and (f) biosynthetic pathways leading to amino acids, purines, pyrimidines, coenzymes, and other key molecules in metabolism (cf. Lazcano, 1995).

Recent results have confirmed the above conclusions. These traits are far to numerous and complex to assume that they evolved independently or that they are the result of massive multidirectional horizontal transfer events which took place before the earliest speciation events recorded in each of the three lineages. Their presence suggests that the cenancestral population was not a direct, immediate descendant of the RNA world, a protocell or any other pre-life progenitor system (Lazcano, 1995). Very likely, the LCA was already a complex organism, much akin to extant bacteria, and must be considered the last of a long line of simpler earlier cells for which no modern equivalent is known. Moreover, the universal distribution of the same essential features of genome replication, gene expression, basic anabolic reactions, and membrane-associated ATPase-mediated energy production in all known organisms not only provide direct evidence of the monophyletic origin of all extant forms of life, but also imply that the sets of genes encoding the components of these complex traits were frozen a long time ago, i. e., major changes in them are very strongly selected against.

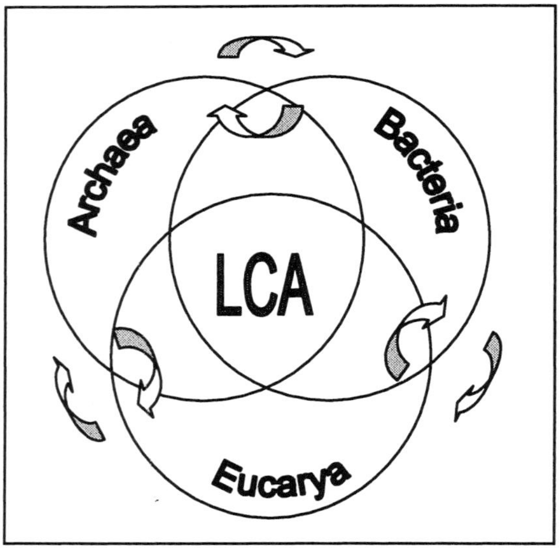

Figure 1. The gene complement of the LCA is defined by the intersection of the complete genomes of the three domains. The arrows represent the horizontal gene transfer between cellular domains.

While trees based on whole genome information have confirmed at a broad level the rRNA-based phylogenies (Snel et al., 1999; Tekaia et al., 1999), it is also true that the congruence between rRNA genes and other molecules is not always ideal. A large variety of phylogenetic trees constructed from DNA and RNA polymerases, elongation factors, F-type ATPase subunits, heat-shock and ribosomal proteins, and an increasingly large set of genes encoding enzymes involved in biosynthetic pathways, have confirmed the existence of the three primary cellular lines of evolutionary descent (Doolittle and Brown, 1994), but there is evidence of extensive horizontal transfer events that have taken place in the past (Doolittle, 1999). In fact, in addition to lateral gene transfer (Figure 1), insights into cenancestral states can be strongly hindered by inadequate biodiversity sampling, polyphyletic gene losses, unequal rates of molecular evolution, convergence, polyphyly, and secondary loss of organelles. These factors clearly limit our ability to recognize the extant molecular relics of the cenancestor.

1.1 THE SEARCH FOR THE ANCESTRAL NUCLEIC ACID POLYMERASE

Replication of genetic material must have been one of the oldest functions to evolve (Figure 2). Ideally, abiotic laboratory polymerization of nucleotides should provide insights into the transition from the prebiotic broth to the extant enzyme-mediated replication of nucleic acids.

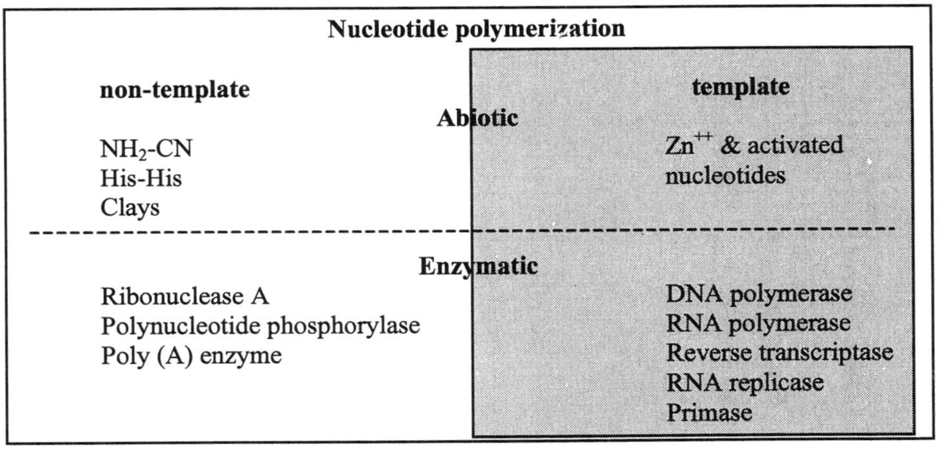

Figure 2. The abiotic and enzymatic polymerization of nucleotides.

In principle, this could also explain the evolutionary development of polymerases, an issue directly related to the chemical composition of the cenancestral genome. Since all extant cells are endowed with DNA genomes, the most parsimonious conclusion is that such genomes were already present in the cenancestral population. However, this hypothesis has been contested by suggestions of an RNA- (Mushegian and Koonin, 1996) or even a mixed DNA-RNA genome for the LCA (Leipe et al., 1999). These proposals are based, at least in part, on the low level of conservation of the primary structure of DNA polymerases (Olsen and Woese, 1996; Edgell and Doolittle, 1997), as well as on the striking differences in their phylogenetic distribution compared with rRNAs, aminoacyl-tRNA-synthetases, and other molecules involved in transcription and translation. This has led to suggestions that DNA genomes, together with the corresponding polymerases, may have been invented independently in the different cell domains (Mushegian and Koonin, 1996; Leipe, *et al.* 1999).

Evolution of enzymes in biological systems often involves the acquisition of new catalytic or binding properties by an existing protein scaffold. However, identification of several non-homologous classes of nucleic acid polymerases (primase, reverse transcriptase (RT), RNA polymerase and DNA polymerase) shows that this is not the case for these enzymes, and demonstrates the polyphyletic origin of template-dependent enzyme-mediated synthesis of phosphodiester bonds (Steitz, 1999).

Based on sequence similarity and crystal structure analysis (Steitz, 1999) DNA polymerases have been classified into five families (Table 1). Three dimensional structures are available for the DNA polymerase families defined by the DNA pol I, DNA pol α, RT, and rat DNA pol β prototypes.

Family	**Representatives**
DNA polymerase I family (A polymerase family)	- Klenow fragment of *Escherichia coli* DNA polymerase I - Klenow fragment of *Bacillus* DNA polymerase I - *Thermus aquaticus* DNA polymerase - T7 RNA and DNA polymerases
DNA polymerase α (B family DNA polymerase or family II)	- All eukaryotic replicating DNA polymerases (α,δ,ε) - Phage T4 DNA polymerase - RB69 Phage polymerase
Reverse transcriptase family	- HIV reverse transcriptase - RNA-dependent RNA polymerase - Telomerase
Rat DNA polymerase β	- DNA polymerase β (rat)
Bacterial DNA polymerase III	- Bacterial DNA polymerase III, on the basis of amino acid sequence comparisons.

Table 1. Classification of DNA polymerases into five families according to sequence similarity and tertiary structure criteria (cf. Steitz, 1999).

All DNA polymerases whose tertiary structure has been determinated appear to share a common overall architectural feature comparable to a right hand shape. This structure is not so evident, however, in the case of rat DNA pol β and its homologues. The structure of the other polymerases has been described as consisting of "thumb", "palm", and "finger" domains (Kohlstaedt, et al, 1992). Detailed analysis of the three dimensional structure of DNA polymerases from the pol I, pol α, and RT families suggest that their palm sub-domain has a single origin, i.e., it is homologous in all of them, while the fingers and the thumb sub-domains are different in all four of the families for which structures are known (Brautigam and Steitz, 1998). The complex evolutionary history of nucleic acid polymerases, combined with the wide sequence space explored by these enzymes during biological evolution, strongly hinders the identification of the ancestral polymerase.

As argued here, the three-dimensional homology between the palm domains of DNA polymerase I and DNA polymerases B, which includes all eukaryotic replicating DNA polymerases (Steitz, 1999), can be extended to suggest that such domain, which catalyses the phosphodiester bond, was already present in the cenancestor. As shown here, the structural multiple alignment of the palm sub-domain of DNA polymerases belonging to the pol I and pol α families from the tree cellular domains of life strongly suggests that this sub-domain is the most ancient protein segment found within these enzymes and could have been present in the LCA.

2. Material and Methods

The crystal structures from the following DNA polymerases sequences were downloaded from Protein Data Bank (www.rcsb.org/pdb/): DNA polymerases A family: 1KLN, *Escherichia coli*; 1TAQ, *Thermus aquaticus*; 1XWL *Bacillus stearothermophilus*; and from DNA polymerases B family: 1TGO, *Thermococcus gorgonarius*; 1D5A, and *Desulfurococcus sp*. Tok;.

The palm sub-domains of all of them, following the classification of CATH database (www.biochem.ucl.ac.uk/bsm/cath_new/index.html), were aligned manually using the program SPDBV (Guex, and Peitsch, 1997) (www.expasy.ch/spdbv/text/refs.htm) to construct a structural multiple alignment.

The sequence of the palm domain from *T. gorgonarius* (1TGO) was used as a query against the SwissProt database in the NCBI server (www.ncbi.nlm.nih.gov/BLAST/), using Blast (Altschul, et al, 1997). Sequences from eukaryoticDNA polymerases thus identified using this metod, were added to the structural multiple alignment using the program ClustalX v1.81 (Thompson, et al, 1997).

The multiple structural alignment was performed by first aligning the two archaeal and the three bacterial palm sub-domains separately, in order to identify the conserved residues in each of the families. This was followed by the manual alignment of all structures looking for the 3-dimensional conserved residues identified before.

3. Results

The multiple structural alignment of the primary structure of the different palm sub-domains in, is shown in Figure 3.

The Blast search found eight eucaryotic DNA polymerases: DPOD_HUMAN DNA polymerase delta catalytic subunit (Expect = 4e-04); DPOD_BOVIN DNA polymerase delta catalytic subunit (Expect = 5e-04); DPOD_MESAU DNA polymerase delta catalytic subunit (Expect = 7e-04); DPOD_RAT DNA polymerase delta catalytic subunit (Expect = 8e-04); DPOZ_HUMAN DNA polymerase zeta catalytic subunit (Expect = 9e-04); DPOD_MOUSE DNA polymerase delta catalytic subunit (Expect = 0.001); DPOZ_MOUSE DNA polymerase zeta catalytic subunit (Expect = 0.001); DPOD_SOYBN DNA polymerase delta catalytic subunit (Expect = 001).

```
                 hhhhhhhhhhhhhhhhhhhhhhhhhhhhhhhhhh__sssssss_ssss
>1KLN            aaaeraainapmqgtaadiikramiavdawlqaeqprvrmimqvhdelv

              hhhhhhhhhhhhhhhhhhhhhhhhhhhhhhhhhhhhh______sssss_ssss
>1XWL         fnvrsfaermamntpiqgsaadiikkamidlnarlkeerlqahlllqvhdeli

                          hhhhhhhhhhhhhhhhhhhhhhhhh  h__sssss_ssss
>1TAQ                     mpvqgtaadlmklamvklfprlee--mgarmllqvhdelv

                         hhhhhhhhhhhhhhhhhhhhhhhhhh   _____sssssssssss
>1TGO                    kecaesvtawgrqyiettireieek--fgfkvlyadtdgff

                         hhhhhhhhhhhhhhhhhhhhhhhhhh   _____sssss_ssss
>1D5A                    recaesvtawgrqyiettmreieek--fgfkvlyadtdgff
------------------------------------------------------------------
                    sssss___hhhhhhhhhh__________    ___sssssss__hhhh
>1KLN               fevhkddvdavakqihqlmenctrld----vpllvevgsgenwdqa

                    sssss___hhhhhhhhhhhhh______    ___sssssss_______
>1XWL               leapkeemerlcrlvpevmeqavtlr----vplkvdyhygstwyda

                    sssss__hhhhhhhhhhhhh_______    ___sssssssssshhh
>1TAQ               leapkeraeavarlakevmegvypla----vplevevgigedwlsake

                    ss_____ _hhhhhhhhhhhhhhhhh______sssssssss
>1TGO               atipga-daetvkkkakefldyinaklpglleleyegfykr

                    ss_____ _hhhhhhhhhhh___hhhh______sssssss
>1D5A               atipga-daetvknkakeflnyinprlpglleleyegf
```

Figure 3. Multiple structural alignment of the palm subdomains. Enclosed in boxes are the residues which are structurally homologous in all the domains studied here. The sequences are as follows: DNA polymerases A family: 1KLN, *Escherichia coli*; 1XWL *Bacillus stearothermophilus;* 1TAQ, *Thermus aquaticus*; and from DNA polymerases B family: 1TGO, *Thermococcus gorgonarius*; 1D5A, and *Desulfurococcus sp*. Tok. Above each sequence the secondary structure is shown: h, α-helix, and s, for β-strand.

4. Discusion and conclusions

Although the availability of completely sequenced cellular genomes has enhanced the likehood of more accurate reconstructions of ancestral states, horizontal gene transfer can strongly hinder our ability to understand the characteristics of the last common ancestor. As shown by the current discussions on the chemical nature of the LCA genome (Mushegian and Koonin, 1996; Leipe, *et al.*, 1999), our attempts to understand the distant past can also be limited by the polyphyletic origin of DNA polymerases, whose classification into different families reflects a case of convergence. Nevertheless, the evidence presented here clearly shows that the palm subdomains of the I and II

DNA polymerase families, which are found in all three cell lineages, have a common origin that has conserved the same tertiary structure and is thus an indication of the monophyletic origin of these enzymes. The lack of a crystalized eukaryotic replicative DNA polymerase has not allowed the recognition of the common origin of these polymerases. As shown here however, their monophyletic origin is recognizable even at the primary structure level (Blast search). The evolutionary conservation of this subdomain, which is involved in the catalysis of the phosphoribosyl transfer reaction (Steitz, 1999), is probably due to the central role it plays in the synthesis of polynucleotides.

On the other hand, the lack of homology between the other subdomains (i.e., the thumb and finger) indicates the easyness by which nucleotide-binding motifs can evolve. A possible evolutionary sequence of nucleotide polymerization agents, starting from the prebiotic synthesis of phosphodiester bonds (and omitting the existence of possible preRNA worlds) is shown in Figure 4. This scheme is based on Steitz (1999) suggestion of a stepwise-emergence of functional peptides in an ribozymic replicase, and on the evolution of polymerization agents discussed elsewhere (Lazcano et al., 1988).

A possible evolutionary sequence of nucleotide polymerization

Prebiotic polymerization of nucleotide-derivatives mediated by His-His, NH_2CN, Zn^{++} (perhaps in chemically active surfaces?)

▼

Ion-dependent ribozymic RNA replicase

▼

Appearance of protein synthesis leads to the development of RNA-binding peptides

▼

Stepwise substitution of the ribozymic RNA scaffold by peptides

▼

Appearance of an ion-dependent proteinic ancestral polymerase (palm-subdomain?)

▼

Polyphyletic origin of extant polymerases

Figure 4. Possible evolutionary sequence of nucleotide polymerization agents, starting from the prebiotic synthesis of phosphodiester bonds (and omitting the existence of possible preRNA worlds).

Given the lack of absolute chemical specificity that polymerases exhibit for both template and substrate (Lazcano et al., 1988), it is quite possible that the conserved ion-dependent palm-subdomain discussed here was part of an ancestral replicase and transcriptase during the RNA/protein world stage (Figure 4). This possibility is supported by the homology between the viral T7 RNA and DNA polymerase. However, the highly conserved sequences of the β and β' subunits of the DNA-dependent RNA polymerase which are found in all three cellular domains, indicate that by the time the LCA had evolved, a modern type of oligomeric RNA polymerase had already evolved. Why polymerases have originated independently several times and why the level of divergence within each family of DNA polymerases is so high, are still open questions that deserve further attention.

Acknowledgements

Support from DGAPA-UNAM/PAPIIT IN213598 project is gratefully acknowledged.

5. References

Altschul, S.F., Madden, T.L., Schäffer, A.A., Zhang, J., Zhang, Z., Miller, W. and Lipman, D.J. (1997) Gapped BLAST and PSI-BLAST: a new generation of protein database search programs, *Nucleic Acids Res.* **25**, 3389-3402

Brautigam, C.A., and Steitz, T.A. (1998) Structural and functional insights provided by crystal structures of DNA polymerases and their substrate complexes, *Curr. Opin. Struct. Biol.*, **8**, 54-63

Doolittle, W.F. (1999) Phylogeny classification and the universal tree, *Science*, **284**, 2124-2128

Doolittle, W. F. and Brown, J. R. (1994) Tempo, mode, the progenote and the universal root, *Proc. Natl. Acad. Sci. USA* **91**, 6721-6728

Edgell, R.D. and Doolittle, W.F. (1997) Archaea and the origin(s) of DNA replication proteins, *Cell*, **89**, 995-998

Fitch, W.M., Upper, K. (1987) The phylogeny of tRNA sequences provides evidence for ambiguity reduction in the origin of the genetic code, *Cold Spring Harb Symp Quant Biol.* **52**, 759-767

Guex, N. and Peitsch, M.C. (1997) SWISS-MODEL and the Swiss-Pdb Viewer: An environment for comparative protein modeling, *Electrophoresis,* **18**, 2714-2723

Kohlstaedt, L.A., Wang, J., Friedman, J.M., Rice,, P.A., and Steitz, T.A. (1992) Crystal structure at 3.5 Å of HIV-1 reverse transcriptase complexed with an inhibitor, *Science*, **264**, 1781-1790

Lazcano, A., Fastag, J., Gariglio, P., Ramírez, C. and Oró, J. (1988) On the early evolution of RNA polymerase, *J. Mol. Evol.* **27**, 365-376

Lazcano, A., Fox, G. E., and Oró, J. (1992) Life before DNA: the origin and evolution of early Archean cells, in R. P. Mortlock (ed), *The Evolution of Metabolic Function*, CRC Press, Boca Raton, pp. 237-295

Lazcano, A. (1995) Cellular evolution during the Early Archaean: what happened between the progenote and the cenancestor? *Microbiologia SEM*, **11**, 185-198

Leipe, D.D., Aravind, L., Koonin, E.V. (1999) Did DNA replication evolve twice independently? *Nucleic. Acids. Res.* **27**, 3389-3401

Mushegian, A.R., and Koonin, E.V. (1996) A minimal gene set for cellular life derived by comparison of complete bacterial genomes, *Proc. Natl. Acad. Sci. U S A,* **93**, 10268-10273

Olsen, G.J. and Woese, C.R. (1996) Lessons from an Archaeal genome: what are we learning from *Methanococcus jannaschii*? *Trends. Genet.* **12**, 377-379

Snel, B., Bork, P., and Huynen, M. A. (1999) Genome phylogeny based on gene content, *Nature Genetics* **21**, 108-110

Steitz, T.A. (1999) DNA polymerases: structural diversity and common mechanisms, *J. Biol. Chem.* **274**, 17395-17398

Tekaia, F., Lazcano, A., and Dujon, B. (1999) The genomic tree as revealed from whole proteome comparisons, *Genome Research* **9**, 550-557

Thompson,J.D., Gibson, T.J., Plewniak,F., Jeanmougin,F. and Higgins,D.G. (1997) The ClustalX windows interface: flexible strategies for multiple sequence alignment aided by quality analysis tools, *Nucleic Acids Res.* **24**, 4876-4882

Woese, C. R. (1987) Bacterial evolution, *Microbiol. Reviews* **51**, 221-271

Woese, C.R. and Fox, G.E. (1977) The concept of cellular evolution, *Jour. Mol. Evol.* **10**, 1-6

Woese, C. R., Kandler, O., and Wheelis, M. L. (1990) Towards a natural system of organisms, proposal for the domains Archaea, Bacteria, and Eucarya, *Proc. Natl. Acad. Sci. USA* **87**, 4576-4579

LIKELIHOOD OF TRANSPORT OF LIFE BETWEEN THE PLANETS OF OUR SOLAR SYSTEM

GERDA HORNECK
German Aerospace Center DLR, Institute of Aerospace Medicine,
D 51170 Cologne, Germany

Abstract
During a hypothetical interplanetary transfer, living organisms have to cope with three major challenges: (i) the escape process, (ii) the long duration exposure to space; and (iii) the entering process. A comprehensive experimental and theoretical study of the probabilities of microorganisms surviving the complex interplay of the parameters of space (e.g., vacuum, UV- and ionizing radiation, temperature extremes) concludes that radiation-resistant microbes could survive a journey from one planet to another in our solar system if they were shielded by a layer of meteorite material.

1. Introduction

The idea that microscopic forms of life, for example spores, can be propagated in space and thereby seeding life from one planet to another first appeared in 1865 (Richter, 1865) and was more precisely formulated later on by Arrhenius (1903) as theory of Panspermia. In the time following, the theory has been criticized with arguments such as it cannot be experimentally tested, it shunts aside the question of the origin of life, and spores will not survive long-time exposure to the hostile environment of space, especially vacuum and radiation (reviewed in Horneck 1995). Hence, Panspermia was no further considered a likely route to the appearance of life on Earth.

However, recent observations have provided support to revisit Panspermia (reviewed in Horneck 1995). These are: (i) the detection of meteorites, some of lunar and some of Martian origin; (ii) the detection of organics and supposition of microbial fossils in one of the Martian meteorites; (iii) the probability of small particles of diameters between 0.5 μm and 1 cm or even boulder-sized rocks reaching escape velocities by the impact of large meteorites on a planet, e.g. on Earth or Mars (Vickery and Melosh 1987); (iv) the ability of bacterial spores to survive the shock waves of a simulated meteorite impact to a certain extent; (v) the high UV-resistance of micro-organisms at the low temperatures of deep space; (vi) the high survival of bacterial spores over extended periods in space; (vii) the paleogeochemical evidence of a very early appearance of life on Earth in the form of metabolically advanced microbial prokaryotic ecosystems leaving not more than approximately 0.4 Ga for the evolution of life from the simple

J. Chela-Flores et al. (eds.), First Steps in the Origin of Life in the Universe, 231–236.

precursor molecules to the level of a prokaryotic, photoautotrophic cell; (viii) the biochemical evidence of a common ancestor for all life forms on Earth; and (ix) the feasibility of artificial or directed transport by probes sent to other planets. Although we do not know whether Panspermia is likely to have occurred in the history of the solar system, or whether it is feasible at all, with space technology a new tool is available for testing experimentally whether microorganisms are capable of surviving a hypothetical journey from one planet to another.

The hypothetical scenario of interplanetary transfer of life would involve the following steps: (i) escape process, i.e. removal to space of biological material which has survived being lifted from the surface to high altitudes; (ii) interim state in space, i.e. survival of the biological material over time scales comparable with the interplanetary or interstellar passage; (iii) entry process, i.e. non-destructive deposition of the biological material on another planet. The chances of microbes to survive these steps will be discussed in the following.

2. Escape Process

Hypervelocity impacts of large objects such as meteorites, asteroids, or comets are considered as the most plausible process capable of ejecting microbe-bearing surface material from a planet or moon into space. The peak shock pressure received by the 15 Martian meteorites, detected on Earth, range from about 20 GPa to about 45 GPa with estimated post-shock temperatures between about 100°C at 20 GPa and about 600 °C at 45 GPa. Although these impacts are very energetic, a certain fraction of ejecta is not heated above 100 °C. These low temperature fragments are ejected from the so-called spall zone, i.e. the surface layer of the target, where the resulting shock is considerably reduced by superimposition of the reflected shock wave on the direct one (Melosh, 1984). These 15 Martian meteorites, so far detected on Earth, represent probably only a diminishing fraction of those imported from Mars within Earth's history. It has been calculated that within the last 4 Ga more than 10^9 fragments of a diameter of ≥2 m and temperatures ≤100°C were ejected from Mars, of which about 5 % arrived on Earth after a journey in space of ≤8 Ma. The corresponding numbers for a transfer from Earth to Mars are about 10^8 fragments ejected from the Earth with about 0.1% arriving on Mars within 8 Ma (Mileikowsky et al. 2000). During the preceding period of "heavy bombardment" even 10 times higher numbers are estimated.

In experiments, simulating impacts comparable to those experienced by the Martian meteorites, the survival of microbes was tested after subjecting spores of *Bacillus subtilis* to accelerations, jerks or shock waves. In shock recovery experiments with an explosive set-up, the survival of spores of *B. subtilis* was studied to a shock treatment in the pressure range which some Martian meteorites have experienced (Horneck et al. 2001). It was found that the a substantial fraction of spores (up to 10^{-4}) were able to survive a peak shock pressure of 32 ± 1 GPa and a post-shock temperature of about 250°C. These data support the hypothesis that bacterial spores may survive an impact-induced escape process in a scenario of interplanetary transfer of life. Assuming a mean spore density of 10^8 spores/g in e.g., desert soil or rock, a 1 kg rock would accommodate approximately 10^{11} spores, of which up to 10^7 could survive even extremely high shock pressures occurring during a meteorite impact. At more moderate shock waves as they would occur in the spall zone of an impact, even substantial higher survival rates are expected.

3. Interim State in Space

The environment, microorganisms in space have to cope with, is characterized by a high vacuum, an intense radiation climate of galactic and solar origin and extreme temperatures (Table 1).

TABLE 1. Parameters of the environment of interplanetary space relevant for studies on microbial responses to space (modified from Horneck, 1999)

Space parameter	Interplanetary space	Earth orbit	Simulation facility
Space vacuum			
Pressure (Pa)*)	10^{-14}	10^{-7}-10^{-4}	10^{-7}-10^{-4}
Residual gas (part/cm^{-3})	1H	10^4-10^5 H 10^4-10^6 He 10^3-10^6 N 10^3-10^7 O	Different values[a)]
Solar electromagnetic radiation			
Irradiance (W/m²)	Different values[b)]	1380	Different values[c)]
Spectral range (nm)	Continuum	Continuum	Different spectra[c)]
Cosmic ionizing radiation			
Dose (Gy/a)	≤ 0.1[d)]	400 -10 000[e)]	wide range[c)]
Temperature (K)	> 4[b)]	wide range[b)]	wide range[c)]
Microgravity (g)	$< 10^{-6}$	10^{-3}-10^{-6}	0-1000

*) 1 Pa = 10^{-5} bar
a) Depending on pumping system and requirements of the experimenter
b) Depending on orientation and distance to sun
c) Depending on the radiation source and filter system
d) Depending on shielding, highest values at mass shielding of 0.15 g/cm²
e) Depending on altitude and shielding, highest values at high altitudes and shield of 0.15 g/cm²

Because of its extreme dehydrating effect, space vacuum has been considered to be one of the factors that may prevent interplanetary transfer of life (Nussinov and Lysenko, 1983). However, space experiments have shown that up to 70 % of bacterial and fungal spores survive short-term (e.g., 10 days) exposure to space vacuum, even without any protection (Horneck, 1993a). The chances of survival in space are increased if the spores are embedded in chemical protectants such as sugars, or salt crystals, or if they are exposed in thick layers. For example, 30 % of *B. subtilis* spores survived nearly 6 years of exposure to space vacuum, if embedded in salt crystals, whereas 80 % survived in the presence of glucose (Horneck et al. 1994). It has been suggested that sugars and polyalcohols stabilize the structure of the cellular macromolecules, especially during vacuum-induced dehydration, leading to increased rates of survival.

Solar UV radiation has been found to be the most deleterious factor of space, as tested with dried preparations of viruses, and of bacterial and fungal spores (Horneck, 1993a). The reason for this high biological efficiency is the highly energetic UV-C (190-280 nm) and vacuum UV radiation (< 190 nm) that is directly absorbed by the DNA, as demonstrated by action spectroscopy in space (Horneck *et al.* 1995). This action spectrum follows the same wavelength dependence as the absorption spectrum of pure DNA. It is important to note that in addition to the well known peak around 260 nm, the DNA absorption shows a second maximum towards shorter wavelengths. In the vacuum UV region, absorption of UV by the sugar phosphate backbone of the DNA becomes decisive. The full spectrum of extraterrestrial UV radiation kills unprotected spores of *B. subtilis* within seconds (Horneck and Brack 1992). These harmful UV ranges do not reach the surface of the Earth because they are effectively absorbed by the Earth's atmosphere. Compared to the average UV radiation climate at the surface of the Earth, the extraterrestrial UV radiation is about three orders of magnitude more efficient in killing spores of *B. subtilis* (Horneck et al. 1996).

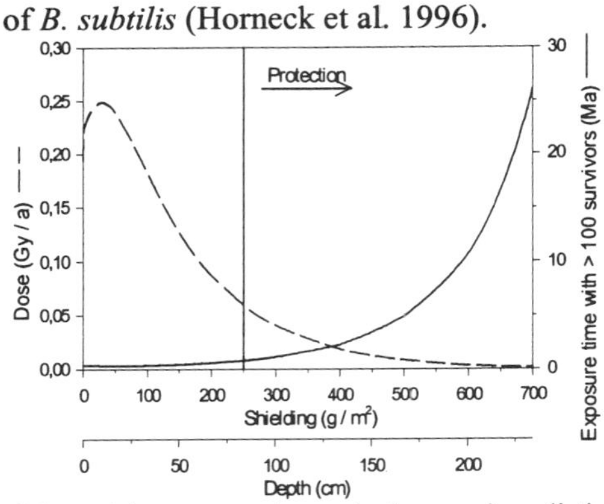

FIGURE 1. Shielding of bacterial spores against galactic cosmic radiation by meteorite material (data from Mileikowsky et al. 2000)

Among the ionizing components of radiation in space the heavy primaries, the so-called HZE particles, are the most biologically effective species (reviewed in Horneck, 1993b; Kiefer et al. 1996). Because these particles interact with the shielding material by creating secondary radiation, with increasing shielding thickness the dose rates go through a maximum. It reaches its maximum behind a shielding layer of about 10 cm, behind about 130 cm the value is the same as obtained without any shielding and only for higher shielding thickness the dose rate reduces significantly (Mileikowsky et al. 2000). The calculations also show that even after 25 million years in space, a substantial fraction of a spore population (10^{-6}) would survive the exposure to cosmic radiation if shielded by 2 to 3 m of meteorite material. The same surviving fraction would be reached after about 500 thousand years without any shielding, and after 350 thousand years after 1 m of shielding (Fig. 1).

4. Entry Process

When captured by a planet with an atmosphere, most meteorites are subjected to very high temperatures during landing. However, because the fall through the atmosphere takes a few seconds only, the outermost layers form a kind of heat shield and the heat does not reach the inner parts of the meteorite. During entry, the fate of the meteorite strongly depends on its size: large meteorites may brake into pieces, however these may be still large enough to remain cool inside until hitting the surface of the planet; medium sized meteorites may obtain a melted crust, whereas the inner part still remains cool; micrometeorites of a few μm in size may tumble through the atmosphere without being heated at all above 100 °C. Therefore, it is quite possible that a substantial number of microbes can survive the landing process on a planet.

5. Conclusions

Although it will be difficult to prove that life was transported through our solar system, experiments in space allow to estimate the chances of resistant microbial forms to survive exposure to the complex matrix of the space environment. This will provide basic data on whether such travel is possible. Experiments in space, which were performed on free platforms since the Apollo era and more sophisticated on the external platform of Spacelab and on free flying satellites, such as LDEF, EURECA, and FOTON, have provided insight into responses of bacterial spores and other extremophile microorganisms to the parameters of space. They support the supposition that space – although it is very hostile to terrestrial life – may not be an absolute barrier for cross-fertilization in the solar system. The various effects on the DNA of the space travelers might even serve evolutionary purposes. Hence, the transport of microbes inside rocks through the solar system might be a feasible process. If protected against solar UV and galactic cosmic radiation, spores may survive inside meteorites over extended periods of time. Several questions remain which are still unanswered, such as: Can spores survive the ejection process? Can spores survive the landing process? What are the most realistic test organisms for Panspermia? What kind of protection, chemical of physical, could be provided by the meteorite material or intrinsically by the organisms?

Relevant studies to tackle these questions will be performed in future studies in space and at ground based facilities (Horneck, 1999; Horneck et al. 2000). For future research on bacterial spores and other microorganisms in space, the European Space Agency ESA is developing the EXPOSE facility that is to be attached to the External Pallet of the truss structure of the International Space Station for 1.5 years during the early utilization period of the Space Station. EXPOSE will support long-term *in situ* studies of microbes in artificial meteorites, as well as of microbial communities from special ecological niches, such as endolithic and endoevaporitic ecosystems (Horneck et al. 1999). These experiments on the Responses of Organisms to the Space Environment (ROSE) include the study of photobiological processes in simulated radiation climates of planets (e.g. early Earth, early and present Mars, and the role of the ozone layer in protecting the biosphere from harmful UV-B radiation), as well as studies of the probabilities and limitations for life to be distributed beyond its planet of origin. Here-to-fore, the results from the EXPOSE experiments will eventually provide clues to a better understanding of the processes regulating the interactions of life with its environment.

6. References

Arrhenius, S. (1903) Die Verbreitung des Lebens im Weltenraum, *Die Umschau* **7**, 481-485.

Horneck, G. (1993a) Responses of *Bacillus subtilis* spores to space environment: results from experiments in space, *Origins Life Evol. Biosphere* **23**, 37-52.

Horneck, G. (1993b) The Biostack concept and its application in space and at accelerators: studies on *Bacillus subtilis* spores, in C.E. Swenberg, G. Horneck, and E.G. Stassinopoulos (eds.) *Biological Effects and Physics of Solar and Galactic Cosmic Radiation,* Part A, Plenum, New York, 99-115.

Horneck, G. (1995) Exobiology, the study of the origin, evolution and distribution of life within the context of cosmic evolution: a review, *Planet. Space Sci.* **43**, 189-217.

Horneck, G. (1999) Astrobiology studies of microbes in simulated interplanetary space, in P. Ehrenfreund, C. Krafft, H. Kochan, and V. Pirronello (eds.) Laboratory *Astrophysics and Space Research*. Kluwer, Dordrecht, pp. 667-685.

Horneck, G. and Brack, A. (1992) Study of the origin, evolution and distribution of life with emphasis on exobiology experiments in Earth orbit, in S.L. Bonting (ed.) *Advances in Space Biology and Medicine*, Vol. 2, JAI Press, Greenwich, CT, pp. 229-262.

Horneck, G., Bücker, H., and Reitz, G. (1994) Long-term survival of bacterial spores in space, *Adv. Space Res.* **14**, (10)41-(10)45.

Horneck, G., Eschweiler, U., Reitz, G., Wehner, J., Willimek, R., and Strauch, K. (1995) Biological responses to space: results of the experiment „Exobiological Unit" of ERA on EURECA I, *Adv. Space Res.* **16**, (8)105-(8)111.

Horneck, G., Reitz, G., Rettberg, P., Schuber, M., Kochan, H., Möhlmann, D., Richter, L., and Seidlitz, H. (2000) A ground-based program for exobiological experiments on the International Space Station, *Planet. Space Sci.* **48**, 507-513.

Horneck, G., Rettberg, P., Rabbow, E., Strauch, W., Seckmeyer, G., Facius, R., Reitz, G., Strauch, K., and Schott, J.U. (1996). Biological dosimetry of solar radiation for different simulated ozone column thickness, *J. Photochem. Photobiol. B: Biol.* **32,** 189-196.

Horneck, G., Stöffler, D., Eschweiler, U., and Hornemann, U. (2001) Bacterial spores survive simulated meteorite impact, *Icarus*, **149**, 285-290.

Horneck, G., Wynn-Williams, D.D., Mancinelli, R., Cadet, J., Munakata, N., Rontó, G., Edwards, H.G.M., Hock, B., Wänke, H., Reitz, G., Dachev, T., Häder, D.P., and Brillouet, C. (1999) Biological experiments on the EXPOSE facility of the International Space Station. *Proc. 2nd Europ. Symp. on Utilisation of the Internat. Space Station*, ESTEC, Noordwijk, ESA SP-433, pp. 459-468.

Kiefer, J., Kost, M., and Schenk-Meuser, K. (1996) Radiation biology, in D. Moore, P. Bie, and H. Oser (eds.) *Biological and Medical Research in Space*. Springer, Berlin, 300-367.

Melosh, H.J. (1984) Impact ejection, spallation and the origin of meteorites, *Icarus,* **59**, 234-260.

Mileikowsky, C., Cucinotta, F.A., Wilson, J.W., Gladman, B., Horneck, G., Lindegren, L., Melosh, J., Rickman, H., Valtonen, M., and Zheng, J.Q. (2000) Natural transfer of viable microbes in space. Part 1. From Mars to Earth and Earth to Mars. *Icarus* **145**, 391-427

Nussinov, M.D. and Lysenko, S.V. (1983) Cosmic vacuum prevents Radiopanspermia, *Origins Life Evol. Biosphere* **13**, 153-164.

Richter, H. (1865) Zur Darwinschen Lehre, *Schmidts Jahrbuch Ges. Med.* **126**, 243-249.

THE NUCLEOTIDE AND NUCLEOTIDE-LIKE COENZYMES IN PRIMITIVE METABOLISM, PHOTOBIOLOGY AND EVOLUTION

M.S. KRITSKY, M.P. KOLESNIKOV, T.A. LYUDNIKOVA,
T.A. TELEGINA, V.A. OTROSHCHENKO, and A.G. MALYGIN
A.N. Bach Institute of Biochemistry, Russian Academy of Sciences
Leninsky Prospekt 33, Moscow, 117071 - Russian Federation
E-mail <inbio@online.ru>, fax (+7-095) 9542732

1. Introduction

The evolutionary link between hypothetical primordial RNA world [1] and the modern R(D)NA/Protein biosphere can hardly be revealed by only searching for chemical roots of primitive genetic coding. The analysis of this problem bears numerous questions regarding the structure and function of the RNA world constituents. One of the questions arisen is how competent are the RNA world's molecules as metabolic catalysts?

After the discovery of catalytic activity in polyribonucleotides, the RNA world protobionts could be seemingly treated as the systems using chemically uniform molecules both to store genetic information and catalyse metabolic reactions. The closer inspection shows that a rather narrow assortment of catalytic activities exhibited by modern ribosymes is mostly related to reactions within the polynucleotide chain itself [2]. The RNA world apparently lacks catalysts for the electron and proton transfer reactions underlying transformation of energy in biosphere. Here we presume a participation in early evolution of molecules capable to store genetic information as well as to (photo)catalyse chemical reactions. These may have been polynucleotides bearing among their bases the non-canonical heterocycles similar to some modern coenzymes.

2. How Compatible Are Coenzymes with the RNA World?

Coenzymes as adenosine-5′-triphosphate or uridine diphosphate sugars are the derivatives of RNA monomers, and most others have much in common with them. Biogenic isoalloxasines and pteridines known as flavin and pterin 'nucleotides' such as riboflavin-5'-phosphate (FMN) and molybdopterin phosphate (MPP) [3] are build after the nucleotide mode: <nitrogen base – carbohydrate (or its derivative) – phosphate>. In a broader sense, to the same category fall folates which glutamate residue(s), not phosphate serves as acidic tail. Nicotinamide adenine dinucleotide (NAD), flavin adenine dinucleotide (FAD) and the MPP-containing dinucleotides mimic the RNA structure, except the 5′, 5′-pyrophosphate, not 3′, 5′-phosphodiester bound, links their monomers.

How closely these coenzyme molecules stand to the RNA world? At the first

J. Chela-Flores et al. (eds.), First Steps in the Origin of Life in the Universe, 237–240.

glance, the set of ribonucleic bases is strictly limited by an AUGC quartet, and the enzymatic transcription is so highly conservative, that enzymatic binding of the coenzyme to RNA is excluded. A more thorough analysis shows, however, that the closeness of some coenzyme to RNA world is sufficient for their molecules were not rejected by the enzymes, catalysing the polynucleotide formation. For instance, NAD and FAD can substitute ATP in the initiation of the template-dependent RNA synthesis catalysed by prokaryotic RNA polymerase [4]. Thus, the binding of coenzymes to polynucleotide chain is not chemically or biochemically forbidden and the molecular chimeras can exist, in which nicotinamide, isoalloxasine or pteridine derivatives are covalently bound to RNA. Moreover, the pterin base structure permits its hydrogen bonding with cytosine partner base in the opposite chain. To a lesser extent, flavin and nicotinamide bases are competent to form hygrogen bonds correspondingly with adenine and cytosine.

3. The Coenzymes Active in Photocatalysis

Photoexcitation dramaticaly increases catalytic activity that flavins and pterins show even in the absence of apoenzyme. Heterocycle plays a key role in this process, but its side groups can modify excitation parameters and the photosensitising activity [5]. The chromophore's interaction with a neighbouring base, for instance, the flavin-to-adenine staking in FAD decreases the excitation efficiency of the molecule, but does not totally deprive it of photochemical activity. The flavins'- and pterins' excited molecules are active in the energy and electron transfer processes. Strongly electronophilic, they subtract electrons from various donors including those with a highly positive E'_O values. The reactions involve free radical mechanism [5] and bring to a formation of active biological reductants: dihydroflavins and the dihydro- and tetrahydropterins [5,6]. The excited reduced coenzymes such as $FADH_2$ and dihydropterins are also active in electron phototransfer [6,7] under anaerobic and aerobic condition. The presence of O_2 sharply decreases a viability of their reduced molecules, and $^1O_2$ generated by the triplet coenzyme changes the mechanism of electron transfer reaction [8]. These reactions, bringing to the free energy accumulation in products, proceed in solution, or, in case of lipofilic flavins, the redox equivalents can be translocated across lipid membrane [5]. Important participants of chemical evolution such as amino acids or the PNA backbone's repeating unit ethylenediamine monoacetate and its precursor ethylene diamine diacetate [9] are electron donors for reactions photocatalysed by coenzymes [5,6,10]. Besides, the capability of the flavins' excitation and relaxation cycle to translocate protons across membrane has been suggested [5].

In organisms apoprotein-bound coenzymes participate in catalytic reactions of electron and group transfer and act as regulators for enzyme proteins. Besides, flavins and pterins are chromophores for protein photoreceptors through which the UV-A and short-wave visible radiation govern development of organisms. Like chlorophyll and in contrast to biological photoregulators such as rhodopsin and phytochrome, the excited coenzyme photosensors have been shown (or in some cases strongly suspected) to photosensitise electron transfer. The coenzyme-binding photosensors are members of several structurally distant protein families. The first group includes DNA photolyases, the enzymes repairing the UV-damaged nucleic molecule, and their structural homologue, cryptochromes, which are photoregulators for plant and, probably, the animal development [9]. Each of these proteins contains two chromophores: in photolyase the first one is $FADH_2$ in the enzyme's active centre, and the second, a light-harvester, is 5,10-N-methenyl-tetrahydrofolate (MTHF), or in some organisms, 7,8-didemethyl-10-hydroxy-5-dezazariboflavin. Similarly, cryptochrome binds FAD and MTHF [8,11]. The second family consists of polypeptides containing the so-called PAS and/or LOV regulator domains capable to bind FMN [12,13]. These polypeptides, which conformation and, correspondingly, the metabolic regulation function is driven by the excited flavin reactions, have likely descended from redox sensing proteins. Like in PAS/LOV-domain containing proteins, the activity of the third group, eukaryotic nitrate reductases, depends on their redox condition, which can be controlled by excited FAD chromophore [14].

4. Coenzymes in Primitive RNA: A Hypothesis

Chemical modelling has demonstrated that isoalloxasines, pteridines and nicotinamides could arise abiotically, and thermal processes of chemical evolution are known to be a source for photochemically active flavins which, bound to proteinoids, can drive photophosphorylation of ADP to ATP [15-18]. The evolutionary age of flavin-binding proteins estimated as 3.5÷4.0 Gy also points to the coenzymes's antiquity [19], and they have been suggested to be primitive organic biocatalysts [20]. The presence in all living kingdoms along with abiotic availability of coenzymes suggested a possible origin of coenzyme photosensors from early (pre-chlorophilic) photosystems, which has shifted in evolution from the solar energy utilisation to cell regulation [10].

Considering the closeness of modern coenzymes to RNA nucleotides we can asume the presence of a vaster set of heterocycle bases in primitive abiogenic polynucleotides than in modern RNA. The coenzyme-related bases such as pteridine, isoalloxasine and nicotinamide present in RNA sufficiently expanded the catalytic spectrum of the RNA world. Their localisation in polynucleotide termini seems more probable, but the non-terminal position of such odd bases should not be excluded, too.

There was a natural selection process that has brought to a limitation of the RNA bases by a modern AUGC quartet. A lesser sterical fitness of flavins and pterins as compared to the canonical RNA bases could be a selection character favouring elimination of these bases from genetic material. Important for negative selection is also a wastefulness of flavins or pterins biosyntheses, which needs several additional enzymatic steps after their precursor guanine nucleotides' have been formed. Moreover, a presence of these coenzymes increased its photodegradation chance for polymer. Both the canonical and coenzyme bases are more or less equal before the deteriorating action of solar radiation with $\lambda \leq 300$ nm. The prevented access of this spectral area due to the ozone shield or a thick water layer absorption has resulted in a decreased degradation risk only for AUGC bases, not for pteridines or, especially, isoalloxasines, which molecules absorb a more long-wave light as well.

It seems probable that rejected from polynucleotides, the coenzyme-related molecules have changed their host polymer to survive in evolution. The selection of primitive genetically ordered polypeptides by their affinity to coenzymes could give rise to the catalytically potent coenzyme-apoprotein complexes and strongly favour the development of the genotypes that has encoded a synthesis of such affine polypeptides.

Supported by Russian Foundation for Basic Research Grant 98-04-48328a.

5. References

1. Gilbert, W.: The RNA World, *Nature (London)*, **319** (1986), 618.
2. Venkstern, T.V.: *The Ribozymes and Small Ribosomal RNAs Genes*, Itogi Nauki I Tehniki, Seria: Molekularnaya Biologia, vol. **27**, VINITI, Moscow (1989), 284 p. (in Russian).
3. Kisker, C., Schindelin, H., and Rees, D.C.: Molybdenum-cofactor-containing enzymes: structure and mechanism, *Annu. Rev. Biochem.*, **66** (1997), 233-267.
4. Malygin, A.G. and Shemyakin, M.F.: Adenosine, NAD and FAD can initiate template dependent RNA synthesis catalyzed by *Escherichia coli* RNA polymerase, *FEBS Letts.*, **102** (1979), 51-54.
5. Presti, D.E.: The photobiology of carotenes and flavins, in D.E. Cosens and D. Vince-Price (eds.), *The Biology of Photoreception,* Soc. Exp. Biol. (1983), pp. 133-180.
6. Kritsky, M.S., Lyudnikova, T.A., Mironov, E.A., and Moskaleva, I.V.: The UV radiation-driven reduction of pterins in aqueous solution, *J. Photochem. Photobiol., **B**: Biol.*, **48** (1997), 43-48.
7. Sancar, A.: Structure and function of DNA photolyase, *Biochemistry*, **33** (1994), 2-9.
8. Krasnovskii, A.A., Jr., Chernysheva, E.K., and Kritskii, M.S.: Investigation of the role of active forms of oxygen in flavin-photosensitized oxidation of NADH, *Biochemistry (Moscow),* **52** (1987), 1273-1282.
9. Nelson, K.E., Levy, M., Miller, S.L.: Peptide nucleic acids rather than RNA may have been the first genetic molecule, *Proc. Natl. Acad. Sci.,* **97** (2000), 3868-3871.
10. Kritsky, M.S., Lyudnikova, T.A., Mironov, E.A., and Neverov, K.V.: The evolutionary roots of coenzyme photobiochemistry, in J. Chela-Flores and F. Raulin (eds.): *Exobiology: Matter, Energy and Information in the Origin and Evolution of Life in the Universe*. Kluwer Acad. Press, Dordrecht, etc. (1998), pp. 187-190.
11. Ahmad, M. and Cashmore, A.R.: Seeing blue: the discovery of cryptochrome, *Plant Mol. Biol.*, **30** (1996), 851-861.
12. Ballario, P. and Macino, G.: White collar proteins: PASsing the light signal in *Neurospora crassa, Trends in Microbiol.,* **5** (1997), 458-462.
13. Christie, J.M., Salomon, M., Nozue, K., Wada, M., and Briggs, W.R.: LOV (light, oxygen, or voltage) domains of the bluelight photoreceptor phototropin (nph1): Binding sites for chromophore flavin mononucleotide, *Proc. Natl. Acad. Sci.,* **96** (1999), 8779-8783.
14. De la Rosa, M.A., Roncel, M., and Navarro, J.A.: Flavin-mediated photoregulation of nitrate reductase. A key point of control in inorganic nitrogen photosynthetic metabolism, *Bioelectrochem. Bioenerg.*, **27** (1989), 355-364.*25.*
15. Ferris, J.F., Kuder, J.E., and Catalano, A.W.: Photochemical reactions and the chemical evolution of purines and nicotinamide derivatives, *Science*, **166** (1969), 765.
16. Heinz, B., Ried, W., and Dose, K.: Thermische Erzeugung von Pteridinen and Flavinen aus Aminosäueregemischen, *Angew. Chem.*, **91** (1979), 510-511.
17. Kolesnikov, M.P. and Kritsky, M.S.: Investigation of chemical structure and the photochemial activity of abiogenic flavin pigment, *J. Evol. Bioch. Physiol.*, in press.
18. Kolesnikov, M.P.: Proteinoid microspheres and the process of prebiological photophosphorylation, *Origins of Life and Evolution of the Biosphere*, **21** (1991), 31-37.
19. Rossman, M.G., Moras, D., and Olsen K.W.: Chemical and biological evolution of a nucleotide-binding protein, *Nature (London),* **250** (1974), 194-199.
20. White, H.B., III: Coenzymes as metabolic fossils, *J. Mol. Evol.*, **7**, () 101-104.

FROM EXTREMOPHILES TO ASTROBIOLOGY

JOSEPH SECKBACH[1] and AHARON OREN[2]
1P.O. Box 1132, Efrat 90435, Israel, and 2Division of Microbial and Molecular Ecology, The Institute of Life Sciences, and The Moshe Shilo Minerva Center for Marine Biogeochemistry, The Hebrew University of Jerusalem, Jerusalem 91904 Israel

1. Introduction to Extremophiles

The 3.5 – 3.8 billion years during which Prokaryotes have lived on Earth have been sufficient time to evolve the characteristics necessary to colonize every habitat compatible with the stability of biomolecules. It has now been well established that diverse communities of microorganisms are able to thrive in very extreme niches on and below the global surface, habitats hostile to most other forms of life. Nearly all the extremophiles are Prokaryotes, small-sized microorganisms that have a very high rate of reproduction and a high degree of physiological flexibility. These characteristics enable their fast adaptation to various extreme habitats. Extremophiles grow at the various frontiers of life: at extreme levels of temperature (thermophiles and psychrophiles), pH (acidophiles and alkaliphiles), pressure (barophiles), dryness (xerophiles), hypersaline environments (halophiles) and high levels of radiation; many can live in the absence of molecular oxygen. These extremophiles provide us with excellent models for the study of biodiversity on Earth, for the formulation of theories on the origin of life, and for speculations on the possible existence of extraterrestrial microbes. More information on the extremophiles may be found in Horikoshi and Grant (1998), Madigan (in Seckbach 2000), Ollivier et al. (in Seckbach 2000), Oren (in Seckbach 2000), Rainey and Ward-Rainey (in Seckbach 2000), Seckbach (1997, 2000) and Seckbach and Oren (in Seckbach 2000). Below are a few examples of microorganisms living on the edge of life.

2. Thermophiles and Acidophiles

Temperature is one of the most important environmental factors governing survival and growth of microorganisms. Every organism has its characteristic cardinal temperatures: minimum, optimum, and maximum. Microorganisms may live at temperatures ranging from subzero (as for example in the Arctic and Antarctic zones) to the boiling point of water and above (in hot springs or in submarine super-heated hydrothermal vents). Microorganisms may exist between –50 and 1130C. Organisms may be classified according to their temperature requirements: psychrophiles grow from -20 to <200C, mesophiles prefer temperatures between 200 and 400C, and

J. Chela-Flores et al. (eds.), First Steps in the Origin of Life in the Universe, 241–245.

thermophiles and hyperthermophiles thrive at temperatures above 450, and up to 1130C. The thermophiles (temperature range between 450 and 750C) include many bacteria and eukaryotic protists. Thermophilic cyanobacteria may thrive at up to 750C, which designates the upper temperature limit for photosynthesis. *Cyanidium caldarium*, a red alga and its cohorts (Seckbach, 1994) and certain fungi live in acidic environments at an upper temperature of 55-600C (Brock, 1978).

Many heterotrophic and chemoautotrophic bacteria even grow at higher temperature levels. Hyperthermophiles, which grow above 750C, are anaerobes and are found in hot environments such as compost piles, hot springs, hydrothermal vents, oil wells, volcanic lakes, etc. The archaeon *Pyrolobus fumarii* is currently the most thermophilic of all known Prokaryotes; it grows up to a temperature of 1130C (Stetter, 1998). This species is able to survive even in an autoclave at 1210C for over one hour (Madigan and Oren, 1999). Hyperthermophilic Prokaryotes have been detected deep within the earth at upper temperatures around 1100C, where geothermal heating supports their lifestyle (Onstott et al., in Seckbach 1999). Hydrothermal vents also provide environments for non-acidophilic species such as *Thermococcus*, which has a temperature range of 56-1030C, or the methanogenic species *Methanopyrus* that lives in hydrothermal vents and can grow over the temperature range of 84-1100C. Bacteria and Archaea may reach areas deeper than 4000 meters below the continental crust. Any greater depth is presumably too hot for life, which is not known to survive temperatures much above 1130C.

Microbial activity of all three domains of life – Bacteria, Archaea, and Eukarya - may occur at both edges of the pH scale, from close to pH 0.7 to above pH 11. Some microorganisms grow near the lower end of the pH scale, growing optimally below pH 3, and even down to pH 0. Thermophilic (growing in the range of 45-700C) acidophilic bacteria (optimum pH around 3) have been isolated from a variety of thermal acidic environments (Darland and Brock, 1971). Several species of hyperthermophilic Archaea grow at low pH values (Stetter, 1998). Among the Eukaryotes, which grow at pH values around zero are genera of fungi such as *Acontium, Cephalosporium*, and *Trichosporon* (Schleper et al. 1995). Some acidothermophilic algae (family Cyanidiaceae) thrive at high temperatures (45-570C) in very acidic solutions (pH 1–4) (Allen 1959; Seckbach, 1994, 1997). *Cyanidium caldarium* was even reported to grow in the laboratory in 1 N H2SO4 (Allen, 1959). The green (halophilic) alga *Dunaliella acidophila* grows at pH values as low as 0.5–3.0 with an optimum at pH 1.0 (Pick, in Seckbach 1999). Accumulation of sulfuric acid has been reported in the vacuoles of brown algae (Sasaki et al., 1999). Recently, acid-tolerant filamentous cyanobacteria have been demonstrated in acid lakes in Bavaria (Germany) at a pH of 2.9 (Steinberg et al., 1998). The archaeon *Thermoplasma acidophilum* grows at ~600C and slightly below pH 0, while the hyperacidophilic Archaea *Picrophilus oshimae* and *Picrophilus torridus* grow down to pH -0.06, which is the lowest pH known to date to support growth (Schleper et al., 1995). The acidophilic cells maintain their intracellular pH near neutral (see e.g. Beardall and Entwisle, 1984), which is achieved by the special properties of the membranes, possessing strong proton pumping activity and selective permeability properties preventing the uncontrolled entrance of protons.

3. Barophiles and Psychrophiles

Barophiles are weight-loving organisms adapted to high pressures, while the psychrophiles are cold-loving organisms. These microbes inhabit deep environments, thousands of meters beneath the surface of the oceans (Deming and Huston, in Seckbach 2000), and occur also in sedimentary rocks or within igneous formations below the ground surface (Onstott et al., in Seckbach 1999). Some are also hyperthemophiles. Such barophilic microorganisms grow at a pressure range from a minimum of >0.1 MPa to maximum values of over 100 MPa (equivalent to 1000 atmospheres), with optimum values between 10-50 MPa. Some barophilic thermophilic microorganisms grow close to the submarine hydrothermal vents in the ocean depths. Similar microorganisms are found very deep in the ground, where they grow at high temperatures.

The oceans have an average temperature of 1 to 50C. Cryohabitats such those that as exist in Antarctica, the Arctic or the permafrost of Siberia have microbes adapted to severe cold. The freezing point determines the lower temperature for microbial growth; when water solidifies, growth is suspended. Extensive microbial growth of a diverse community may occur in pockets of liquid water in permanently frozen areas. In such zones, frozen lake ice and sea ice thaw during the summer period. Several types of eukaryotic algae occur in cold environments, including in snow (Ling, 1996). Some of these algae, including the green algae *Chlamydomonas* and *Chloromonas* and certain dinoflagellates (Hoham and Duval, 2001; Roberts, in Seckbach 1999), color the snow green, orange, or red through their pigmentation. Among the psychrophilic Prokaryotes we may mention *Polaromonas vacuolata*, which grows between 0-12 0C with an optimum at 40C. It has been proposed that some of psychrophilic microbes produce antifreeze proteins, which interfere with the crystallization process of water, thus preventing the formation of ice crystals within their cells.

4. Astrobiology

Microorganisms adapted to various extreme environments may serve as potential analogues for extraterrestrial life. Since extremophiles are able to survive far below the ground surface and deep in the oceans under extremes of temperature and pH, it is possible that life on Earth could have started under such harsh conditions (Davies, 1999). It may even be possible that life still exists at present in some celestial bodies.

Some extremophiles have been "prisoners" in their sealed habitats for a long period of time dormancy (as spores) and can be revived upon a change in their environment. Vreeland et al. (2000) claim to have revived halophiles after a period of 250 million years of dormancy in deep salt mines. The microbes discovered in cores recovered from the depths of Vostok station (Antarctica) over the hidden deep lake (Priscu et al., 1999; Karl et al., 1999) may serve as a good analogue for the possibility of life in the permafrost subsurface of Mars or Jupiter's moon, Europa. This Jovian satellite could currently possess an "ocean" of liquid water beneath its surface (Phillips et al., 2000). There is strong evidence that these two bodies may have had (or still have) liquid water below their surface. Since liquid water is the vital solvent for life and its presence is an absolute requirement for growth, there may be some chance to find some living traces in those heavenly bodies.

The hyperthermophiles and extreme acidophiles occurring in very hot and acidic niches represent life forms that may exist on acidic hot planets such as Venus. Likewise, the extreme halophiles may represent life forms that might be present elsewhere in the Universe where high concentrations of salt prevail. Some microbes adapted to grow anaerobically on various atmospheres, including under pure CO2, (Seckbach, 1994) or ammonia (Siegel, in Seckbach 1999) could survive and grow on celestial bodies containing similar atmospheres. The reviving of dormant microorganisms exposed for a long time to dryness (Davis, 1972) or to harsh habitats such as the lunar surface (Mitchell and Ellis, 1971) or buried in underground salt mines (Vreeland et al., 2000) indicates the ability of microbial life to cope with harsh conditions. These facts suggest that life may perhaps exist in the subsurface of extraterrestrial bodies. The debate about the authenticity of the Martian "nanobes" (Brack and Pillinger, 1998) observed in meteorite AHL84001 recovered from Antarctica (McKay et al., 1996) has greatly stimulated the interest in the possible existence of ancient Martian life. For the time being, however, the question of the existence of extraterrestrial life, at present or in the past, must remain unanswered.

5. References

Allen, M.B. (1959). *Arch. Mikrobiol.* **32**, 270-277.

Beardall, J. and Entwisle, L. (1984) *Phycologia* **23**, 397-399.

Brack, A. and Pillinger, C.T. (1998) *Extremophiles* **2**, 313-319.

Brock, T.D. (1978) *Thermophilic Microorganisms and Life at High Temperatures.* Springer-Verlag, New York.

Darland, G. and Brock, T.D. (1971) *J. Gen. Microbiol.* **67**, 9-15.

Davies, P. (1999) *The Fifth Miracle*, Simon & Schuster, New York.

Davis, J.S. (1972) *The Biologist* **54**: 52-93.

Hoham, R.W. and Duval, B. (2001) In: H.G. Jones, J.W. Pomeroy, D.A. Walker and R.W. Hoham (eds.) *Snow Ecology: An Interdisciplinary Examination of Snow-Covered Ecosystems*, Cambridge University Press, UK, pp. 168-228.

Horikoshi, K. and Grant, W.D. (1998) *Extremophiles, Microbial Life in Extreme Environments*, Wiley-Liss Publishers, New York.

Karl, D.M., Bird, D.F., Björkman, K., Houlihan, T., Shackelford, R. and Tupas, L. (1999) *Science* **286**: 2144-2147.

Ling, H.U. (1996) Hydrobiologia **336**, 99-106.

Madigan, M.T. and Oren, A. (1999) *Curr. Opinion Microbiol.* **2**: 265-269.

McKay, D.S., Gibson, E.K., Thomas-Keprta, K.L., Vali, H., Romanek, C.S., Clemett, S.J., Chillier, X.D.F., Maechling, C.N., and Zare, R.N. (1996) *Science* **273**: 924-930.

Mitchell, F.J. and Ellis, W.L. (1971) In: A.A. Levinson (ed.) *Proceedings of the Second Lunar Science Conference*, vol. **3**. The MIT press, Cambridge, MA. pp. 2721-2733.

Phillips, C.B., McEwen, A.S., Hoppa, G.V., Fagents, S.A., Greeley, R., Klemaszewski, J.E., Pappalardo, R.T., Klaasen, K.P., and Breneman, H.H. (2000) *J. Geophys. Res. Planets* **105**, 22579-22597.

Priscu, J.C., Adams, E.E., Lyons, W.B., Voytek, M.A., Mogk, D.W., Brown, R.L., McKay, C.P., Takacs, C.D., Welch, K.A., Wolf, C.F., Kirshtein, J.D., and Avci, R. (1999) *Science* **286**, 2141-2144.

Sasaki, H., Kataoka, H., Kamiya, M. and Kawai, H. (1999) *J. Phycol.* **35**, 732-739.

Schleper, C., Pühler, G, Kühlmorgen, B. and Zilllig, W. (1995) *Nature* **375**, 741-742.

Seckbach, J. (1994) (ed.) *Evolutionary Pathways and Enigmatic Algae: Cyanidium caldarium (Rhodophyta) and Related Cells.* Kluwer Academic Publishers, Dordrecht.

Seckbach, J. (1997) In: C.B. Cosmovici, S. Bowyer and D. Wertheimer (eds.) *Astronomical and Biochemical Origins and the Search for Life in the Universe.* Editrice Compositori, Bologna, pp. 511-523.

Seckbach, J. (1999) (ed.) *Enigmatic Microorganisms and Life in Extreme Environments.* Kluwer Academic Publishers, Dordrecht.

Seckbach, J. (2000) (ed.) *Journey to Diverse Microbial Worlds: Adaptation to Exotic Environments.* Kluwer Academic Publishers, Dordrecht.

Steinberg, C.E.W., Schäfer and Beisker, W. (1998) *Acta Hydrochim. Hydrobiol.* **26**, 13-19.

Stetter, K.O. (1998) In: K. Horikoshi and W.D. Grant (eds.) *Extremophiles: Microbial Life in Extreme Environments.* Wiley-Liss, Inc., New York, pp. 1-24.

Vreeland, R.H., Rosenzweig, W.D. and Powers, D.W. (2000) *Nature* **407**, 897-900.

ADAPTATIVE RESPONSE OF A FUNGUS SPECIES TO HYDROCARBON ENVIRONMENTS - RESULTS FROM EXPERIMENTS IN LABORATORY

VICENTE MARCANO[1], PEDRO BENITEZ[1] AND ERNESTO PALACIOS-PRÜ[1,2] *
[1] *Electron Microscopy Center, Evolutionary Biology and Chemistry Laboratory, University of the Andes, P. O. Box 163, Mérida, Venezuela.* prupal@ula.ve
[2] *Internacional Institute of Advanced Studies, USB, Caracas, Venezuela.*

1. Introduction

The growth of microorganisms under extreme environmental conditions such as exposure to UVC and UVB radiations, low or high temperatures, absence of oxygen, minimum water supply, and very low pH has been known for many years. This fact motivated the development of studies under *in vitro* conditions similar to those of the early Earth and extraterrestrial environments. Currently, it is well known the capacity of some microorganisms to support extreme conditions in exobiological experiments of spatial simulation (Dose & Klein 1996) and out of the Earth´s atmosphere (Horneck 1993). These microorganisms are even capable to metabolize extraterrestrial organics present in the Murchison meteorite (Mautner et al., 1995), and abiotic 'tholins', say, laboratory materials that provide good analogues to meteoritic and cometary organics (Chyba et al., 1990). However, most of these studies have been done using prokaryotes, viz. *Bacillus, Pseudomonas, Flavobacterium, Methanobacterium.* Most recently, Marcano et al. (1999) demonstrated the capacities of extreme adaptation of an eukaryotic organism that grows in hydrocarbon media, viz. *Fusarium alkanophillum* V. Marcano & Palacios-Prü. We showed here additional data relative to the extreme adaptation of this fungus species.

2. Experimental section

Fungus samples were grown during 3-900 days at all the temperatures between 15-35°C on culture media that included potato dextrose agar PDA, Czapek CZA modified and malt extract-agar MA having light and heavy hydrocarbons, viz. aliphatic, aromatic. Other cultures were assayed in several simulation (stainless steel and Pyrex) chambers having variable concentrations of creosote (0.5-5.0 %), $HgCl_2$ (0.1-2.0 %), H_2O_2 (30%, 0.2-2.0 μl/ml), including UVC and UVB radiation exposures (253.7 and 354 nm lamps, 2200 μW cm^{-2}), low pressure (1 atm), and CO_2 and NH_3 vapours. In all the experiments,

* To whom correspondence should be mailed.

J. Chela-Flores et al. (eds.), First Steps in the Origin of Life in the Universe, 247–250.

the cell morphology was evaluated by EM. Samples were fixed with 3% paraformaldehyde and 3% glutaraldehyde mixed in 0.1 M sodium cacodylate buffer, pH 6.3. After 2 h, tissue plugs were postfixed in 1% osmium tetraoxide. The samples were dehydrated and finally embedded in Epon. Thick sections of resin-embedded specimens were treated with 2% paraphenylenediamine and the thin sections were contrasted with uranyl acetate and lead citrate. Specimens were examined in a Hitachi H-7000 TEM. Aqueous extracts of the fungus from several media were separated on silica gel Merck G plates. After application of the extract and elution standards, the TLC plates were developed with a mixture of butanol:glacial acetic acid:water (100:10:30). The dried plates were then exposed to UV (354 nm) to display the bands which were scraped off. The fractions were eluted with methanol, identified by UV spectra, and additionally were analyzed by FTIR spectroscopy. Protein extracts were obtained with PBS pH 7.0, filtered through a 0.22 μm Millipore membrane and quantified using the modified Lowry assay. The molecular weights of the protein products were estimated by SDS-PAGE, using 20% gel concentration. Aqueous extracts from several cultures having hydrocarbons were analyzed for lectins following standard methods. The extracts were tested against group A, B and O cells. Inhibition tests were performed using equal volumes of lectin and 0.1 M solutions of several carbohydrates. Hydrolysis of the protein extracts was performed in the vapor phase, using 6N HCl/7% thioglycolic acid, under vacuum, at 110°C. After 24 hr hydrolysis, the tubes were dried in vacuum at 40°C over sodium hydroxide pellets for 30 min and then derivatized with dansyl chloride. Separation of dansyl amino acids was carried out on a C18 column (Bio-Sil ODS 5 μm reverse-phase, 4 x 250 mm, Bio-Rad, Richmond, CA) thermostated at 40°C. For the separation by HPLC, a gradient was formed with two mobile phases: solvent A, 30 mM sodium phosphate, pH 6.5, 5% acetonitrile; and solvent B, 30 mM sodium phosphate, pH 6.5, 55% acetonitrile. The flow rate was 1 ml/min. Dansyl amino acids were detected at 254 nm by an LKB 2141 absorbance detector. Quantification of chromatographic peaks and data treatment were done using an EZChrom™ chromatography Data System. Fungal respiration was measured using a Gilson model 5/6 H oxygraph and a "Clarke" type calibrated oxygen electrode. Electrode was polarized at a voltage of 0.8 V and operated at constant temperature (25°C). Fungus samples growing in several hydrocarbon media were removed from petri capsules, pestled and then centrifuged for 10 minutes at 5°C. Each 100 mg cells was suspended in 1.5 ml of Milli-Q water and enclosed in a chamber of 3 ml connected to the oxygen electrode.

3. Results and Discussion

Fusarium alkanophillum showed a mean growth rate in hydrocarbon and agar media of 1.9 mm/hour. *Fusarium alkanophillum* was able to grow by very long term and degrade several hydrocarbons with minimum water (0.25-1 ml/500 ml hydrocarbon) and oxygen (1.26 - 4.24 μM/hour, or < 5 %/hour) requirements, producing between 0.5-3.6 g (dry weight) of biomass. These hydrocarbons included kerosene, gasoil, mineral oil, paraffin and asphalt. The growth of *F. alkanophillum* was inhibited in media containing < C_{10} or

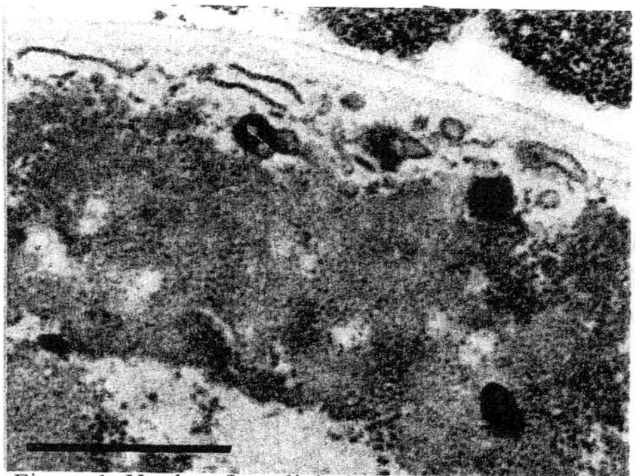

Figure 1. Hyphae from light hydrocarbons showing absence of plasmalemma, nuclear envelope and organelle membranes; scale = 555 nm.

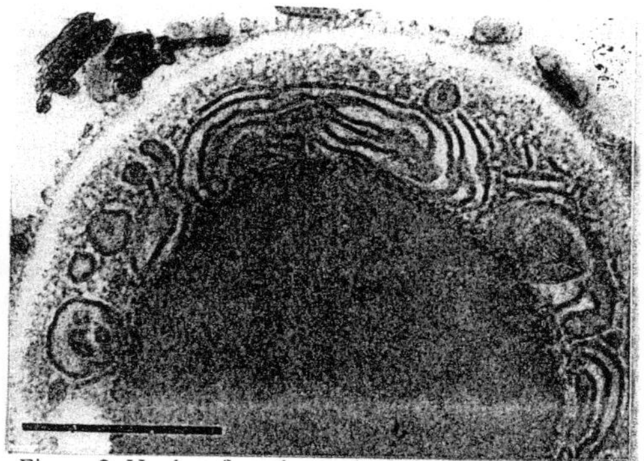

Figure 2. Hyphae from heavy hydrocarbons exhibiting multivesicles and multivacuolar bodies and absence of plasmalemma and nuclear envelopes; scale= 444 nm.

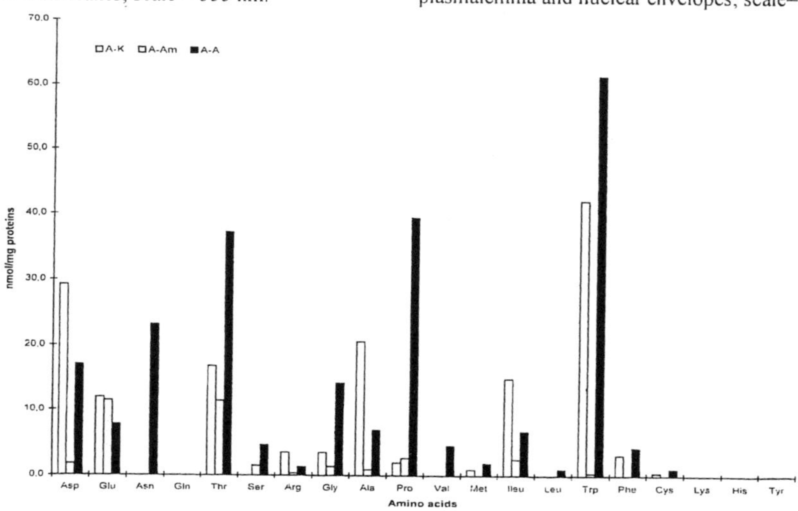

Figure 3. Amino acids present in hydrolyzed aqueous extracts from fungal cells grown in several media; A, agar; K, kerosen; AM, mineral oil .

aromatic hydrocarbons, viz. dioxane, hexane, toluene, benzene and xylene. It was able to grow aerobically in PDA, MA and CZA culture media. *Fusarium alkanophillum* showed no growth on several hydrocarbon media with creosote or $HgCl_2$. No growth was observed in several synthetic media containing urea, ammonium carbonate and potasium ferrocyanide as nitrogen sources in the presence of hydrocarbons. This organism is proteolytic, since albumin serum, glycoprotein and gammaglobulin serum served as a good source of carbon and nitrogen in cultures with hydrocarbons, but in media containing other proteins that lack sulphur linkages no growth was observed, e.g. casein, papaine, trypsin. In cultures with kerosene and PDA or sulphur-containing proteins having H_2O_2 between 0.2-0.6 μl/ml, a notable growth (1.5-2.5 mm/hour) was observed, whereas in higher concentrations, the growth was inhibited. In simulation experiments under NH_3 atmosphere having UVB and UVC radiations and mineral oil in PDA or sulphur-proteins, *F. alkanophillum* showed a retarded growth (≈ 0.5 mm/hour). In other experiments under CO_2 atmosphere having UVB and UVC radiations and mineral oil in PDA or sulphur-proteins, *F. alkanophillum* showed enhanced growth (≈ 2.8 mm/hour). Spore formation was stimulated markedly by exposure to 354 nm UV-radiation. Optimal growth occurs between 15-25°C. Immersion in mineral oil at all temperatures, from –20 to 55°C, during 15 days had no deleterious effects on the subsequent growth of the fungus from germinated spores.

Analysis by TEM of *F. alkanophillum* grown in light hydrocarbon media revealed the absence of plasmalemma, nuclear membranes and other cytomembranes (Figure 1). On the other hand, in heavier hydrocarbon media, vesicles, vacuoles and groups of parallel membranes resembling smooth endoplasmic reticulum were seen (Figure 2). Mitochondria were not observed in both cases. Samples from heavy and light hydrocarbon media revealed disperse and irregularlly disposed nuclear material, associated to free ribosomes. Aqueous extracts of secreted material from several hydrocarbon-cultures separated by TLC and analyzed by FTIR spectroscopy revealed the presence of fatty acids, UV-protective indole pigments, and showed a positive reaction to the Lowry test for protein, including a positive hemagglutinating activity and precipitation of glycoconjugates, suggesting the presence of lectins. HPLC analysis of aqueous extracts of hydrolyzed cells showed high proportions of Trp, Ala, Thr, Glu and Asp, and the occurrence of amino acids having sulphur (Figure 3).

In summary, *F. alkanophillum* showed diverse typical features of an extremophile organism, such as facultative anaerobism, capacity to grow under oxidizing or reducing atmospheres, minimum water requirement, growth and sporulation stimulated by exposure under UVC and UVB radiations or dark, enhanced development in several enriched media in non-aromatic hydrocarbons and sulphur, and varied morphological and physiological adaptive changes. The characteristics of the electron-transport system developed in non-aqueous media are not known. The efficient adaptation of *F. alkanophillum* to grow in hydrocarbon media and the high requirements of sulphur suggest a long evolutionary process that would have occurred in hydrocarbon natural zones, viz. near hot springs and hydrothermal vents, asphalt lakes, ponds.

The absence of lipidic membranes in *F. alkanophillum* constitute an important finding in the cell biology of the terrestrial species that deserve careful evaluation. Moreover, the study of the physiological mechanisms involved in anhydrous conditions must be of interest in the exobiology of planetary bodies (Heinrich, 1972) having hydrocarbon potential niches, viz. Titan (Clarkes & Ferris, 1997) because it offers a different vision related with the search of life in the universe.

4. References

Chyba, C. F., Thomas, P. J., Brookshaw, L. and Sagan, C.: 1990, *Science* **249**, 336.

Clarke, D. W. and Ferris, J. P.: 1997, *Origins of life and Evol.Biosphere* **27**: 225.

Dose, K. and Klein, A.: 1996, *Origins of life and Evol.Biosphere* **26**: 47.

Heinrich, M. R.: 1972, in (eds.) D. L. Rohlfing and A. I Oparin, *Molecular Evolution, Prebiological and Biological*, Plenum Press, New York, pp. 331-339.

Horneck G. 1993. *Origins of life and Evol.Biosphere* **23**: 37.

Marcano, V, Benitez, P. and Palacios-Prü E.: 1999, in *Abstracts of the 12th International Conference on the Origin of Life, ISSOL'99*, held in San Diego (USA), p. 112.

Mautner, M. N., Leonard, R. L. and Deamer, D. W.: 1995, *Planetary and Space Science* **43**, 139.

ORIGINS OF LIFE IN THE UNIVERSE AND EARLIEST PROKARYOTIC MICROORGANISMS ON EARTH

V.C.TEWARI

Wadia Institute of Himalayan Geology Dehra Dun 248001, India and
The Abdus Salam InternationalCentre for Theoretical Physics34136,Trieste,Italy

Abstract : The discovery of fossilised life forms or nannobacteria from Martian meteorite ALH 84001 supports the hypothesis that the early evolution of life may have taken place on Mars and earliest prokaryotic microorganisms and first stromatolites similar to the planet Earth around 3.5 billion years might have proliferated on Mars. The eukaryotic algae from the Deoban and the Krol Group of the Himalaya, India has been recorded. There is evidence of vegetative reproduction during Neoproterozoic. Akinites have been reported from Deoban cherts indicative of cell division and the formation of thick cellular wall. Akinite germination have been observed in which they underwent cell divisions to produce a cellular hormogone in Deoban cells.

1. Introduction

The origins of life on Earth is explained by chemical evolution hypothesis by pioneer researchers (Oparin, 1924;Miller,1953). It is quite important that Murchison meteorite contains the same amino acids as Miller (1953) obtained in his famous experiment. Carbon isotope ratios of the oldest sediments from the Isua Supracrustal rocks indicate indirect evidence for the existence of life on Earth at 3800 Ma. The extraterrestrial theory suggest that bacteria exists in outer space and the microbial life is transported to the Earth by meteorites, asteroids and comets. McKay *et al.*(1996) reported nannobacteria (20-200nm) and large bacteria from Martian meteorite ALH-84001 from Antarctica. In the present paper, the earliest occurrences of prokaryotes (cyanobacteria and stromatolites) on Earth and early evolution of life in the Lesser Himalaya, India has been discussed.

2. Earliest Prokaryotic Microorganisms on Earth

Schopf (1993) recorded well preserved filamentous cynobacteria from the Apex Formation in the Warrawoona Group. The important taxa are ***Primaevifilum laticellulosum*, *Archaeoscillatoriopsis grandis***. Silicified stromatolite has also been recorded from Warrawoona Group. Tewari (1998) has compared some Archaean cyno-bacteria with those of Proterozoic genus ***Oscillatoriopsis media*** described from the Deoban Group of the Lesser Himalaya, India (Tewari,1989). Raha *et al.*(2000) have described possible Archaean microfossils in the stromatolitic Banded Iron Formation

J. Chela-Flores et al. (eds.), First Steps in the Origin of Life in the Universe, 251–254.

identified as ***Eobacterium bailadilensis*** (Figure 1e reproduced by author's permission)

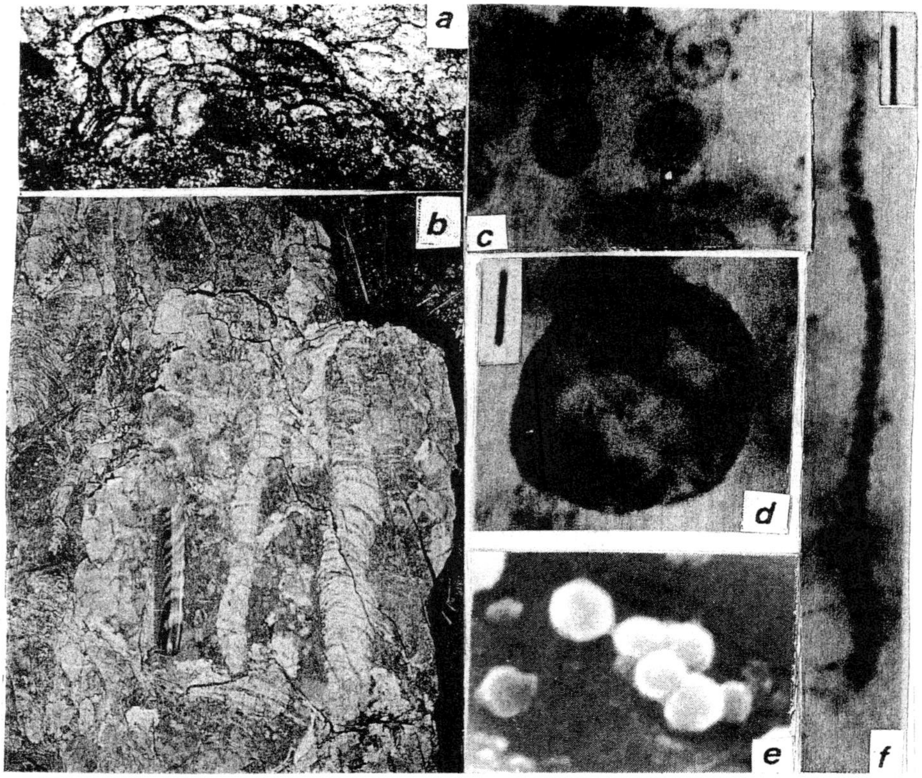

Figure 1.

a. Microstromatolites (1mm) in the Menga Limestone, Arunachal Pradesh.
b. Neoproterozoic microbial buildups of large columnar and branching stromatolites. 1 m high. From the Upper Krol Formation, Mussoorie Syncline.
c. Coccoid microfossils containing eccentrically placed possible nuclei in Deoban cells. (8-12 μm in diameter).
d. ***Kildinosphaera,*** planktonic eukaryotic algae (up to 260 μm in diameter, Deoban Cells.
e. Spheroidal microfossils less than 1 μm showing binary reproduction by fission in Archaean Banded Iron Formation, India. (Photo P.K. Raha.)
f. ***Gunflinta minuta*** (350 μm), Deoban chert.

3. Early Evolution and Diversification of Life in the Lesser Himalaya India

The Deoban-Gangolihat carbonate belt of the Lesser Himalaya show good development of Meso to Neopro-terozoic stromatolite assemblage: (Tewari, 1989;1993). Recently, ***Microstromatolites*** and filamentous bacteria have been discovered from the Menga Limestone of the Arunachal Pradesh in the Eastern Himalaya (Figure 1a). Well preserved organic walled microfossils (OWM's) have been recorded from the Deoban cherts and stromatolites (Shukla *et al.,*1986;Tewari,1989). Isolated hexactinellid and monaxon sponge spicules with cyanobacterial filaments have been more recently discovered from the Gangolihat Dolomite (Tewari *et al.*, 2000). The prokaryotic and eukaryotic biota include ***Eomycetopsis robusta, Oscillatoriopsis media, Gunflintia minuta*** (Figure 1f***), Myxococcoides minor, Sphaerophycus parvam, Archaeotrichion sp, Biocatenoides*** and ***Kildinosphaera sp.*** (Figure1c,d). These coccoid microbiota contain black to amber coloured eccentrically placed possible fossilised nuclei or pyrenoid like bodies /cell organelles commonly found in eukaryotic algae (Figure 1c). ***Kildinosphaera*** (Figure1d) is a eukaryotic algae representing reproductive cyst forming stage. There is also evidence of vegetative reproduction in Deoban cells. ***Akinites*** present in the Deoban cherts are indicative of cell division and the formation of thick cellular wall (Figure1a) .***Akinite*** germination underwent cell divisions to produce a cellular hormogone in cells.The eukaryotic phytoplankton radiation was initiated about 1100-1000 Ma.ago and the lower part of the Deoban-Gangolihat belt indicate the same age. Radiation of new acantho-morphic acritarchs, stromatolites (Figure1b), multicellular brown algae ***Vendotaenids, Tyrosotaenia, Krolotaenia*** and Ediacaran metazoans like ***Cyclomedusa, Charniodiscus, Dickinsonia, Spriggina Pteridinium*** etc.was recorded from Australia, Krol Group of the Lesser Himalaya India, China, Europe, Africa, Siberia and Canada (Narbonne and Hoffman,1987; Tewari, 1993; 1999).

4. Discussion and Conclusion

Prokaryotic life (cyanobacteria and stromatolites) are the oldest known record of life on planet Earth. It is possible that the early stages of the origin of life around 3.8-4 Ga.and early evolutionary processes like genetic aspect of eukaryogenesis might have preserved on Mars (Chela Flores, 1998). There is a strong possibility of preservation of micro-stromatolites, stromatolites, prokaryotic to eukaryotic algae and the Ediacaran life on Mars (Tewari, 1998; 1999). Comets may have been a potential source of organic compounds on the early Earth (Oro *et al.*, 1971). Jupiter's satellite Europa has been identified as a potential site for the extra terrestrial search for life in the solar system. The prokaryotic to euraryotic evolution of life on Earth, palaeoclimatic events of Neoproterozoic snow ball Earth and the extinction and further emergence of highly organised life after glaciation can be used as a possible model for the search of extraterrestrial life in the solar system and an important tool for future astrobioligical research.

5. Acknowledgements

The author is grateful to Professor Julian Chela Flores, Abdus Salam Internatioal Centre for Theoretical Physics,Trieste, Italy for encouragement, discussions and reviewing the manuscript. He is thankful to Ms. Rosalba Bonaccorsi, Dipar-timento di Scienze Geologiche, Ambientale Marine, University of Trieste for critically going through the paper and valueable suggestions. Dr. B.R.Kerur is thanked for going through the earlier draft of the paper. The Directors of the Wadia Institute of Himalayan Geology Dehra Dun, India and Abdus Salam I.C.T.P., Trieste, Italy are thanked for providing facilities.

6. References

Barghoorn, E.S.and Schopf, J.W.(1966) Microorganisms three billion years old from Precambrian of South Africa, *Science,* **152,** 758-763

Chela Flores, J.(1998) A search for extraterrestrial eukaryotes; physical and palaeontological aspects. *Origins of Life and Evolution of the Biosphere* **28,** 583-585

Mc Kay, C.P.(1997) The search for life on Mars.*Origins of Life and Evolution of the Biosphere,27,* 263-289.

Mc Kay , D.S., Gibson, E.K. *et al.*(1996) Search for past life on Mars : Possible relic biogenic Activity in Martian Meteorite ALH 84001,*Science* **273**, 924-930.

Miller, S.L. (1953) A production of amino acids under possible primitive earth conditions, *Science* **117**, 528-529.

Narbonne, Guy M. and Hoffman , H.J. (1987) Ediacaran biota of the Werenecke Mountains, Yukon, Canada,*Palaeontology* **30 (4),** 647-676.

Oparin, A.I. (1924) *Proiskhozhdenite Zhizni (Origin of Life).*Moscow,Izd. Moscovskii Rabochii (in Russian).

Oro, J., Gilbert, J., Lichtenstein, H., Wikstrom , S. and Flory., D.A.(1971) Amino acids, aliphatic and aromatic hydrocarbons in the Murchison meteorite, *Nature* **230,** 105-106.

Raha, P.K., *et al.* (2000) Possible microfossils from the Archaean Banded Iron Formation (Bailadila Group), Madhya Pradesh, India,*Journal Geological Society of India* **55**, 663-673.

Schopf, J.W. (1983) *Earth's Earliest Biosphere.*Princeton University Press,New Jersey.

Schopf, J.W. (1993) Microfossils of the Early Archaean Apex Chert : New evidence of the Antiquity of life,*Science* **260**, 640-646.

Shukla, M.,Tewari, V.C. and Yadav,V.K. (1986) Late Precambrian microfossils from the Deoban Limestone Formation, Lesser Himalaya,*Palaeobotanist,* Himalaya,*Palaeobotanist,* **35 (3),** 347-356.

Tewari,V.C. (1989) Upper Proterozoic-Lower Cambrian stromatolites and Indian Stratigraphy *Himalayan Geology* **13,** 119-146.

Tewari,V.C. (1993) Precambrian and Lower Cambrian stromatolites of the Lesser Himalaya, India *Geophytology Spl. Pub.* **23 (1),** 19-39

Tewrai,V.C. (1998) Earliest microbes on Earth and possible occurrence of stromatolites on Mars.in: J. Chela Flores and R. Raulin (eds.) *Exobiology: Matter,Energy and Information in the Origin and Evolution of Life in the Universe,* Kluwer Academic Publishers, Netherlands.pp.261-265.

Tewari,V.C. (1999) *Vendotaenids* :Earliest megascopic multicellular algae on Earth, *Geoscience Journal,* **20 (1),** 77-85.

Tewari, Meera,Pant, C.C.,Tewari, V.C. (2000) Neoproterozoic sponge spicules and organic walled microfossils from the Gangolihat Dolomite, Lesser Himalaya, India, *Current* Science **79 (5),** 651-654.

PERSISTENCE OF LIVING PLANKTONIC FORAMINIFERA (*Neogloboquadrina pachyderma*) IN ANTARCTIC SEA-ICE INFERRED FROM A STUDY OF A SEDIMENT CORE (ROSS SEA CONTINENTAL MARGIN)

R. BONACCORSI AND R. MELIS

Dip. Scienze Geologiche, Ambientali e Marine (DiSGAM)
University of Trieste, via E. Weiss 2, 34127 Trieste, Italy
e-mail: bonaccor@univ.trieste.it

Abstract: This work attempts to constrain and discuss the persistence of living planktonic foraminifera *Neogloboquadrina pachyderma* (Ehrenberg, 1861) in Antarctic sea-ice (overwintering strategy) during the Last Glacial Maximum (LGM). Samples taken from Level 12 (216.5-238 cm-depth) of a gravity sediment core ANTA95-89C (74° 29.100'S – 175° 34.059' W; 2058 m-depth) have radiocarbon dates of 28.2-17.2 yr BP. During the LGM, the Ross Sea Ice Shelf and associated sea ice extended far seaward of the position of this core. From 15 up to 60-80% of *N. pachyderma* were found in a three-layer interval referred as Level 12 and associated with up to 10.8-65.7 % of sand to gravel-sized Iceberg Rafted Debris (IBRD). This suggests that foraminifera lived in wide areas of Ice Shelf/sea-ice during the LGM. Results of such palaeoclimate/palaeoenvironmental investigation are integrated into an Astrobiology Roadmap science framework.

1. Introduction

One of the general principles of the astrobiology, a relatively "new field" of science related to the genesis, evolution, and persistence of life in a broad variety of "extreme" environments, is to combine classical fields of the science (e.g., Life, Earth and Planetary sciences) in a non-traditional way (Principle 1).

This work aims to i) constrain the persistence of living foraminifera in the past sea-ice by using a combination of multi-component analyses, micropalaeontological observation, and AMS ^{14}C dating on a sediment core, and ii) integrate different approaches within an astrobiological framework.

The Antarctic Ice Shelf and its seasonal sea-ice are examples of "extreme" marine environments, and the investigation of life-forms capable of exploiting such settings might provide insight into the possibilities and limits of life in other worlds' ecosystems (e.g., Irion, 2000), as outlined in the science scheme (i.e., Goal 8) of the NASA Astrobiology Roadmap (Morrison and Schmidt, 1999). High latitude environments (e.g., Ice sheets, subglacial lakes, sea-ice microhabitats, soils from the Dry valleys and other ultra-cold deserts) represent important analogues for planetary bodies where ice/water interfaces are expected (i.e., Jupiter's moons Europa/?Callisto),

J. Chela-Flores et al. (eds.), First Steps in the Origin of Life in the Universe, 255–260.

or where liquid/frozen water might either presently occur (e.g., Martian permafrost and ?groundwater) or have occurred (Mars's ancient Ocean).

The occurrence in Antarctic sea-ice of living specimens of (left-coiling) *Neogloboquadrina pachyderma* (Ehrenberg, 1861) (cold-water form) was emphasised by several authors (Lipps and Krebbs, 1974, Spindler and Dieckmann, 1986, Spindler *et al.*, 1990, Dieckmann *et al.*, 1991, Palmisano and Garrison, 1993). This species is one example of marine life in extreme settings (e.g., brine channels, pockets, and pores of granular-frazil ice), which can be characterised by hyper-saline water (up to 83-150‰) and temperatures down to –9.6°C (Spindler and Dieckmann, 1986). Hence, investigation on past, Recent and present sedimentary records to constrain limits of life in extreme environments (Objective 7 of Goal 5) satisfy at least one fundamental question ("Does life exist elsewhere in the Universe?"), two goals (i.e., Goals 5 and 8) and one objective, as listed in the Astrobiology Roadmap.

Background. Accumulation of polar biogenic carbonates in deep-sea sediments is affected by biological (such as primary productivity) and chemical-physical factors of the water column, and/or dilution from terrigenous inputs (Berger, 1970, Rao, 1996). Glacial conditions are characterized by lowering of the Carbonate Compensation Depth (CCD) as boundary for $CaCO_3$ preservation, while during interglacials Antarctic deep sediments, normally remaining below the CCD, and lack carbonates (Grobe and Mackensen, 1992). In the Ross Sea, levels of *N. pachyderma* were recognised in two deep-sea cores (Barbieri *et al.*, 1999, Quaia and Cespuglio, 2000) including the LGM (21-18 kyr BP, Martinson *et al.*, 1987). Biogenic "cold" carbonates were also reported in Recent deep-sea sediments (Taviani *et al.*, 1993), and near-shore Pleistocene stratigraphic units from the Cape Roberts Project (CRP-1) (Webb and Strong, 1998).

ANTA95-89C, a gravity core retrieved from the Ross Sea continental margin (74 29.100'S-175 34.059'W; 2058 m-depth (Fig. 1) consists of Unit A (0-238 cm depth) and Unit B (238-404 cm depth) and laminated alternating to coarse-grained lithofacies, for a total of 32 levels (Fig. 1d). Among these, we will focus on Level 12 (216.5-238 cm-depth, Fig. 2), a glacialmarine diamicton, almost entirely dominated by *N. pachyderma* and benthic species (Bonaccorsi, in prep), which was interpreted as related with dramatic pulses in Ross Ice Shelf extent (Bonaccorsi *et al.*, 2000).

2. Materials and methods

ANTA95-89C was collected in the framework of the Programma Nazionale di Ricerche in Antartide (PNRA). Comparisons between ultra high-resolution logs of physical (e.g., Wet Bulk Density, and Magnetic Susceptibility) and whole-core X-ray photographs allowed the recognition of lithological boundaries within levels (Bonaccorsi *et al.*, 2000) and the high resolution sampling of Level 12. Sedimentological methods (e.g., TOC by Perkin Elmer 2400 CHN, and water content determinations) are described elsewhere (Bonaccorsi, in prep.). Gravel (> 2mm), sand (2000-62.5μm) and mud (<62.5μm) fractions were determined on oven-dried (T < 50°C) samples. The percent abundance of foraminifera was evaluated semi-quantitatively and a Scanning Electron Microscope (SEM) was used to assess their degree of preservation (Figs. 3a-c). Three accelerator mass spectrometer (AMS) ^{14}C ages were made on well preserved >150μm-size tests for Levels 12a and 12b at Woods Hole/NOSAMS facility

3. Results and discussion

The X-ray three-layered anatomy of Level 12, and sublevels of this 22 cm-thick glaciomarine diamicton are shown in Fig. 2. Changes in physical properties (Bonaccorsi *et al.*, 2000) mark both the lithology and composition of Levels 12c, 12b, and 12a, and its chronostratigraphy (Tab. 1 and Tab. 2).

Level	Depth (cm)	Lab. Code	AMS-^{14}C age (yr BP, uncorrected)
12a	216.5-218	0S-24143	17,200 ± 190
12c	229-230	0S-27146	24,300 ± 130
12c	237-238	0S-24144	28,200 ± 380

Tab. 1 – AMS (uncorrected) ^{14}C ages in Levels *12a* and *12c*.

Level	Depth (cm)	*N. pachy* %	Gravel >2 mm	Sand (2000-62.5μm)	Clay (<62.5μm)
12a	216.5-223	25-35	11.3 - 42.7	7.21 - 30.7	39.4 - 69.3
			25.9 ± 13.4	21.3 ± 9.3	52.4 ± 13.1
12b	223-229	0-15	2.5 - 23.2	29.9 - 39.9	46.9 - 58.3
			9.3 ± 8.3	37.4 ± 4.2	53.4 ± 4.3
12c	229-238	60-75	0.0 - 4.0	10.8 - 65.7	32.3 - 89.2
			2.2 ± 1.5	50.5 ± 22.5	47.3 ± 23.7

Tab. 2 – Percentages of *N. pachyderma* and summary statistic for gravel and total sand content (wt %).

In Level 12, the up core decreasing sand-size fraction is associated with an increasing abundance of gravel (Fig. 2). Water content and TOC trends covary in phase with the gravel fraction and underline the textural diversity among Levels 12a, 12b, and 12c, which are also characterized by changes in the abundance of *N. pachyderma* (Tab. 2). The recovered planktonic assemblage is likely not reworked since a) no turbidity flows are evident from different depositional ^{14}C ages, physical logs, and lithology (fine sands at bottom, gravel and pebbles at top; Fig. 2); b) large adults and tiny juveniles are present (figs 3a-c); and c) they are generally well-preserved (not oxidized or fragmented), likely because of good preservation during glacials as also suggested by Quaia and Cespuglio (2000) for similar levels from two deep-sea cores. A direct source from sea-ice can be inferred by the occurrence of large encrusted adult forms and not-encrusted (juveniles) (figs. 3a-c). The latter ("tiny form with four to four and one half chambers") were found both alive in sea-ice (Spindler and Dieckmann, 1986) and near-shore older units (Webb and Strong, 1998). The abundance of *N. pachyderma* (from 15 up to 75-80% in Levels 12a, 12b, and 12c) was associated to inputs of IBRD (Figs. 2), which are proxies for climate-induced calving and melting of ice masses (Kellogg and Kellogg, 1988, Anderson *et al.*, 1991) nearby the ice marginal zone. In core 89C biogenic carbonates layers were deposited between 28.2 and 17.2 kyr BP. They can be interpreted as three different pulses of Ross Ice cover over a period of time bracketing the LGM, and followed by a rapid retreat of the Ross Ice Shelf to a southernmost position. Similar results from two deep-sea cores (Brambati *et al.*, in press, Quaia and Cespuglio, 2000) suggest that sea-ice, and other types of floating ice masses (Anderson *et al.*, 1991), were synchronously persistent along the western and central sides of the Antarctic margin (latitudes 70° - 74° S). *N. pachyderma*-rich levels in ANTA95-89C might be consistent with the abundance of living juvenile and adult specimens in the annual sea-ice (Spindler and Dieckmann, 1986, Spindler *et al.*, 1990). In fact, during the Antarctic winter, this planktonic species can be incorporated by the fast-forming sea-ice in brine channels and ice pores where a sustainable food web allows the individuals to survive and grow (Dieckmann *et al.*, 1991). Then, when spring ice melting occurs, individuals return to the water column and can exploit newly available food resources. At the end of their life-cycle their tests may become part of sedimentary records.

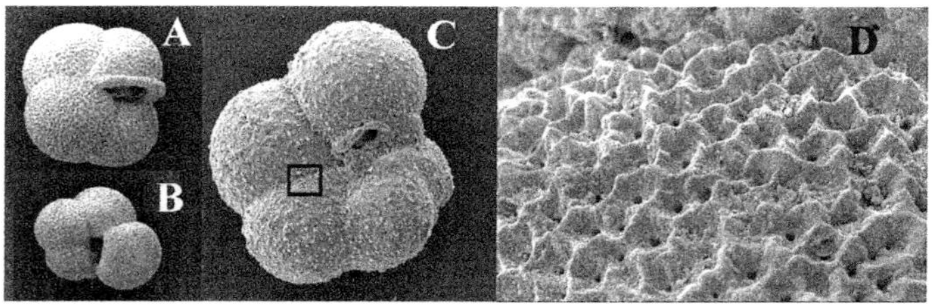

Fig. 3 - SEM photographs of Level 12a (216.5-218 cm) *N. pachyderma* (left-coiling), Umbilical view. Adult (A, 70×); non-encrusted juvenile specimen (sea-ice forms) (B, 70× and C, 125×) and test's features from specimen C (D, 380 ×).

4. Conclusions

N. pachyderma-rich layers were extracted from core ANTA95-89C and yielded radiocarbon dates of 28.2-17.2 yr BP. The abundance, accumulation and preservation of this species in deep-sea glaciomarine sediments was likely allowed by relationships with sea ice settings, and pre-post depositional mechanisms.

When recognized in recent and older sediments, the abundance of *N. pachyderma* and its morphological variants can be both a proxy for deposition under prevailing of annual sea-ice conditions and preservation of "cold" carbonates during glacials. Also, the non-encrusted form, if typical of live specimens testifies to an overwintering strategy that might have been conservative over the time.

Constraining paleo-environmental models, according to the present-day natural history of such extremophile organisms from the sedimentary record, fulfils the general principles, goals, and objectives of the Astrobiology Roadmap.

5. Acknowledgements

This work is funded by the Italian Programma Nazionale di Ricerche in Antartide (PNRA) and a grant/cooperative agreement from the National Oceanic and Atmospheric Administration. The authors are grateful to A. Brambati (DiSGAM), who sponsored this work, R.F. Anderson for supporting the AMS ^{14}C, and J. Linch-Stieglitz for the hospitality in her lab at Lamont-Doherty Earth Observatory (L-DEO). Also, we wish to thank I. Premoli Silva, M.B. Cita (University of Milan), and N. Pugliese (DiSGAM) for valuable suggestions, A. Piotrowski (L-DEO) and J.A. Al-Saadi (NASA Langley Research Center) for their helpful proofread and critical review of this manuscript; P. Fanzutti and T. Quaia (DiSGAM) for useful discussion, and T. Ubaldini for assistance with SEM. The views expressed herein are those of the authors and do not necessarily reflect the views of NOAA or any of its sub-agencies.

6. References

Anderson, J.B., Kennedy, D.S., Smith, M.J, and Domack, E.W. (1991) Sedimentary facies associated with Antarctica's floating ice masses, in J.B Anderson and G.M Ashley, (eds.), *Glacial Marine Sedimentation; Paleoclimatic significance*, Boulder, Colorado, GSA Special Paper 261.

Berger, W.H. (1970) Planktonic foraminifera: selective dissolution and the lysocline. Mar. Geol. 8, 11-138.

Barbieri, R., D'Onofrio, S., Melis, R., and Westall, F. (1999) r-Selected benthic foraminifera with associated bacterial colonies in Upper Pleistocene sediments of the Ross Sea (Antarctica): implication for calcium carbonate preservation, Palaeogeogr., Palaeoclim., Palaeoecol. 149, 41-57.

Bonaccorsi, R., Brambati, A., Busetti, M., and Fanzutti, G.P. (2000) Relationship Among X-Ray Lithofacies, Magnetic Susceptibility, P-Wave Velocity and Bulk Density in Core ANTA95-89C (Ross Sea, Antarctica): First Results, *Terra Antarctica Reports* **4**, 185-198.

Bonaccorsi, R. (in prep). Evolution of the Ice Sheet-Ice Shelf-Continental Margin System of the Ross Sea Area (Antarctica) from late Quaternary sediment records, *Ph.D. thesis*, University of Trieste.

Brambati A., Melis R., Quaia T. and Salvi, G. (in press) Late Quaternary climatic changes in the Ross Sea area (Antarctica), in J.A. Gamble, D.N.B. Skinner, S. Henrys and R. Lynch (eds.), *Antarctic Earth Sciences at the Close of a Millenium*, Royal Society of New Zealand, Bulletin 35

Dieckmann, G.S., Spindler, M., Lange, M.A., Ackley, S.F., and Eicken, H. (1991) Antarctic sea ice: a habitat for the foraminifer *Neogloboquadrina pachyderma*, *J. For. Res.* **21**(2), 182-189.

Garrison, D.L. (1991) Antarctic sea-ice biota, *Am. Zool.* **31**, 17-33.

Grobe, H., and Makensen A. (1992) Late Quaternary climatic cycles as recorded in sediments from the Antarctic continental margins, *Ant. Res. Ser.* **56**, 349-376.

Irion, R. (2000) The Science of Astrobiology takes shapes, *Science* **288**, 603-605.

Kellogg, T.B., and Kellogg, D.E. (1988) Antarctic cryogenic sediments: Biotic and inorganic facies of ice Shelf and marine-based ice sheet environments, *Palaeogeogr., Palaeoclimatol., Palaeoecol.* **67**, 51-74.

Lipps, J.H., and Krebbs, W.N. (1974) Planktonic Foraminifera associated with Antarctic sea ice, *J. For. Res.* **4**, 80-85.

Martinson, D.G., Pisias, N.G., Hays, J.D., Imbrie, J., Moore, T.C. jr., and Shackleton, N.J. (1987) Age dating and the orbital theory of the Ice Ages: development of a high-resolution 0 to 300,000-year chronostratigraphy, *Quat. Res.* **27**, 1-29.

Morrison, D., and Schmidt, G.K. (1999) *NASA Astrobiology Roadmap*, D. Morrison and G.K. Schmidt (eds.), Ames Research Center, Moffet Field, California, 44 pp.

Palmisano, A.C., and Garrison, D.L, (1993) Microorganisms in Antarctic sea ice, in E.I. Friedmann, (ed.), *Antarctic Microbiology*, Wiley-Liss Inc., N.Y, pp. 167-218.

Rao, C.P. (1996). Modern Carbonates, tropical, temperate, polar. *Arts of Tasmania*, Hobart, Tasmania, 206 pp.

Quaia, T., and Cespuglio, G. (2000). Stable Isotopes Records from the Western Ross Sea Continental Slope (Antarctica). Considerations on Carbonate Preservation, *Terra Antartica Reports* **4**, 199-210.

Spindler, M., and Dieckmann, G.S. (1986) Distribution and abundance of the planktic foraminifer *N. pachyderma* in the sea ice of the Weddell Sea (Antarctica), *Polar Biology*, **5**, 185-191.

Spindler, M., Dieckmann, G.S., and Lange, M.A. (1990) Seasonal and geographic variations in sea ice community structure of the Weddell Sea, In K.R. Kerry and G. Hempel (eds.), *Antarctic Ecosystem, Ecological Change and Conservation*. Spinger-Verlag, Berlin, 130-135.

Taviani, M., Reid, D.E., and Anderson, J.B. (1993) Skeletal and isotopic composition and paleoclimatic significance of late Pleistocene carbonates, Ross Sea, Antarctica, *J. Sedim. Petrol.* **63**, 84-90.

Webb, P.N, and Strong, C.P. (1998) Occurrence, Stratigraphy distribution and Palaeoecology of Quaternary Foraminifera from CRP-1, *Terra Antarctica* **5**(2), 455-472.

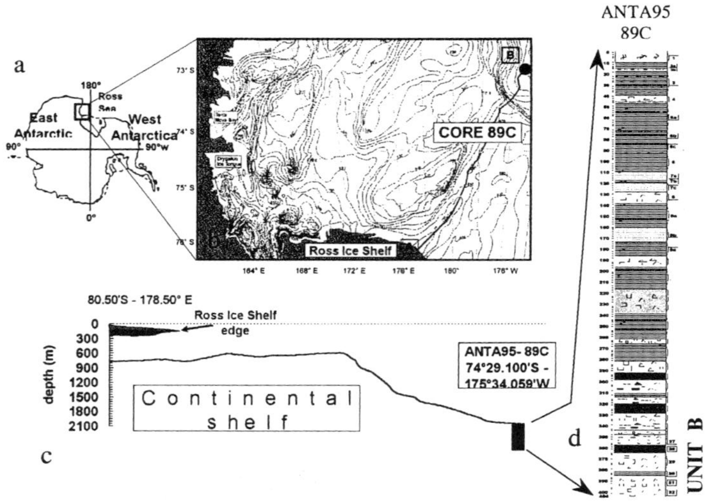

Fig. 1 - a) Study Area (central Ross Sea); b) bathymetry and core location; c) profile (A-B) along the Glomar Challenger basin and off the adjacent slope; and d) X-ray core stratigraphy: (1) diamictons and (2) laminated levels (modified after Bonaccorsi *et al.*, 2000).

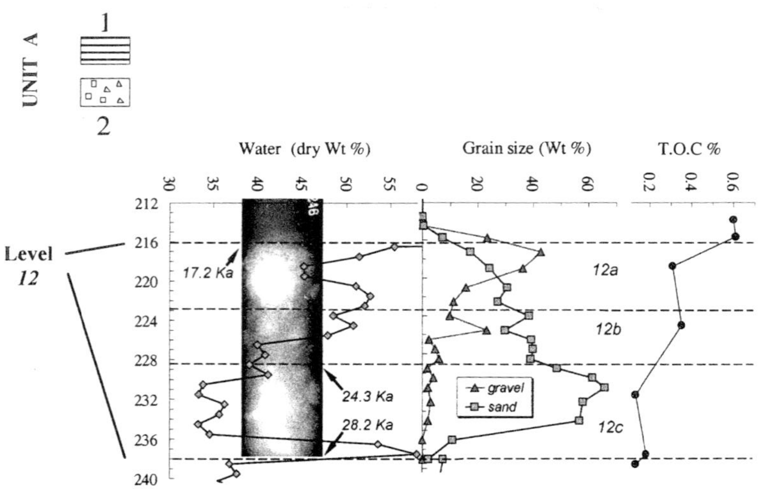

Fig. 2 - Core ANTA95-89C. X-ray anatomy and sedimentary features (water, sand/gravel fractions and total organic carbon percentages, TOC) of the *N. pachyderma*-rich levels (12a, 12b, and 12c). No gravity flows are suggested from the downcore trend of sand/gravel-size fractions and AMS ^{14}C dates (inorganic carbon fraction).

TESTING EVOLUTIONARY CONVERGENCE:

From RNA World to Intelligence

K. TAHIR SHAH
Dipartimento di Elettrotecnica Elettronica Informatica
Università di Trieste, Trieste, Italy
email: shah@spin.it

1. Introduction

The major exobiological issue today is the ontogenesis and evolution of life elsewhere in the universe. The issue is equivalent to the problem of repetition of terrestrial life and its biological diversity "if the tape were played twice", to use Gould's metaphor [Gould 1989]. Various authors [e.g., Mayr 1985 & Monod 1972] argued against evolution of terrestrial-like life and intelligence elsewhere in the universe - life and intelligence, especially the evolution of humanoid as only once in a universe's life time show. The thrust of arguments has been, not that intelligent life cannot exist, but that it has extremely low probability of occurrence elsewhere in the universe or repetition on Earth, given that an extremely large number of evolutionary paths available. This school of thought assigns a significant role to contingency and contends that natural selection is not the only cause of evolutionary change [Gould 1994]. Whether we accept the role of contingency or not, we should make a careful assessment of such notions as evolutionary convergence in the light of such criticism and especially the domain of its validity. On the other hand, a large majority of biologists believe in what has been called evolutionary convergence. The most optimistic view among others of evolution of intelligent life is represented by Minsky [1985], who argues that all intelligent problem-solvers are subject to the same ultimate constraints, limitations on space, time, and materials and thus similar to us. His argument derive strength from two basic principles which he called 'economics' and 'sparseness. The idea behind these principles is that optimality somehow constraints evolution to converge both at biological and psychological levels. Minsky's arguments is at a level of abstraction different from the molecular biology, the cognitive level, which is different from arguing for evolutionary convergence leading to complex brain such as that of *homo sapiens*. But, how closely related are, in general, Darwinian evolution based on natural selection/adaptation and the optimization process, and the selection of cognitive processes to optimize computational resources of a nervous system in particular? This is a point at issue which is normally dealt in evolutionary psychological studies [see, e.g., Barkow et al. 1992].

Other authors enthusiastic of evolutionary convergence believe that, "the reality of convergence suggests that tape of life can be run as many times as we like and in principle intelligence will surely emerge." [Connway-Morris 1998]. But the fact remains that at present there is no hard evidence to support any such contention except through extrapolation of a rather vague and local notion of morphological convergence. To be an

J. Chela-Flores et al. (eds.), First Steps in the Origin of Life in the Universe, 261–266.

acceptable hypothesis, evolutionary convergence must be testable beyond anecdotal evidence and with least amount of extrapolation. An essential first step should therefore be to clarify basic issues by defining the notion of evolutionary convergence with a higher degree of precision so as to narrow down search for hard evidence to support one or the other of the two views. A similar stand was also taken by R.F. Doolittle [1994] by pointing out that care is needed in the use of the term 'convergence' in molecular evolution. This note summarizes our study of this issue, the details of which are to published elsewhere.

2. Testing Evolutionary Convergence in Stages

Many authors [e.g., Austin 1998 & references therein] define evolutionary convergence simply as the acquisition of morphologically similar traits between distinctly related organisms. A large number of cases of morphological similarity in various organism led to the hypothesis of convergence which was eventually extended indiscriminately to all situations. For example, as one can see easily that convergence in morphology does in no way imply convergence in the emergence and development of the central nervous system and cognitive capabilities. Nevertheless, some authors deduce from this hypothesis the inevitability of the emergence of not only particular biological properties but higher cognitive faculties too [e.g., Chela-Flores 1999 and reference therein]. For example, it is stated that if life started somewhere else it would be constrained to take an evolutionary pathway to eukaryogenesis and multicellularity [Seckbach et al. 1998], thus asserting convergence at one of the major transition period. However, the issue is much more complex than either the re-appearance of the same morphology in some cases or the assertion that certain transitions (say, from RNA world to RNA-DNA-Protein world) are inevitable. There were numerous events, a change in their temporal order of occurrence, for example, would have changed the course of further evolution. It is then necessary not only to test convergence for major evolutionary transition periods but also with respect to function, structure, mechanism, sequence (abstract ordering of nucleotides or amino acids), morphology and cognitive universals. Moreover, we should concentrate on the basic life processes rather than specific traits. Thus extending the idea of Doolittle & Brown [1994] we divide the evolutionary history into a multistage evolution of life's basic machinery and major transition periods as follows.

Pre-Darwinian (chemical) evolution: This covers from prebiotic chemistry to oligonucleotides and peptides. A basic question requiring an answer during this epoch is why nature has selected specific prebiotic chemistry among many alternatives - a pentose and not a hexose and eventually ribofuranose and not ribopyranose for its nucleic acids? Chemical etiology is experimental study to narrow down the number of possible answers to the above question [Eschenmoser 1999]. The aim is not only to show how life on Earth could have originated, but to provide experimental evidence for evolutionary convergence hypothesis which implies that we must be able to produce again in the laboratory a transition from the pre-RNA world with many alternatives to the RNA world under similar conditions. Or, we must find such an evidence in exobiological experiments.

Possible End Product: 1-Biopolymer Life Form (Self-replicating RNAs and RNA cell).

Transition Period of Macromolecular Symbiosis: The first kind of symbiotic event that led to 2-biopolymer life form. Transition occurred from RNA genomes and ribozymes to primitive translation and eventually to ribosomal RNAs, transcription, ribosomal proteins, gene regulation, DNA genome and a full translation system. The basic questions are: 1) Will information processing machines for transcription, RNA processing and translation rise again out of RNA world? 2) Has it evolved elsewhere in the universe? 3) Is it the only alternative towards complex life?

End Product: Emergence of Breakthrough Organism (Benner et al. 1989).

Progressive-Darwinian (cellular) evolution: First fully functional cell (Last Universal Common Ancestor) and divergence to eubacteria and archaea.

Transition Period of Cellular (Bacterial) Symbiosis: Emergence of eukaryotes.

Postprogressive-Darwinian evolution: Multicellular life, its diversification, the emergence and task-directed (activation-dependent) evolution of the brain. Evolution of perceptual capabilities, sensorial interfaces, and cognitive capabilities. Relevant issues in this epoch are: 1) The question asked by Gould [1994] whether biological diversity would be same if it is to start all over again and 2) A clear indication of which convergence types (as described in the sequel) are observed in nature and their interrelationship.

Period of Transition to Higher Cognitive Functions: Evolution of social intelligence and tool use.

For progressive and post-progressive periods we further divide convergence into the following types.

Functional Convergence: It implies the emergence of some molecular functionality (e.g., enzyme that catalyze the same reaction) independently and more than once in the evolutionary history. Examples include many kinds of proteases, superoxide dismutases, ribonucleotide reductases and sugar kinases. However, we do not know whether some of the cases of convergence are simply due to horizontal transfer of relevant genes to a completely unrelated species.

Mechanistic Convergence: In this case, the same geometry of protein from completely different folds and sequence arrangements is obtained.

Structural Convergence: When three-dimensional structures of proteins are significantly similar but for which no clear sequence resemblance exists.

Sequence Convergence: When different structural problems are solved by evolving similar sequences from unrelated ancestors. This is quite different from the parallel evolution of similar sequences that are similar from the beginning. This type of convergence is yet to be observed.

Morphological Convergence: This implies acquisition of morphologically similar traits between distinctly related organisms.

Cognitive Reality Convergence: This refers to re-emergence of the same cognitive universals [Shah 1998] and the homologous perceptual and higher cognitive functional circuits in the brain during evolutionary history [Shah 2000]. Most scientists and mathematicians believe in realism and ignore the cognitive system and ultimately its hardware-dependence of perceived reality. Thus they expect to find the alien view of the world as ours. Even if one take for granted all of the convergence types, the matter is not settled there. What is required for inter-species communication, say between us and an intelligent alien species, is *reality convergence*, meaning that their and our notions of reality coincide. Mathematics, like natural language, is a species-specific and domain-specific capacity which shapes our scientific thinking to a large extent. Using mathematics for communication with extraterrestrials is perhaps on the same footing as using any terrestrial language. Thus the problem of inter-species communication requires both biological and psychological convergences.

3. Unavoidable Regularities Due to Mathematical and Chemical Constraints on Macromolecules

Both Van der Waerdon's and Ramsey theorems in combinatorics assert that sequences formed by a finite set of symbols will eventually have subsequences with regularity properties (see e.g., Lothaire, 1983). This is known as the *principle of unavoidable regularities*. To apply these theorems we interpret four nucleotides as four abstract symbols and macromolecules as sequences of these symbols. This means that purely random sequences of certain length simply cannot exist if only four symbols are used to form them. In particular, studies in ergodic theory of combinatorics show that Morse-Thue type and other recurrent sequences are a consequence of such regularities [Furstenberg 1987]. Such sequences contain as their subsequences either palindromes or inverse palindromes - the so called hair-pin structure. However, the principle of unavoidable regularities does in no way implies that regularities of RNA world species had exactly these kind of mathematical regularities. Other factors, such as chemical constraints played their role to give them their final shape.

Crick-Watson base pairing and free-energy changes impose strong constraints on the stability of macromolecular sequences. For instance, some tetramers are more stable than the others and hence likely to survive longer than others under thermal variations. For instance, at 37°C, GCCGGC is more stable than CCGGC, and ACCGGC is more stable than CCGGC. Helixes without loops GCGGCG, GCGGCGA, GCAACCA, GCGGUCA are the most stable as compared to helixes with bulge and helix with hairpin loops [Turner & Bevilacqua 1993, Turner et al, 1987]. Hairpins represent the dominant secondary structure elements in RNA. Such sequences are found with exceptionally high frequency in many RNAs and are characterized by high thermodynamics stability. Such RNA hairpins define nucleation sites for folding, determine tertiary structure in ribozymes, protect mRNA from degradation and recognized by RNA binding proteins.

Using techniques from computability theories to analyse sequence spaces studies on artificial life also help us to discover generalization about evolving biological systems. Example are abstract descriptions of prebiotic chemistry using Lambda Calculus, molecular automaton to mimic the "molecular logic of the living state", rewrite rules model, cellular automata, and computer programs that self-replicate, mutate, and adapt by natural selection. Results from both unavoidable regularities and those in artificial life studies suggest that certain structures will emerge due to laws of mathematics and chemistry independent of environmental constraints and evolutionary path, thus supporting partially the hypothesis of convergence in pre-Darwinian epoch.

4. Evolution Ex Vivo of RNA World

Szostak & Ellington [1993], Szostak [1993], Lorsch & Szostak [1994], Bartel & Szostak [1994], and Wilson & Szostak [1995] developed an iterative method for the isolation of nucleic acid sequences with specific ligand-binding (or catalytic) properties from a very large pool of "random" sequences (see, e.g., Joyce 2000, for his earlier work and comments) . The i*n vitro* selection cycle is used to enrich the RNA pool with desired species. Gold et al.

[1993] developed another method known as SELEX (or Systematic Evolution of Ligands by EXponential enrichment), which is used to test the concept of the RNA world by assessing the catalytic breadth of RNA - by studying the catalytic and binding properties of small RNA molecules. The conclusion is that RNA apparently has a huge distribution of sizes and shapes, including even a collection of rather short RNA molecules, which could have been used for an enormous variety of chemistry prior to the invention of proteins in the RNA world. Both kinds of experiments tackled a major issue, that of catalytic molecules essential for virtually all processes in RNA world. Such experiments, therefore, by demonstrating reproducibility of catalytic agent necessary for life processes give strong support to convergence hypothesis. Given this series of evidence we seriously contemplate the thought of considering the existence of an RNA cell, a self-replicating unit inside a membrane. It is likely that elsewhere in our solar system the life processes never had a transition from RNA world to RNA-DNA-Protein world. We should consider search for a RNA cell as an alternative to the terrestrial prokaryotes either on Mars or Europa.

5. Convergence in Progressive Darwinian Era

To test evolutionary convergence hypothesis we need to find answers to the following questions: 1) Can we find *in vitro* (or find in an extraterrestrial world) a mechanism of high fidelity that faithfully reproduces genetic material (whatever it is) so that it is inherited generation after generation? 2) Can we produce *in vitro* (or find in an extraterrestrial world) 2-biopolymer life form with mechanisms such as translation to transfer information from genetic material to specify construction of many other types of structures that constitute a living organism? The relationship between location and function is called "positional information" that is both spatial and temporal. Two related questions are then: 3) Can we produce *in vitro* such positional arrangements? That is, producing simply *de novo* all macromolecular components is not enough. They are to be arranged is a specific way representing specific algorithms. 4) There are other structures such as introns and molecular chaperons which are not a part of the mature structure. Can we produce such arrangements *in vitro*? As far as transition from RNA World to RNA-DNA-Protein World in concerned, there are no *in vitro* evolution experiments to confirm the possibility of *de novo* formation of an RNA-DNA-Protein world with all its complex information processing mechanisms. Of course, any evidence from extraterrestrial source, even unicellular but similar to terrestrial life, will add support to evolutionary convergence in a significant and decisive way. Similarly, a definite evidence for the re-emergence of eukaryogenesis is required either in the laboratory or in extraterrestrial context as suggested by Chela-Flores [1999] for Europa mission, should it confirm the existence of eukaryogenesis, shall be a milestone in the search for life in the universe and a step in the proof of strong convergence.

6. Conclusion

In a broad sense, studies in evolutionary convergence seek to answer the following questions: 1) How abundant is primitive life (terrestrial type or otherwise) in the universe? 2) Are there examples of eukaryotic and multi-cellular complex life in the universe? 3) What evidence we have of intelligent life in the universe (SETI studies)? The outcome of our study is that at early stages of Earth's history (pre-Darwinian epoch), the chance hypothesis is not tenable in the light of theoretical and experimental studies. Although there is only partial support for the convergence hypothesis, nevertheless, it seems that some form of life is bound to re-emerge. However, virtually there is no evidence to support the repetition of complex information processing mechanisms (replication with DNA repair, transcription, RNA processing and translation). The same can be said about eukaryogenesis.

7. References

Austin, D.F. (1998) Parallel and Convergent Evolution in the Convolvulaceae, in *Diversity and Taxonomy of Tropical Flowering Plants*, eds., P. Mathews and M. Sivadasan, Mentor Books, Calicut, India, pp. 201-234.

Barkow, J.,Cosmides, L. and Tobby, J. (1992) *The Adapted Mind: Evolutionary psychology and the generation of culture*, Oxford University Press, New York.

Bartel, D.P. and Szostak, J.W. (1994) Isolation of New Ribozymes from a Large Pool of random Sequences, *Science* **261**, 1411-1417.

Chela-Flores, J. (1999) Eukaryogenesis: The Search for an Evolutionary Transition Towards Intelligence in an Extreme Environmental Habitat of the Outer Solar System, in *Enigmatic Microorganisms and Life in Extreme Environments*, ed., J. Seckbach, Kluwer Academic Publishers, pp. 63-71.

Conway-Morris, S. (1998) *The crucible of Creation*, Oxford University Press.

Doolittle, W.F. and Brown, J.R. (1994) Tempo, Mode, the Progenote, and the Universal Root, *Proc. Natl. Acad. Sci. USA* **91**, 6721-6728.

Doolittle, R. F. (1994) Convergent evolution: The need to be explicit, *TIBS*, January issue, pp. 15-18.

Eschenmoser, A. (1999) Chemical Etiology of Nucleic Acid Structure, *Science* **284**, 2118-2124.

Furstenberg, H. (1987) *Recuurence in Ergodic Theory and Combinatorial Number Theory*, Princeton University Press, NJ, USA.

Gold, L., et al. (1993) RNA: The Shape of Things to Come, in *The RNA World*, eds. R.F. Gasteland and J.F. Atkins, Cold Spring Harbor Laboratory Press, pp. 497-509.

Gould, S.J. (1989) *This Wonderful Life*, Norton, New York.

Gould, S.J. (1994) *Scientific American*, October issue.

Joyce, J.F. (2000) Ribozyme Evolution at the Crossroads, *Science* **289**, 401-402.

Lorsch, J.R. and Szostak, J.W. (1994) In vitro evolution of new ribozymes with polynucleotide kinase activity, *Nature* **371**, 31-36.

Lothaire, M. (1983) *Combinatorics on Words*, Addison-Wesley, Reading, MA.

Mayr, E. (1985) The Probability of Extraterrestrial Intelligent Life in *Extraterrestrial Science nd Alien Intelligence*, ed. E. Regis Jr., Cambridge University Press, UK, pp. 23-30.

Minsky, M. (1985) Why intelligent aliens will be intelligible in *Extraterrestrials, Science and Alien Intelligence*, ed.: E. Regis Jr., Cambridge University Press, Cambridge, UK, pp. 117-128.

Monod, J. (1972) *Chance and Necessity*, Collins, London.

Seckbach, J., Westall, F., and Chela-Flores, J., Introduction to Astrobiology: Origin, Evolution, Distribution and Destiny of Life in the Universe, in *Journey to Diverse Microbial Worlds: Adaptation to Exotic Environments*, ed. J. Seckbach, Kluwer Academic Publishers, (in press).

Shah, K. Tahir.(1998) Cognitive Universals: Abstract Psychology of Terrestrial and Extraterrestrial Intelligence, in *Exobiology: Matter, Energy and Information in the Origin and Evolution of Life in t he Universe*, Eds., J. Chela-Flores and F. Raulin, Kluwer Academic Press, pp 161-164.

Shah, K. T. (2000) Constructive Realism: An Introduction to Cognitive Reality, DEEI, Universita di Trieste, Trieste, Italy, (to be published).

Szostak, J.W. and Ellington, A.D. (1993) In Vitro Selection of Functional RNA Sequences, in *The RNA World*, eds. R.F. Gasteland and J.F. Atkins, Cold Spring Harbor Laboratory Press, pp. 511-531.

Szostak, J.W. (1993) Evolution *ex vivo*, *Nature* **361**, 119-120.

Turner, D.H. et al. (1987) Improved Parameters for Prediction of RNA Structure, *Cold Springr Harbor Symposia on Quantitative Biology*, vol. **LII**, pp. 123-132.

Turner, D.H. and Bevilacqua, P.C. (1993) Thermodynamic Considerations for Evolution by RNA, in *The RNA World*, eds. R.F. Gasteland and J.F. Atkins, Cold Spring Harbor Laboratory Press, pp. 447-463.

Wilson, C. and Szostak, J.W. (1995) *In vitro* evolution of a self-alkylating ribozyme, *Nature* **374**, 777-782.

Section 6:

Is there indigenous organic material on Mars?

THE SEARCH FOR A SECOND GENESIS OF LIFE IN OUR SOLAR SYSTEM

CHRISTOPHER P. MCKAY

Space Science Division
NASA Ames Research Center
Moffett Field, CA 94035 USA

Abstract

One of the key goals of Astrobiology is to search for another example of life. In the near term, Mars and Europa are the likely targets for a search for a second genesis. On Mars the search is likely to be conducted in the frozen permafrost. On Europa active life may be found in the ocean beneath the ice cover.

Introduction

One of the deepest results of modern biology is the demonstration of the chemical and genetic unity of life on Earth. All life on Earth is constructed from the same basic biochemical building blocks consisting of 20 amino acids with left handed symmetry, five nucleotides, a few sugars of right handed symmetry and some lipids (Lehninger, 1975). Using the metaphor of computers this is equivalent to saying that all life shares the same hardware. Beyond hardware similarity, it is now known that all life has fundamentally the same software. The genetic code of life is common in organisms. This was first elucidated by Carl Woese using studies of ribisomal RNA (Woese, 1987).

One of the goals of Astrobiology is to understand life on a fundamental level. This would include understanding how life originated on Earth, if there is life beyond the Earth, and in what conditions life can originate. A broad understanding of the fundamental nature of life is difficult to achieve with only the one example of life on Earth. The many theories posited for the origin of life are shown in Fig. 1.

Some have argued that the "hammer of evolution is heavy" (Norm Pace, personal communication) and life anywhere is likely to be composed of identical biochemical and genetic patterns. In this view, everything about biochemistry and genetics, including the structure of DNA, the triplex coding that specifies each amino acid and the amino acids used has been optimized by evolutionary selection and should

J. Chela-Flores et al. (eds.), First Steps in the Origin of Life in the Universe, 269–277.

therefore be the same for life anywhere. However the suggestion that the details of biochemistry are optimal and not contingent on historical details is not compelling. Indeed, even if it happens that the optimization of biochemistry will inexorably lead to the same amino acids and sugars, there is no reason why the left handed amino acids and the right handed sugars would be selected. In only this one aspect there are four permutations of biochemical systems that are functionally and energetically identical to Earth life. More broadly, we do not know enough possible biochemistries to argue that there is but one global optimum in the possible phase space of biochemical function. In a system as complex as biochemistry it is likely that there are numerous local optima and the details of the optimum found by evolutionary selection on another world would likely depend on the initial conditions and random developments in the early biological history on that world.

Table 1. Three approaches to a Second Example of Life.

Construct alternative living systems in the laboratory
Communicate with intelligent extraterrestrials
Find intact lifeforms on Mars or Europa

How can the deep questions about the fundamental nature of life and possible alternative biologies be answered? It may be that extensive study of the one example of biochemistry that we have on Earth will show what is necessary for life and what is contingent. Perhaps this will come from experimental construction of alternative living systems in the laboratory. An alternative search for a second example of life is based on SETI (Search for Extraterrestrial Intelligence). Clearly communication with intelligent aliens will allow us to inquire, *inter alia*, as to the nature of their biochemical and genetic makeup. This paper focuses on a third approach to detecting a new biochemistry: the search for life on Mars and Europa.

Fossils are Not Enough

The search for life beyond the Earth is usually associated with a search for fossils (Fig 2). Indeed there is some speculative evidence for fossil remains in the martian meteorites (McKay et al., 1996; Friedmann et al., 2001, Thomas-Keptra et al., 2001). However fossils alone, even clearly biogenic fossils, will not address the question of a second example of biochemistry.

Ideally a second example of life would be expressed as a functioning, reproducing microbial ecosystem. In this case the structure and function of the biomolecules could be observed. The role of the genetic material in reproduction and development could be studied. In addition, community interactions among different species could be monitored. These observations could provide key comparisons and contrasts to life on Earth.

Failing to find extant life, the next most desirable possibility is discovery of dead but still intact organisms; possibly dead from radiation or thermal shock. From such micro-corpses it would be possible to determine the structure of the alien biochemistry and the range of biomolecules used. It may also be possible to develop sufficient knowledge of the genetics of the alien life form to be able to revive the organisms using sophisticated genetic repair techniques customized to the alien biochemistry. This would likely be a large and complex undertaking requiring a level of study of the alien biochemistry comparable to the level of study of Earth-based biochemistry. However, the task is worth the reward if a truly second genesis can be reconstructed.

Even fragmented organic remains of lifeforms that do not constitute an entire organism may provide evidence of a second genesis and clues as to the nature of the alien biochemistry. Organic material composed of clearly non-uniform distributions of organic molecules could be indicative of biochemistry. For example, if organic material was discovered on another world and it contained large amounts of right handed amino acids and virtually no left handed amino acids that would be both an indication of life and proof of a second biochemistry distinct from Earth life.

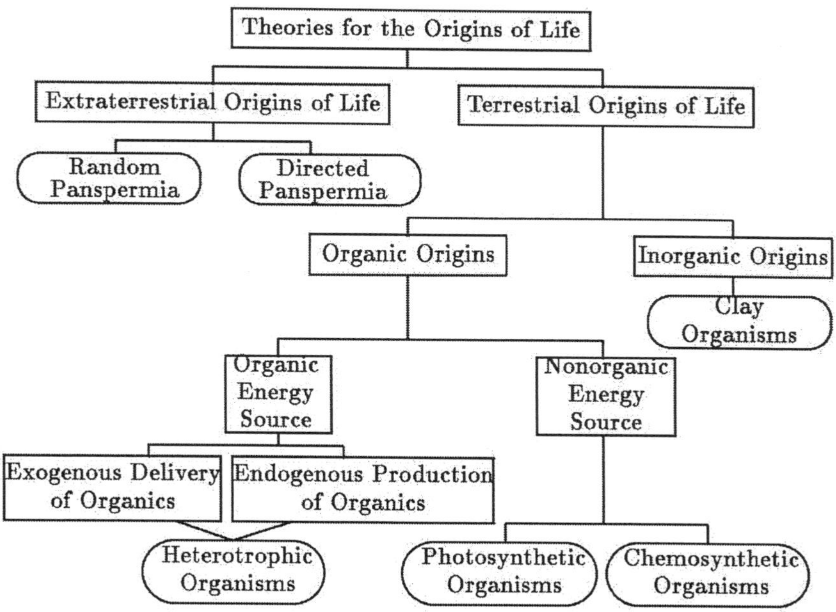

Figure 1. Diagram of the theories for the origin of life on Earth as organized by Davis and McKay (1996). All environments that were present on the early Earth would have been present on the early Mars. Thus the possible origin of life on Mars is based directly on analogy with Earth. For Europa the options are more restrictive and only a chemosynthetic origin of life would be plausible there.

Mars

Mars is the most promising candidate for a near-term search for a second example of life (see e.g., McKay, 1997). There is considerable evidence that Mars had liquid water early in its history and liquid water habitats may have persisted on Mars for a time period long enough for the origin of life (McKay and Davis, 1991).

But even if there was life on Mars it was not necessarily a second genesis. There are more than a dozen meteorites found on the Earth that originated on Mars (McSween, 1994). Weiss et al. (2000) have shown that during ejection from Mars, transfer to Earth, and entry into the Earth's atmosphere the internal temperature of these Mars rocks did not exceed 40°C. Thus microorganisms could be transferred from Mars to Earth and presumably, from Earth to Mars (Mileikowsky et al., 2000). Indeed, Sleep and Zahnle (1998) have argued that it is more likely that life originated on Mars and was subsequently carried to Earth.

Because of the possibility of interplanetary cross-fertilization between Earth and Mars the question of a second genesis cannot be assumed and must be determined directly. Table 2 lists possible locations of a second biochemistry on Mars.

Table 2. Possible Locations of Biochemistry on Mars.

Dormant cells in the soil	Inconsistent with Viking results
Dormant cells in amber or salt	Amber unlikely; salt preservation unproven
Subsurface life	Possible but improbable due to insufficient water
Dormant cells in permafrost	Most likely to succeed

In the most optimistic scenario there may be dormant organisms present in the martian soil. This result has been suggested based on the results of the Labeled Release experiment on the Viking landers (Levin and Straat, 1981). However a biological explanation for the reactivity of the martian soil is at odds with the other instruments on the Viking landers, and chemical explanations have been proposed (eg. Yen et al., 2000; Zent and McKay, 1994).

On Earth the oldest preserved viable organisms are found encased in amber and salt crystals. Cano and Borucki (1995) retrieved viable microorganisms (*Bascilus*) from 25 Myr old amber and microorganisms have subsequently been retrieved from amber as old as 90 Myr -- the upper limits apparently set by the age of amber. Vreeland et al. (2001) have reported viable spores (*Bascilus*) in salt deposits that are 250 Myr old. Unfortunately salt, unlike amber, may be permeable to small water droplets over long timescales; the genetic sequencing did not indicate, nor did it refute, a long age for the recovered microorganisms. Amber is produced by trees and is unlikely to have been produced on Mars. Salt deposits, in contrast, are very probable on Mars and if the preservation of microorganisms in salt over geological times is confirmed this could be a promising direction for future Mars exploration.

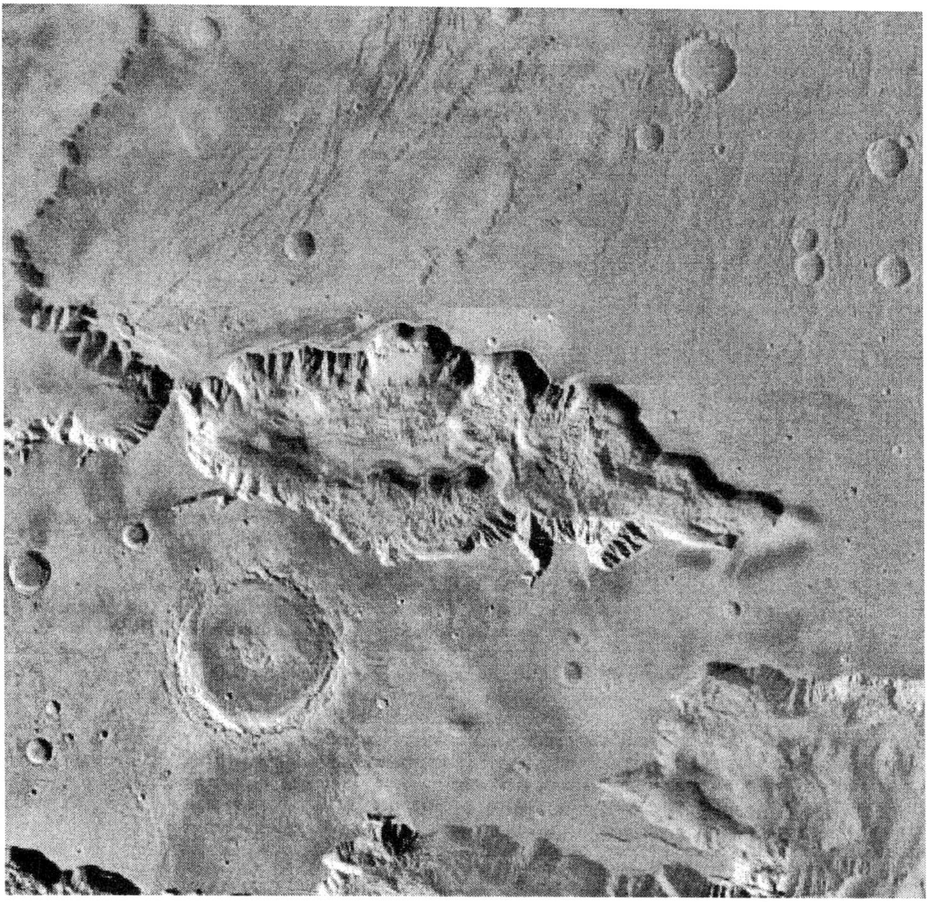

Figure 2. Hebes Chasma on Mars, a possible target for future Astrobiology missions to Mars, is a box canyon that may contain a plateau of sedimentary material deposited at a time when the canyon was full of water. These deposits may be carbonates (McKay and Nedell, 1985) and may hold fossil evidence of life on Mars. However fossils alone will not address the question of a second genesis on Mars. The canyon is about 100 km wide. (NASA Image)

A final, and possibly the most promising, way to search for intact biochemical remains of life on Mars may be in the martian permafrost. On Earth, viable organisms have been recovered from 3 Myr old Siberian permafrost (Gilichinsky et al., 1992; Rivkina et al., 2000). There are two factors that limit long term dormancy in permafrost: thermal decay and ionizing radiation. Thermal decay is an inevitable thermodynamic process at any finite temperature. This would include racemization of amino acids, breaking of chemical bonds, and other chemical changes. Generally speaking thermal decay is an exponential function of temperature determined by an activation energy. Because martian permafrost is at -70°C or less, thermal decay is not likely to be a cause of loss of viability for frozen martian microorganisms. Given survival in Siberia for 3 Myr at -10°C, and assuming that every 10°C temperature drop increases survival by a factor of 3, then at -70°C survival for 3 Gyr is possible. Survivability in martian permafrost is likely limited by the background ionizing radiation due to U, Th, and K in the martian crust. Assuming, consistent with the elemental analysis of the martian meteorites, that Mars has crustal abundance of these radiogenic elements comparable to Earth levels then the radiation dose in the martian subsurface would be about a few tenths of a rad per year. This would imply a lethal dose for most soil microorganisms in about 10 Myr (longer than the age of the Siberian permafrost) and 100 Myr for even the most radiation resistant microorganisms *Dinococcus radiodurans* (Minton, 1994; Battista, 1997). Even in the case that deep freezing reduces radiation injury by a factor of 5, it would still be the case that microorganisms in the martian permafrost would have accumulated many times the lethal dose over 3.5 Gyr. Without the possibility of repair in the deeply frozen state, the irradiated organisms would likely be dead when warmed. However, their biochemistry and genetic material would be largely intact allowing for direct analysis. Furthermore it is plausible that methods could be developed that would allow for artificially repairing the genetic and cellular damage and thereby restoring the martian microbes to the viable state.

If there was ever life on Mars the prospects for determining if it had a separate origin from life on Earth are good; there are many promising directions for exploration. Unfortunately, from studies of the martian meteorites and given the impact history of the Earth and Mars it is probable that both planets share a common biology. If this is borne out by direct investigation then the search for a second genesis within our solar system must expand beyond Mars to the outer solar system and in particular to Europa.

Europa

Europa is one of the four Galilean moons of Jupiter and is second in distance from Jupiter, after Io. Due to interactions between the moons and with Jupiter, there is a forced eccentricity to the Galilean moons. This results in tidal heating, which on Io is strong enough to produce volcanic eruptions. The surface of Europa is composed of ice and there are many indications that Europa has a subsurface ocean that may be about 100 km deep. The thickness of the ice cover is thought to be about 10 km (Pappalardo et al., 1999), although some interpretations of the spacecraft data suggest

negligible thickness in certain locations (Greenberg et al., 2000). If Europa has an ocean, a search for life within that ocean would follow. And if there is life in the ocean on Europa it is unlikely to have come from Earth or Mars.
Europa, unlike Mars, may have biologically produced organic material at its surface. The linear features on Europa may be cracks in the surface ice through which ocean water has flowed. If this is the case then the surface deposit associated with the cracks may contain organic material produced by any life within the ocean. An impact sampling mission, such as the proposed Europa Ice Clipper, to collect some of this material may provide the first access to alien biochemistry.

Conclusion

The search for second genesis has both philosophical and practical implications. If there is life beyond the Earth this has important implications for our understanding of the universe and our place and future in it. On the other hand, if there is no life beyond the Earth this also has important implications for our understanding of the universe and our place and future in it. The philosophical implications of life beyond the Earth have been the subject of human discourse for centuries. Now, for the first time the question can be address experimentally, by observation, and with investigations by field scientists.

Beyond philosophy, there would likely be important practical benefits from the discovery of a second type of biochemistry. It is arguable that biochemistry is the basic science that has the most influence on the quality of our lives. It is at the root of progress in medicine, agriculture, pest control, food science (eg. chocolate) and even ecological preservation. It may not be an exaggeration to say that our current understanding of biochemistry is limited because it is only based on one data point; one example of how biological systems can be constructed. Expanding the database to two examples of life could bring profound insights that have implications for the applied fields that are rooted in biochemistry. For both the philosophical and practical reasons the search for a second genesis is the keystone of our future exploration of Mars and Europa.

References

Cano R.J., and M.K. Borucki (1995) Revival and identification of bacterial spores in 25- to 40-million-year-old Dominican amber. Science 268, 1060-1064.

Davis, W.L. and C.P. McKay (1996) Origins of life: A comparison of theories and application to Mars. Origins Life Evol. Biosph., 26, 61-73.

Friedmann, E.I., J. Wierzchos, C. Ascaso, and M. Winklhofer (2001) Chains of magnetite crystals in the meteorite ALH84001: Evidence of biological origin. Proc. Nat. Acad. Sci. 98, 2176-2181.

Gilichinsky, D.A., E.A. Vorobyova, L.G. Erokhina, D.G. Fyordorov-Dayvdov, and N.R. Chaikovskaya (1992) Long-term preservation of microbial ecosystems in permafrost. Adv. Space Res. 12 (4) 255-263.

Greenberg, R., P. Geissler, B. Tufts, and G.V. Hoppa (2000) Habitability of Europa's crust: The role of tidal-tectonic processes. J. Geophys. Res. 105, 17551-1756.

Lehninger, A. L. (1975) Biochemistry. Worth, New York.

Levin, G.V. and P.A. Straat (1981) A search for nonbiological explanation of the Viking labeled release life detection experiment, Icarus 45, 494-516.

McKay, C. P. (1997) The search for life on Mars. Origins of Life and Evolution of the Biosphere 27:263-289.

McKay, C.P. and W.L. Davis, (1991) Duration of liquid water habitats on early Mars, Icarus 90, 214-221.

McKay, C.P., and S.S. Nedell (1988) Are there carbonate deposits in Valles Marineris, Mars?, Icarus 73, 142-148.

McKay, D.S., E.K. Gibson, K.L. Thomas-Keprta, H. Vali, C.S. Romanek, S.J. Clement, X.D.F. Chillier, C.R. Maechling, and R.N. Zare (1996) Search for Past Life on Mars: Possible Relic Biogenic Activity in Martian Meteorite ALH84001. Science 273, 924-930.

McSween, H.Y. (1994) What have we learned about Mars from SNC meteorites. Meteoritics 29, 757-779.

Minton, K.W. (1994) DNA repair in the extremely radioresistant bacterium Deinococcus radiodurans. Mol. Microbiol. 13 9-15.

Battista, J.R. (1997) Against all odds: The survival strategies of Deinococcus radiodurans, 51, 203-24.

Mileikowsky, C., F.A. Cucinotta, J.W. Wilson, B. Gladman, G. Horneck, L. Lindegren, J. Melosh, H. Rickman (2000) Natural transfer of viable microbes in space. 1. From Mars to Earth and Earth to Mars. Icarus 145 391-427.

Pappalardo, R. T.; M.J.S. Belton, H.H. Breneman, M.H. Carr, C.R. Chapman, G.C. Collins, T. Denk, S. Fagents, P.E. Geissler, B. Giese, R. Greeley, R. Greenberg, J.W. Head, P. Helfenstein, G. Hoppa, et al. (1999) Does Europa have a subsurface ocean? Evaluation of the geological evidence. J. Geophys. Res. 104, 24,015-24,055.

Rivkina, E.M., E.I. Friedmann, C.P. McKay, and D.A. Gilichinsky (2000) Metabolic activity of permafrost bacteria below the freezing point, Appl. Environ. Microbio. 66, 3230-3233.

Sleep, N.H. and K. Zahnle (1998) Refugia from asteroid impacts on early Mars and the early Earth. J. Geophys. Res. 103, 28,529-28,544.

Thomas-Keprta, K.L., S.J. Clemett, D.A. Bazylinski, J.L. Kirschvink, D.S. McKay, S.J. Wentworth, H. Vali, E.K. Gibson, Jr., M.F. McKay, and C.S. Romanek (2001) Truncated hexa-octahedral magnetite crystals in ALH84001: Presumptive biosignatures. Proc. Nat. Acad. Sci. 98, 2164-2169.

Vreeland, R.H., W.D. Rosenzweig and D.W. Powers (2000) Isolation of a 250 million year old bacterium from primary salt crystals. Nature 407, 897-900.

Weiss B.P., J.L. Kirschvink, F.J. Baudenbacher, H. Vali, N.T. Peters, F.A. Macdonald, and J.P. Wikswo (2000) A Low Temperature Transfer of ALH84001 from Mars to Earth. Science 290, 791-795.

Woese, C.R. (1987) Bacterial evolution. Microbiol. Rev. 51, 221-271.

Yen, A.S., S.S. Kim, M.H. Hecht, M.S. Frant, and B. Murray (2000) Evidence That the Reactivity of the Martian Soil Is Due to Superoxide Ions. Science 289 1909-1912.

Zent, A.P. and C.P. McKay (1994) The chemical reactivity of the martian soil and implications for future missions. Icarus 108, 146-157.

THE PUBLIC LECTURE

ANCIENT LIFE ON MARS AND EARTH

J. WILLIAM SCHOPF

Department of Earth and Space Sciences,
Institute of Geophysics and Planetary Physics
(Center for the Study of Evolution and the Origin of Life),
and Molecular Biology Institute,
University of California, Los Angeles, CA 90095-1567
PHONE: (310) 825-1170
FAX: (310) 825-0097
E-MAIL: Schopf@ess.ucla.edu

ABSTRACT: In the modern epoch, the question of life on Mars dates from the pioneering observations of the great Italian astronomer Giovanni Virginio Schiaparelli (1835-1910) who recorded the presence of linear features on Mars' surface that were thereafter interpreted by Percival Lowell (1855-1916) -- an influential if somewhat eccentric American astronomer -- as artificial canals, aqueducts built by an advanced civilization to link the planet's icy water-rich poles to its arid equatorial deserts. By 1924, belief in "life on Mars" had become so widespread, so notably pervasive, that in August of that year -- when the orbits of Earth and Mars were especially close -- governments worldwide urged their citizens to turn off their home radios so that supersenstitive receivers could listen during an hour-long period for signals from beings on the Red Planet. Though this attempt, of course, came to naught, and though even to the present day there remains no firm evidence of any type of life on Mars -- past or present, primitive or advanced -- the question of life on the Red Planet has lingered on, lying dormant for years at a time only to rise again in the public mind as it did most

J. Chela-Flores et al. (eds.), First Steps in the Origin of Life in the Universe, 279–280.

recently in August 1996, spurred by NASA's now famous report of possible evidence of "ancient Martian microbes.". Though the report first looked promising, it has since failed to meet the test. The late Carl Sagan phrased it well: "Extraordinary claims require extraordinary evidence" and, plainly put, there is still no clearcut evidence, no "smoking gun," to back this extraordinary claim. But it would be wrong to think that the matter is closed, for in truth the question of life on Mars simply remains unanswered -- no more, no less -- a status captured by another telling Saganism: "Absence of evidence is not evidence of absence!" Understanding of the Red Planet is in its infancy; its climate, its range of settings, its history, are all as yet largely undeciphered, and the hunt for life on Mars remains a scientific quest. The tools are at hand. And the rules for the search are very much the same as those used so successfully in tracing life's history on planet Earth. If there once was or is now life on Mars or elsewhere in the accessible reaches of space, science can and will ferret it out.

SEARCH FOR LIFE ON MARS :
THE MARS EXPRESS BEAGLE-2 LANDER

ANDRE BRACK

Centre de Biophysique Moleculaire,
CNRS, rue Charles Sadron, 45071 Orleans
cedex 02, France,
brack@cnrs-orleans.fr

The ESA Exobiology Science Team directed and supported by ESA Manned Spaceflight and Microgravity Directorate designed a multi-user integrated suite of instruments to optimize the search for evidence of life on Mars in subsurface and rock interior samples. The package includes instruments to study sample mineralogy (composition, macroscopic and microscopic structural and textural features), organics (elemental, molecular, chiral), oxidants and petrology (major and minor element composition). Environmental characteristics will also be measured (atmosphere, radiation budget, temperature, pressure, wind speed and direction, humidity, dust activity).

Beagle 2, the exobiology lander of ESA 2003 Mars Express mission will be the first illustration of such a package. It is designed to provide a qualitative, quantitative and isotopic inventory of the noble gases and of the light elements CHON in their organic, inorganic and atmospheric situations. The material will be sampled from protected sites (subsurface and rock interiors) with a mole and a small surface rock grinder and corer mounted on a robotic arm.

J. Chela-Flores et al. (eds.), First Steps in the Origin of Life in the Universe, 281–282.

References

• Beagle 2 : a proposed exobiology lander for ESA's 2003 Mars Express Mission, M.R. SIMS, C.T. PILLINGER, A. BRACK et 26 co-auteurs, Adv. Space Res. 23, No.11 (1999) 1925-1928.

• An ESA study for the search for life on Mars, F. WESTALL, A. BRACK, B. HOFMANN, G. HORNECK, G. KURAT, J. MAXWELL, G.G. ORI, C. PILLINGER, F. RAULIN, N. THOMAS, B. FITTON, P. CLANCY, D. PRIEUR, D. VASSAUX, Planet. Space Sci. 48 (2000) 181-202.

• Exobiology in the Solar System & the search for life on Mars, A. BRACK, B. FITTON and F. RAULIN, ESA Scientific Publication SP 1231 (1999).

SEARCH FOR ENANTIOMERIC EXCESS IN EXTRATERRESTRIAL SAMPLES. INTEREST FOR MARS SAMPLE RETURN.

O. VANDENABEELE-TRAMBOUZE (1), M. DOBRIJEVIC (2), D. DESPOIS (2), A. COMMEYRAS (1), M. GEFFARD (3), M. ALBERT (4), M.F. GRENIER LOUSTALOT (5).

(1) Université de Montpellier II, UMR 5073, CC 017, Place E. Bataillon, F- 34095 Montpellier, France.
(2) Observatoire de Bordeaux, BP 89, F-33270, Floirac, France.
(3) INSERM, 200 Av. Thouars, Bordeaux, France.
(4) Université Paul Sabatier, IMRCP, UMR 5623, 118 route de Narbonne, 31062 Toulouse, France.
(5) SCA-CNRS, BP 22, Vernaison, France.

Introduction: amino acids on Mars.

Amino acids are simple, stable compounds, which play a fundamental role in terrestrial life as building blocs of proteins and enzymes. Terrestrial life mainly uses 20 of them in proteins, (biogenic amino acids) which are alpha-hydrogenated. In contrast, more than 70 have been identified in the Murchison meteorite, many of which bearing two alkyl groups in alpha position. A remarkable characteristic of proteinic amino acids is their homochirality: only one optical isomer (L) is present in terrestrial life, in contrast with the racemic (L=D) mixture produced by most chemical amino acids production scheme.

Indications exist that physico-chemical conditions on early Mars and early Earth might have been close to each other. Life may thus have appeared on Mars 3.5-4 Gyr ago. If not, records of prebiotic chemistry could have been preserved. The low temperature and humidity on Mars since 3 Gyr provide very good conditions to preserve the amino acids and their chirality [1]. If life appeared, it may be extinct or extant. In the former alternative, amino acids and chirality may have been preserved at least partly. In the later alternative, amino acids may be homochiral (L or D) and widely spread. Harsh conditions (oxidizing environment, absence of efficient shielding against solar UV and cosmic rays) seem to preclude substantial concentration of organic matter on the surface. In conclusion, amino acids may be present on the surface or, more probably, in the sub-surface of Mars.

Goals of this work

Very sensitive analysis may allow to detect and quantify very small amounts of amino acids in Mars samples. Detection of amino acids (biogenic or not) and chirality (L or D) may lead to numerous interpretations about the origin of these molecules. Different sources are expected: (1) local martian sources from prebiotic chemistry processes

J. Chela-Flores et al. (eds.), First Steps in the Origin of Life in the Universe, 283–287.

(present or past), (2) degradation of martian biota (present or past), (3) external source (meteorites and micometeorites). Consequently, preparation of the Mars Sample Return (MSR) mission must be oriented towards two directions. First, we have to increase the sensitivity of the analysis to allow the detection and quantification of small concentration of amino acids in tiny samples (few milligrams). Second, it is necessary to study various extraterrestrial samples (chondritic meteorites, martian meteorites and micrometeorites…) in order to prepare the interpretation of future results.

In particular, the determination of the amino acids enantiomeric excess in meteorites and micrometeorites is important since they may have played an important role in organic mater input on primitive planets. This input could have been continuously active on Mars for 4 Gy leading to a complete cover of the Mars surface by meteoritic material [2].

The first step of our research program dedicated to the MSR mission is to develop new analytical methods to reach a sensitivity in agreement with that required by the mission. These methods will be tested on several samples (Murchison, martian meteorites and micrometeorites).

Amino acids in meteorites and micrometeorites

In the Murchison carbonaceous meteorite, a small enantiomeric excess of amino acids was found in some of them. In particular, [3] found that about 7% of the L enantiomer of 2-amino- 2-3 dimethylpentanoic acid occurs in excess in Murchison. Non ambiguous L enantiomer excess determination only concerned non-biogenic amino acids. Problems of contamination and of racemization upon heating during atmospheric entry of biogenic amino acids generated contradictory interpretations about the origin of the enantiomeric excess measured in the Murchison meteorite [2, 4-7]. So, further studies are required to a better knowledge of the composition and the chirality of amino acids in carbonaceous meteorites and to a better understanding of their origin.

In micrometeorites, the diversity and abundance of amino acids are not well known. Only one analysis of a pool of 30 micrometeorites has been performed at this time [8]. This study confirms the presence of amino acids (mainly α-aminoisobutyric acid: AIB). Micrometeorites represent the major fraction of the extraterrestrial input on Earth and seem to softly enter into the atmosphere: 80% of micrometeorite found on Earth are not melted [9]. Therefore, amino acid enantiomeric excess (if existing) inside such materials may have been preserved after the atmospheric entry. So, it is necessary to develop new analytical methods adapted to determination of amino acids and their enantiomeric ratio in one micrometeorite.

Analytical development

We focus now on the determination of amino acid enantiomeric excess in micrometeorites. Such analysis is rendered difficult by the risk of contamination by terrestrial biogenic amino acids and by the tiny size of these bodies (30 - 500 mm). If amino acid abundances are the same in micrometeorites as in the Murchison meteorite, the sensitivity of the method must reach at least 10^{-13} moles/micrometeorite (considering the most abundant amino acid in a micrometeorite of 30μg). Amino acid analysis in samples as small as one micrometeorite is also interesting in the scope of future analysis of very heterogeneous extraterrestrial matter, or of Mars samples, which are to be expected in only small amounts.

TABLE 1. Summary of some analytical methods and results obtained for Murchison

Sample Amount	Enantiomeric determination	Methods Separation and detection (derivatizing reagent)	Results	Author
10 g Murchison	Direct (a)	GC-MS (N-trifluoroacetyl butyl ester)	7 amino acids identified D/L ≤ 1	[7]
2 g Murchison, Allende, Muray	Indirect (b)	HPLC-UV	Equivalent distribution between Murchison and Muray	[15]
20 g – 60g [1] K/T sediment[1] Geological samples[2]	Indirect and Direct	HPLC-Fluo (OPA-NAC) Confirmation: GC-MS[1] (N-trifluoroacetyl butyl ester)	D/L isovaline ≤ 1 2≤ AIB/ Isov ≤4	[1][16] [2] [17]
Unknown quantity Murchison	Direct	GC-MS (N-pentafluoropropionyl ester) (N-trifluoroacetyl butyl ester)	D/L(2a-2,3dmp) ~0.83 D/L (Isovaline) ~ 0.84	[3]
>450 mg ALH 84001 Murchison (100 mg)	Indirect	HPLC-Fluo (OPA-NAC)	Contamination AIB< LOD (c)	[18]
30-35 Micrometeorites (150 – 300 µg) Murchison (5g-5mg)	Indirect and Direct	HPLC-Fluo (OPA-NAC) Confirmation: GC-MS	Isovaline < LOD AIB micrometeorite >Murchison D/L not determinable	[8]
7.3 g Murchison	Direct	GC/C/IRMS	Confirm an extraterrestrial source for an L-enantiomer excess	[19]

(a) achiral derivatization and chiral separation.
(b) chiral derivatization and achiral separation.
(c) LOD: limit of detection.

A critical survey of classical method used in this field (gas chromatography coupled to mass spectrometry, high performance liquid chromatography after OPA-NAC derivatization) revealed that they are not sufficiently sensitive (see table 1). None of these two methods efficiently prevents contamination since both involve the formation of unstable derivatives at the last step prior to analysis.

We recently developed Chloroethylnitrosoureas (CENU) as new derivatizing reagents for amino acids and other nucleophiles [10], for their quantitative analysis by capillary electrophoresis with laser induced fluorescence (CE-LIF) [11, 12] and immunoenzymatic methods (ELISA). CENUs offer new possibilities to prevent contamination since the involved derivatives are stable [13]; moreover their structure may be modified for optimizing the sensitivity without changing their derivatizating properties [10].

Our first investigations focus on non-biological aminoacids, like α-aminoisobutyric acid (AIB) and isovaline, to reduce the problem of contamination. The separation of AIB, isovaline L and D in CE is performed in a carbonate or a borate buffer with a non ionic surfactant. In this kind of separation the limit of detection is inferior to 10 pM and the detection is linear between nM to pM. The sensitivity obtained in CE-LIF is 100 times better than the usual GC-MS method (method that we have also developed) [14].

The first step to use ELISA methods is to obtain specific antibodies. It is impossible to generate directly antibodies against small molecules (haptens) like amino acids. Carriers ("big molecules") coupled to the haptens must be used to induce an immunological response, and so, to produce antibodies against the haptens-carrier complex. Our strategy is to use several different carriers to generate antibodies which are specific against the constant part of the immuno-conjugate. After 6 months, antibodies were obtained and their specificity is evaluated by a competition method. We have shown that the only compound that change the antibodies response is the immuno-conjugate that have been used to produced the antibodies. The enantiomeric recognition is very good. No cross reactivity is observed between isovaline L and D. So, we have demonstrated that specific antibodies can be obtained to allow enantiomeric recognition in complex samples.

At this stage of our work, the new analytical methods we develop for amino acids determination with CENU derivatization coupled with CE-LIF or ELISA allows us to optimize the sensitivity and/or selectivity of the measurement with respect to classical GC-MS or MS-MS methods. These new methods decrease the risk of contamination as the stable derivatives can be formed earlier in the analytical process. The reduction of the number of contamination steps is an important point concerning terrestrial measurement. The structure of our derivatizing reagent can be changed without any modification of its derivatization properties, giving a high adaptability to the method.

Acknowledgements. This work is supported by the CNES (Centre National d'Etudes Spatiales, France).

References

[1] Bada, J.L. and Mc Donald, G.D. 1995. *Icarus* **114**, 139
[2] Bland, P.A. and Smith, T.B. 2000. *Icarus,* **144**, 21
[3] Cronin, J. R., Pizzarello, S. 1997. *Science* **275**, 951
[4] Engel, M.H., Macko S.A. and Silfer, J.A. 1990. *Nature*, **348**, 47.
[5] Engel M.H. and Nagy, B. 1982. *Nature*, **301**, 494.
[6] Bada, J. L., Cronin, J. R., Ho, M.-S., Kvenvolden, K. A., Lawless, J. G., Miller, S. L., Oro, J. and Steinberg, S. 1983. *Nature* **301**, 494.
[7] Kvenvolden, K.A., Lawless, J., Pering, K., Peterson, E., Flores, J., Ponnamperuma, C., Kaplan, I.R., Moore. C. 1970. *Nature*, **228**, 923.
[8] Brinton, K. L.; Engrand, C.; Glavin, D. P.; Bada, J. L.; Maurette, M. 1998. *Orig. Life Evol. Biosphere*, **28**, 413
[9] Maurette. M. 1998. *Orig. Life Evol. Biosphere*, **28**, 385.
[10] Vandenabeele, O., Garrelly, L., Ghelfenstein, M., Commeyras, A. and Mion. L. 1998. *J. Chromatogr. A.* **795**, 239; Vandenabeele, O., Garrelly, L., Commeyras, A. and Mion. L. Patent PCT/FR98/01984 September, 16 1997.
[11] Vandenabeele-Trambouze, O., Albert, M. Bayle, C., Couderc, F., Commeyras, A., Despois, D., Dobrijevic, M. and Grenier-Loustalot. M.F. 2000. *J. Chromatogr. A.*, **894**, 259.
[12] Trambouze-Vandenabeele, O., Grenier-Loustalot, M-F., Albert, M., Despois, D., Dobrijevic, M., Commeyras, A., Geffard, G., Couderc, F. and Bayle, C. 2000. *J. Biol. Chem. Lum.*, **15**, 91.
[13] Vandenabeele-Trambouze, O., Garrely, L., Mion, L., Boiteau, L. and Commeyras. A. 2000. *Adv. Environ. Res.*, in press.
[14] Rodier, C., Vandenabeele-Trambouze, O., Stenberg, R., Coscia, D., Raulin, F., Vidal-Madjar, C., Grenier-Loustalot, M.F., Despois, D., and Dobrijevic. M. 2000. *Advances in Space Research.* In press.
[15] Cronin, J. R. and Moore, C. B. 1971. *Science,* **172**, 1327.
[16] Zhao, M. and Bada, J.L. 1989. *Nature*, **339**, 463.
[17] Zhao, M. and Bada, J.L. 1995. *J. Chromatogr.*, **690**, 55.
[18] Bada, J. L., Glavin, D. P., McDonald, G. D. and Becker, L. 1998. *Science,* **279**, 362.
[19] Engel M.H. and Macko, S.A. 1997. *Nature*, **389**, 265.

CHANGES ON THE SURFACE OF MARS RESULTING FROM INNER FACTORS AND IMPACTORS

ERNESTO PALACIOS-PRÜ[1,2] *AND VICENTE MARCANO[1]
[1] *Electron Microscopy Center, Evolutionary Biology and Chemistry Laboratory University of the Andes, P. O. Box 163, Mérida, Venezuela.* prupal@ula.ve
[2] *Internacional Institute of Advanced Studies, USB, Caracas, Venezuela.*

1. Introduction

Images of the Martian surface transmitted to Earth by the Mars Global Surveyor spacecraft were recently interpreted as evidence of the presence of liquid water (Malin & Edgett, 2000). This finding would suppose the occurrence of gullies within the walls of a very small number of impact craters, south polar pits, and two of the larger Martian valleys display geomorphic features that could be explained by processes associated with groundwater seepage and surface runoff (Malin & Edgett, 2000). Careful examination of the images acquired by Mars Global Surveyor Mars Orbiter Camera (MOC2-234-245) allowed by JPL/NASA/Malin Space Science Systems, showed features that were far away from those interpretations, because it is difficult to conceive runoff, gullies or outflows as originated from the effects of liquid water running on the current Martian surface. As a result, several explanations are explored in this paper obtained mainly from terrestrial examples.

2. Thermal Oscillation Effects

Mars has a daily wide thermal oscillation showing mean values of 50°C. Thus, during the day, in Chrysis Planitia and in Utopía Planitia surface temperatures up to –30°C have been recorded; during the night, temperature may go down as below as –83°C. Wider oscillations have occurred at the subtropical regions, registering 27°C during the day and –89°C during the night in the soils (Tillman et al., 1993; Jakosky et al., 1995). On Earth, the maximum daily thermal oscillations have been recorded in the Andean Páramos, 28.2°C; Guayana´s Highlands, 26.4°C (Azócar & Monasterio, 1980, Marcano, 1998), and Mojave desert, China Lake, ~ 20°C. In Mars, these temperature changes may give place to fractionations on rocks and soils on cliffs which may be transported down the slope (Figure 1) (Sullivan, 1992), similarly as occur in the desertic Páramos during the

* To whom correspondence should be mailed.

J. Chela-Flores et al. (eds.), First Steps in the Origin of Life in the Universe, 289–292.

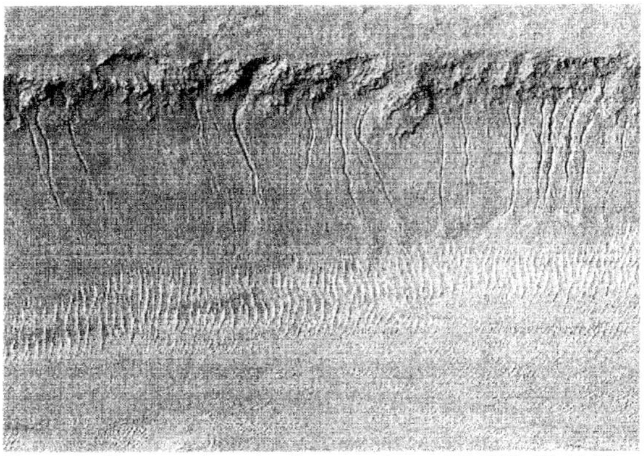

Figure 1. Dunes and erosive lines on slopes, 29.4° S, 39,1° W (MOC2-244, JPL/NASA/Malin Space System)

night (Schubert, 1980), where stripes, and long erosive lines are also formed from freezing-defrosting cycles. Some studies show that weathering and weakening of initially stronger martian slope materials penetrated to depths of ten meters before downslope failure of weathered material occurred, creating the observed avalanche chutes (Sullivan, 1992). These temperature changes, as occurring in Mars, are able to produce cracks on the ground having polymorphic shapes, and thereby outbursting during the warmer days all the hypothetical water retained below ground (permafrost).

3. Pressure Effects

Pressures recorded on Mars correspond to 6.36 mbars with variations of 37% (Owen et al., 1977). This very low pressure, about 100 times less than it is at sea level on Earth (1 bar), can cause water boiling at temperatures higher than 273° K. Furthermore, the atmospheric surface pressure on Mars is remarkably close to the water triple point pressure of 6.1 mbars. Thus, it is very difficult to believe that liquid water run on the surface of Mars forming outflows or gullies without being freezed or evaporated, mainly during a warm afternoon. Further, both low pressure and fragility of the soils by wethearing can cause continuous collapses (Sullivan, 1992) on the surface around the internal margins of craters or valley walls, showing false runoffs or gullies.

4. Wind Effects

Because of the high thermal oscillations being recorded daily in Mars, intense winds are generated mainly toward the equatorial region, having mean velocities of 2 m/sec and gusts up to 15 or 19 m/sec (Murphy et al., 1990). These winds are constant during the whole Martian year and can produce an erosive process that continuously modify the topography giving place to news reliefs and formations such as dunes, monticles,

Figure 2. Dunes, erosive lines and stripes, 29.7° S, 38.6° W (MOC2-240, JPL/NASA/Malin Space System).

channels, polygons, erosive lines (Thomas & Gierasch, 1995), which may be mistaken for and interpreted as erosions produced by liquid water. In Earth, desertic Páramos > 4500 m, show winds with mean velocities of 2.8 m/sec (Azócar & Monasterio, 1980), and gusts with velocities up to 25 m/sec, as occurring on the higher peaks, e.g. Bolívar, Espejo (Marcano, 1994) in Venezuela. These winds may occasion modelling processes of granitic rocks on their surfaces (Schubert & Vivas, 1993), including formation of long and deep erosive lines and sorted and non-sorted stripes (Schubert, 1972; Beaty, 1974).

5. Oxidative Effects

Martian atmosphere is mainly composed by 95.32% CO_2, 2.7% molecular nitrogen and 1.6% argon (Owen et al., 1977). High oxidative atmosphere would generate deterioration of the soil mainly by HCl, O_3, NO_x and H_2O_2 (Hartman & McKay, 1995; Encrenaz et al., 1995) and therefore would cause the formation of stripes or erosive lines similar to those formed by liquid water running on the surface. This effect may also occur on cliffs producing weakening on the surface and, thereby creating dry debris avalanche chutes (Sullivan, 1992).

6. Volcanic and Seismic Effects

An intense volcanic activity (Hodges & Moore, 1993) and plate tectonics (Sleep, 1994) would have been demonstrated could be occurring on Mars. These processes could originate ruptures of lands (faults) or volcanic debris channels over very large distances (Postawko & Mouginis-Mark, 1987) and to simulate erosion lines or false runoffs, such as seen in Alba Patera.

7. Residual Effects of Impacts

The residual effects of shock waves, generated by asteroidal impactors having variable mass and diameter (≤ 1.36 x 10^{20} kg, ≤ 440 km) during the Martian history (Leighton et al., 1965; Melosch & Vickery, 1989; Squyres & Kasting, 1994), such as kinetic energy deposition (up to 1.7 x 10^{20} Joules), impact pressure and high temperatures (> 5,000 K) could give place to scars that may be interpreted as gullies and runoffs. Seismic and gravity wave oscillation effects may also produce weakening on the surface around the internal margins of craters (Melosh, 1982), which could be misread as gullies and runoffs. These effects are very frequent on Hellas Planitia, which is considered the longer second impact basin existing in our solar system (about 2000 km in diameter) (Leighton et al., 1965).

In conclusion, the occurrence of water on Mars near polar caps, have been suficiently demonstrated by the Mariner IV, VI and VII, Viking I and II and Pathfinder spacecraft's and more recently by the Mars Global Surveyor. Although there is evidence of the existence of subterranean water (permafrost) in Mars, it appears that solid evidences of liquid water running on the current surface of Mars are still lacking.

8. References

Azócar, A. & Monasterio, M.: 1980, in: *Estudios Ecológicos en los Páramos Andinos*, ed. M. Monasterio, Ediciones de la Universidad de Los Andes, Mérida, p. 207-223.

Beaty, C. B.: 1974, *Geology* **2**, 565-567

Encrenaz, Th., Bezard, B., Crovisier, J., Coustenis, A., Lellouch, E., Gulkis, S., Atreya, S. K.: 1995, *Planetary and Space Science* **43**, 1485-1516.

Jakosky, B. M:, Henderson, B. G. & Mellon. M. T.: 1995, *J. Geophys. Res.* **100**, 1579-1584.

Hartman, H. & McKay, C.: 1995, *Planetary and Space Science* **43**, 123-128.

Hodges, C. A. & Moore, H. J.: 1993, *U.S. Geological Survey Professional Paper* 1534.

Leighton, R. B., Murray, B. C., Sharp, R. P., Allen, J. D. & Sloan, R. K.: 1965, *Science* **149**, 627-630.

Malin, M. C. & Edgett, K. S.: 2000, *Science* **288**, 2330-2335.

Marcano, V. 1994, *Studies in Andean lichens*. Vol. I, Koeltz Scientific Books, Germany, p. 63.

Marcano, V. 1998. Ecophysic study of forested microrrefuges from the Gran Sabana, Venezuela and of their lower plant and fungus communities. *Rev. Ecol. Latinoamericana* **5**, 21-48.

Melosh, H. J.: 1982, *J. Geophys. Res.* **87**, 371-380.

Melosch, H. J. & Vickery, A. M.: 1989, *Nature* **338**, 487-489.

Murphy, J. R., Conway, B. L. & Tillman, J. E.: 1990, *J. Geophys. Res.* **95**, 14555-14576.

Owen, T., Biemann, K., Rushneck, D. R., Biller, J. E., Howarth, D. W. & Lafleur, A. L.: 1977, *J. Geophys. Res.* **82**, 4635-4639.

Postawko, S, E. & Mouginis-Mark, P.: 1987, *Possible origin of some channels on Alba Patera, Mars*, in Reports of Planetary Geology and Geophysics Program, SEE N87-23341 16-91, p. 303-304, NASA, Washington.

Schubert, C. : 1972, *Acta Científica Venezolana* **23**, 108-152.

Schubert, C.: 1980, in: *Estudios Ecológicos en los Páramos Andinos*, ed. M. Monasterio, Ediciones de la Universidad de Los Andes, Mérida, p. 29-46.

Schubert, C. & Vivas, L. 1993. *El cuaternario de la Cordillera de Mérida, Andes Venezolanos*. Universidad de Los Andes y Fundación Polar, Mérida, pp. 345.

Sleep, N. H.: 1994, *J. Geophys. Res.* **99**, 5639-5655.

Squyres, S. W. & Kasting, J. F. 1994, *Science* **265**, 744-749.

Sullivan, R. H.: 1992, *Evolution of steep Martian slopes*, Ph.D. Thesis Arizona State Univ., Tempe.

Tillman, J. E., Johnson, N. C., Guttorp, P. & Percival, D. B.: 1993, *J. Geophys. Res.* **84**, 10963-10971.

Thomas, P. C. & Gierasch, P. J.: 1995, *J. Geophys. Res.* **100**, 5397-5406.

SPATIAL AND TEMPORARY PATTERNS OF SOME CLIMATE PARAMETERS AROUND THE TIMBERLINE OF PICO DE ORIZABA

LUIS CRUZ- KURI
Instituto de Ciencias Básicas. Universidad Veracruzana
Carr. Xalapa-Ver., Km. 3.5, Las Trancas, Xalapa, Ver., MEXICO

CHRISTOPHER P. MCKAY
NASA-Ames Research Center
Moffett Field, CA 94035, USA

RAFAEL NAVARRO-GONZALEZ
Laboratorio de Química de Plasmas y Estudios Planetarios
Instituto de Ciencias Nucleares, UNAM, Circuito Exterior, Ciudad Universitaria, Apartado Postal 70-543, MEXICO

Abstract. Pico de Orizaba, the highest mountain in Mexico, has environmental conditions that probably could have occurred in ancient Mars and may also be relevant to the future introduction of life on Mars. Related to this, we address in this paper the question why life is constrained at high altitudes in our planet; we particularly focus to the search of those factors that determine timberline. Since March 1999 we started to monitor the environmental conditions of the mountain, recording among other measurements, soil temperature at various depths in different locations within, above and below the timberline of Pico de Orizaba. This is a preliminary report of some findings related to the search of spatial as well as temporal patterns of these parameters.

1. Introduction

Pico de Orizaba (19°N) has the highest treeline in the world (4200 meters) and is the subject of an ongoing study by us supported jointly by the National Aeronatics and Space Administration, (NASA), the National Autonomous Universiy of Mexico (UNAM), and the University of Veracruz (UV). This ongoing work is related to the survival of ecosystems in extreme environments with applications to life on Mars. A subsequent project is contemplated which would expand this ongoing work to consider the response of tropical treeline to global warming and to increased human settlement on the mountain.

The treeline in temperate zones appears to be determined by summer temperatures--- trees grow when the warmest summer month has an average temperature of 10°C or above. We hypothesize that tropical treelines, in contrast, are more sensitive to the annual average temperature. In addition our current work indicates that the treeline on Pico de Orizaba reflects the temperature dependence of the nitrogen cycle. Thus nitrogen generated by human activities could effectively fertilize the soil at treeline altering its position. Thus the sharp discontinuity of treeline may provide an important indicator of global warming and human activity.

J. Chela-Flores et al. (eds.), First Steps in the Origin of Life in the Universe, 293–301.

A detailed database is needed to quantitatively understand the current position and elevation of treeline. The present report addresses some aspects pertaining to the above undertaking, mainly from the perspective of soil temperature as well as external temperature at various locations of the mountain where we have established meteorological *stations*. At some selected sites, inner tree-trunk temperature is also being recorded for pines of the domineering species *P. hartweggi*. Other variables such as relative and absolute humidity, light intensity, dew point, etc., have been included too. In this report, the main objective refers to detection of spatial as well as temporary patterns of these parameters. Due to space limitations, our findings are presented in a partial fashion.

2. Methods

Several locations in the mountain were selected above, below and within the timberline of Pico de Orizaba in each one of two faces, namely, the southern face and the northern face. These locations constitute our meteorological stations. The altitude of the stations varied from 2926 m to 4474 m. The coordinates for our stations were measured with a GPS instrument. In each station, one or more loggers were deployed, some of them buried underground, others left on the ground. Several types of loggers were used: with integrated temperature sensors, with light and humidity detectors as well as loggers with external sensors capable of measuring soil temperatures at various depths as well as inner tree-trunk temperature. The sensors of the multiple channel loggers were placed underground at depths of 10 cm, 20 cm and 40 cm, the other type of loggers with integrated sensors were either placed inside closed cases and buried at 10 cm depth or externally with open cases. All types of loggers have the capability of storing information at a selected frequency which can vary from fractions of a second to several hours. For our study, intervals of one hour were chosen; with this, depending of the type of logger, there is a storing capability of three to five or more months. The raw data were retrieved *in situ* approximately every three months either with the use of a laptop computer or with a more specialized instrument available from the maker of the loggers (data shuttle). In each case, a specific purpose software was required for these tasks (BoxCarPro4). This software allows some graphical processing of the data with the possibility of zooming at selected parts of the graphs in order to show more details; the statistical processing are very simple, however, the software permits the exportation of data in the custom mode so one can use other computing packages. The resulting files were transferred to desk top computers for subsequent processing and analyses. For the analyses presented here, a statistical package was used (SPSS) which allows several types of processing (graphical or numerical) including the ones with a multivariate approach, *e.g.*, cross correlations of time series.

3. Results and Discussion

The first meteorological stations were established in the southern face of the mountain on March of 1999, followed by stations on the northern face later the same year. Previous to the time of the most recent retrieval of logger data, on November of

2000, every face had six stations. Immediately afterwards, more stations were added, also the number of sensors have been increased from three to four in the loggers of multiple channels. For the present report, we restrict ourselves to what was available up to November of 2000. Each retrieval of information spanned a period lasting anywhere from two to five months approximately. The resulting data files were analyzed either individually or by fusion of one or more files. Both, the BoxCarPro4 and SPSS software were used for several types of applications. Due to space considerations, for this report we have made a selection of some files in some of the stations and a choice of only a few typical processing. Tables, graphics and comments follow.

In the timberline region, where stations South 4 and North 4 lie, two subdivisions were made for each station, namely, "within the trees", that is, in the zone where trees are present, and "outside the trees", where there are no trees. Figures 1 to 2 below show the temperatures hourly recorded for the period from May to August of 2000 corresponding to station North 4 (altitude 4061 m) in the zone within the trees. Soil temperatures were measured for three depths, 10cm, 20 cm and 40 cm. It becomes apparent from the zoomed part of Fig. 1, consisting of 40 days for a selected range of dates, that the temperatures are modulated by a wave with a period of approximately twenty days; of course there is also the daily periodicity. This suggests performing more precise types of analyses for the corresponding time series.

Temperatures for depths of 20 cm and 40 cm are presented in Figure 2. Notice the narrowing bands.

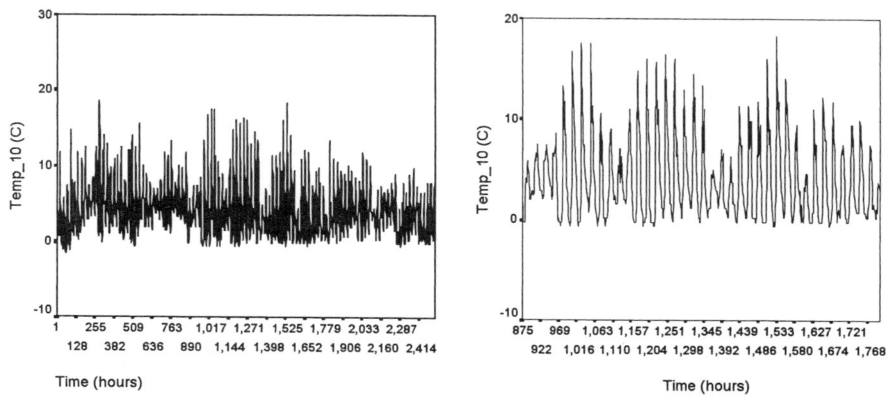

Figure 1. Soil temperatures at 10 cm depth and zoomed, within trees, north station 4

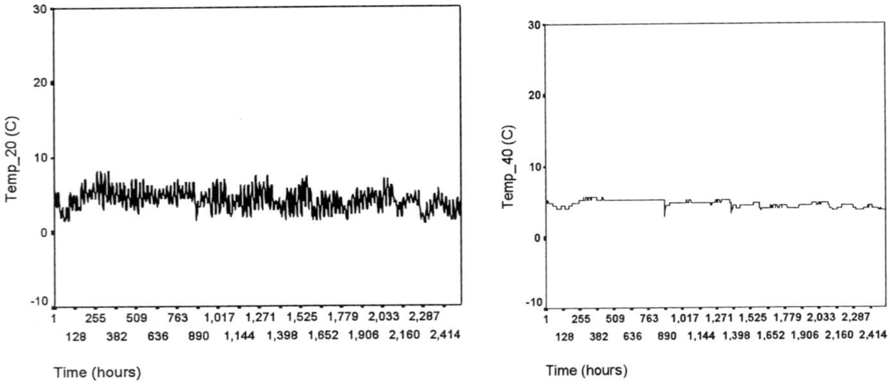

Figure 2. Soil temperatures at 20 cm and 40cm depth, within trees, north station 4

Table 1 below shows descriptive statistics for the same data. As expected, the fluctuations, as measured by standard deviations, decrease with depth of soil; the averages are alike for the three depths, however, herethe relevant influence must be attributed to the difference of variation.

Table 1. Descriptive statistics for soil temperatures at three depths, within trees, north station 4

	N	Minimum	Maximum	Mean	Std. Deviation
Temp_10	2520	-1.51	18.66	4.1311	3.3285
Temp_20	2520	1.17	8.23	4.2742	1.4038
Temp_40	2520	2.89	5.81	4.7940	.5508

Table 2. Descriptive statistics for soil temperatures at three depths, within trees, south station 4

	N	Minimum	Maximum	Mean	Std. Deviation
Temp_10	2555	-1.97	21.71	5.2236	5.2438
Temp_20	2555	3.31	9.42	5.5885	1.4320
Temp_40	2555	4.15	8.23	5.3965	1.1167

Figure 3 below corresponds to north station number 4 and shows the correlograms of two of the three time series for the region within trees. Cross correlations were computed with lags from –7 hours to +7 hours. Likewise Fig 4 shows the counterpart for south station number 4.

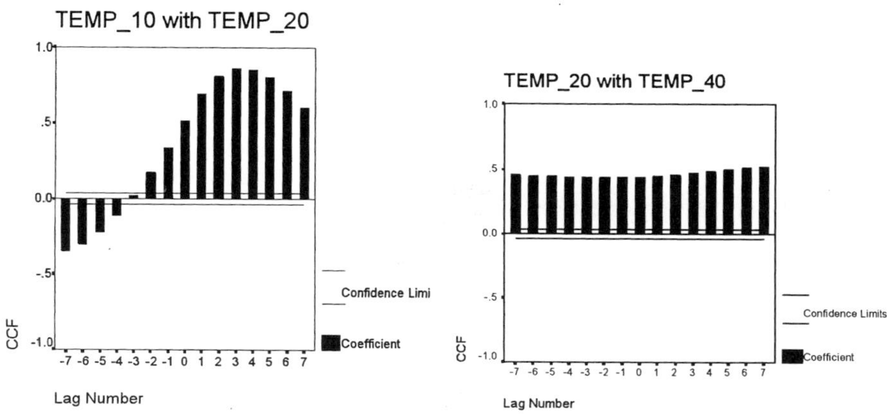

Figure 3. Cross correlations, temperature at 10 cm with temperatures at 20 cm and 20 cm with 40 cm, within trees, north station 4. Lags numbers in units of one hour each

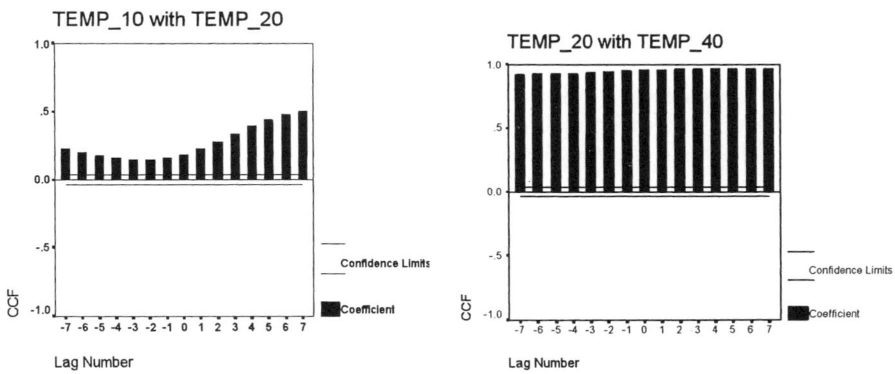

Figure 4. Cross correlations, temperature at 10 cm with temperature at 20 cm, and 20 cm with 40 cm within trees, south station 4

From Figure 3, first part, it can be noticed that the maximum correlation between temperatures at 10 cm and 20 cm occurs when there is a time lag of +3 hours; the correlations being positive for time lags from –3 hours to +7 hours indicate that both temperatures vary with the same pattern, *i.e.*, both temperatures tend to increase together or both tend to decrease together, but not simultaneously. This has to do with the phenomenon of heat conduction within the soil. For lags from –7 hours to –4 hours, the correlations are negative indicating that, when one temperature increases there is a tendency for the other one to decrease, and vice versa. The second part of Fig. 3 shows that heat conduction from a depth of 10 cm is fairly correlated with the one for a depth of 40 cm for any time lag; something similar happens for the correlations between temperatures at 10 cm and 40 cm, although the relationship is weaker.

The first part of Fig. 4 above shows that heat conduction from a depth of 10 cm to 20 cm has a maximum correlation at the time lag of +7 hours, being the weakest correlations at 0 and –1 hours, in agreement with the expected warming and cooling process. The correlations for temperature at 10 cm with temperature at 40 cm fluctuate between 0.255 and 0.343 with same pattern as the one in the first part of Fig. 4. In contrast, the second part of Fig . 4 displays strong correlations ranging from 0.921 to 0.965, which means that knowledge of the temperature at depth 20 cm at a given time almost determines in a linear fashion the temperature at 40 cm for any time within an interval that goes from –7 hours to +7 hours. Suffice to say for the time being that heat flow occurs at a faster rate. This pattern is preserved also to a lesser extent for depths of 10 cm to 40 cm. The graphs provide confidence bands for the correlations; in all cases the correlations are significantly distinct from zero. All these phenomena will be analyzed subsequently.

For the time period spanning from August of 2000 to November of 2000 and for south station 3, another logger recorded external temperature (above the ground) as well as other parameters such as relative humidity, absolute humidity, dew point, luminosity, etc. Two cross correlations are shown below. The first part of Fig. 5 indicates a strong correlation between external temperature and absolute humidity for a time lag of zero hours and for other time lags they decrease in a bell shaped pattern with ranges of 0.471 to 0.458 at the extremes and a standard error of 0.024. Where the highest correlation is estimated as 0.941 with a standard error of 0.024; thus almost 90% (R^2) of the variation of environmental temperature can be accounted in terms of the variation of absolute humidity. The second part of Fig. 5 shows that correlations between external temperature and relative humidity are not as strong ; the maximum correlation of 0.506 occurs at lag zero hours. A similar analysis can be performed for soil temperature with absolute humidity and external temperature.

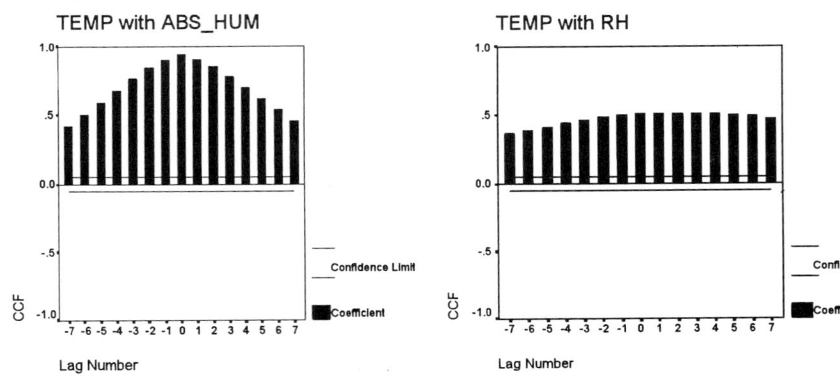

Figure 5. Cross correlations, external temperature with absolute humidity (gm/m^3) and external temperature with relative humidity (%). South station number 3

Similar types of analyses were performed for all the other stations and for different periods of time. An interesting feature is that the corresponding correlograms, for the same variables at the same location, but for different periods, exhibit a variety of patterns ranging from bell-shaped, to U-shaped and also to ones with negative components as well as with positive components. This can be accounted for with the already mentioned phenomenon of heat conduction in the soil, whose properties may vary with soil humidity and other internal conditions, but it may be influenced too by external conditions. All these factors will be taken into consideration in subsequent undertakings.

Finally, one more example with cross correlations is presented; namely, for stations 2 to 5 of the southern face, with data spanning from March of 1999 to August of 2000. Table 3 below shows descriptive statistics of soil temperatures at 10 cm depth.

Table 3. Descriptive statistics for soil temperatures at 10 cm depth, south stations 2, 3, 4 and 5

Temp	N	Minimum	Maximum	Mean	Std. Deviation
Temp_St2	12406	.73	18.66	6.2562	2.7415
Temp_St3	12408	.29	22.09	9.1440	3.1266
Temp_St4	12408	2.03	13.70	6.0341	1.8011
Temp_St5	12408	2.89	12.93	7.0418	1.7355

It can be seen that, for the period considered, station 3 (4210 m) presents the highest average temperature as well as the smallest minimum and the largest maximum; also, this station exhibits the largest temperature variation as measured by the standard deviation. The next largest mean corresponds to station 5 (3923m); also, this station shows the smallest temperature variation. Station 2 (4369 m) has the second largest temperature variation. Figures 6 to 7 below show how the temperatures are distributed for each station.

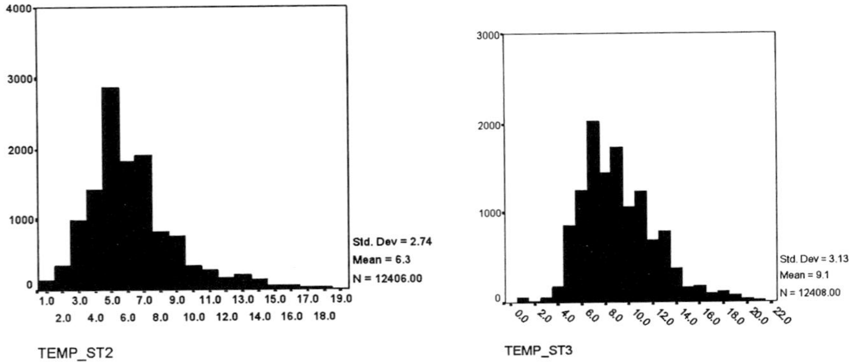

Figure 6. Frequency distributions for soil temperatures at 10 cm depth, south station 2 and 3 respectively

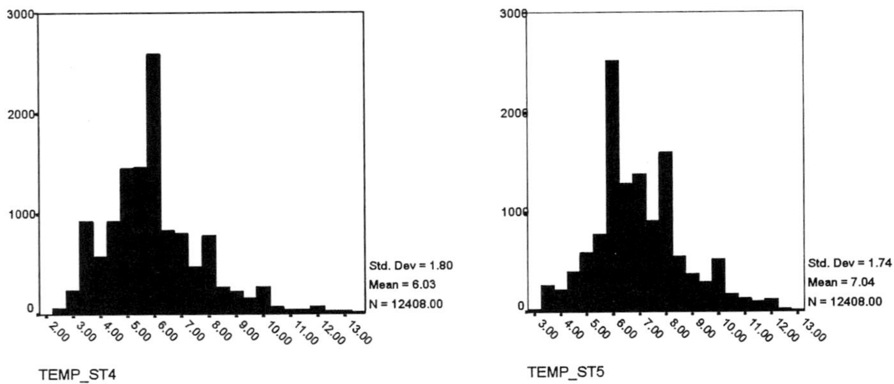

Figure 7. Frequency distributions for soil temperatures at 10 cm depth, south stations 4 and 5 respectively

As a further illustration it is shown below in Fig. 8 cross correlations for two combinations chosen from stations 2 to 5. Station 4 corresponds to the timberline.

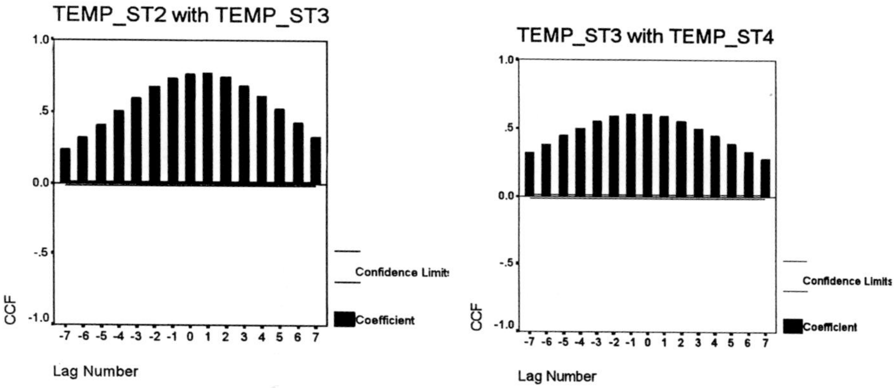

Figure 8. Cross correlations for soil temperature at 10 cm depth at different altitudes, south face.

As a concluding comment it is pertinent to say that one of our next goals is to develop a detailed computer data base of the vegetation and soil type on Pico de Orizaba, Mexico. Particular emphasis will be on the distribution of trees near treeline and the altitude of treeline itself.

Acknowledgements

We are indebted to several persons for a variety of supporting activities. For most valuable logistical support as well as field work at Pico de Orizaba to Roberto Cruz-Kuri and Flavio Leyva-Ruiz. To Rafael Navarro-Aveves, Itzel Pérez-Chávez, Nahúm Castillo-González and José Mora-Domínguez for field work; the last two also for their computing assistance. Raúl Hernández-Montoya contributed substantially for the final version of this document. Two of us (Cruz-Kuri and Navarro-González) warmly acknowledge Prof. Julian Chela-Flores for the generous support to participate in the 2000 Sixth Trieste Conference on Chemical Evolution –FIRST STEPS IN THE ORIGIN OF LIFE IN THE UNIVERSE.

References

Arno S.F. and Hammerly R.P. (1984) Timberline—Mountain and Artic Forest Frontiers. The Mountaineers. Seattle

Pérez-Chávez I., Navarro-González R., McKay C. P. and Cruz-Kuri L. (2000) Tropical Alpine Environments: A Plausible Analog for Ancient and Future Life on Mars, *Astrobiology: Origins from the Big-Bang to Civilisation*, Kluwer Academic Publishers, 297-302.

Körner C. (1998) A Re-assessment of High Elevation Treeline Positions and Their Explanation, *Oecologia* **115**, 445-459

Section 7:

The search for prebiotic and biological indicators in the satellites of the outer solar system

THE SATELLITES OF SATURN

TOBIAS OWEN

Institute for Astronomy
University of Hawaii
Honolulu, HI, USA

Abstract: Saturn's family of satellites includes manifestations of chemical evolution and offers a possible prototype for habitable satellites in planetary systems containing super-giant planets. Titan and the dark side of Iapetus reveal the most obvious evidence of organic chemistry. New observations of Titan in the near infrared have allowed the production of maps of the satellite's surface at three different wavelengths. These maps, together with additional spectroscopic measurements, allow us to place constraints on the location and extent of possible hydrocarbon oceans and large-scale deposits of some types of organic aerosols. We have made a new interpretation of the material covering the dark, leading hemisphere of Iapetus that suggests the presence of nitrogen-rich (N/C > 1) organic material in addition to amorphous carbon and water ice. The origin of this organic material remains obscure, but nearby Titan with its nitrogen-dominated atmosphere and ongoing production of aerosols is an obvious suspect, which can be queried by future observations. The Titan-Saturn pair can be scaled to hypothetical satellites around the newly discovered extrasolar planets to test the idea that satellites of these planets, rather than the planets themselves, might provide habitable environments.

References.

Meier, R., Smith, B.A., Owen, T., and Terrile, R.J. (2000) The Surface of Titan from NICMOS Observations with the Hubble Space Telescope, *Icarus* **147,** 309-316.

Owen, T. (2000) On the Origin of Titan's Atmosphere, *Planet. Space Sci.* **48,** 747-752.

Owen, T., Cruiksnak, D.P., Dalle Ore, C.M., Geballe, T.R., Roush, T.L., de Bergh, C., Meier, R., Pendleton, Y., and Khare, B.N. (2001) Decoding the Domino: The Dark Side of Iapetus, *Icarus* **149**, 160-172.

J. Chela-Flores et al. (eds.), First Steps in the Origin of Life in the Universe, 305.

TITAN : EXOBIOLOGY AND THE CASSINI-HUYGENS MISSION

F. RAULIN[1], P. COLL[1], R. NAVARRO-GONZALEZ[2], S. RAMIREZ[2], Y. BENILAN[1], F. SHINDO[1], V. VUITTON[1] & M-C. GAZEAU[1]
[1]*LISA, CNRS & Universités Paris 12 et Paris 7, Avenue du Général de Gaulle, 94010 Créteil Cedex France*
[2]*Laboratorio de Química de Plasmas y Estudios Planetarios, Instituto de Ciencias Nucleares, UNAM, 04510 México*

1. Introduction

In the solar system, one can consider two classes of extraterrestrial bodies of prime interest for Exobiology. These are planetary bodies where (extinct or extent) life may be present, such as Mars and Europa, and bodies where a complex organic chemistry is taking place. Titan, Saturn's largest satellite is probably, with the comets, one of the most exobiologically interesting bodies of this second kind.

Titan is the only satellite in the Solar system having a dense atmosphere and a surface not visible (due to the presence of atmospheric hazes and aerosol layers). In spite of noticeable differences (specially in size and temperature ranges), several similarities between the environments of Titan and of the primitive Earth can be pointed out (Clarke and Ferris, 1997). Both have a dense N_2 atmosphere with the presence of greenhouse gases and anti-greenhouse species inducing analogous temperature vertical profiles, with a troposphere and a stratosphere. Titan's atmosphere also includes a noticeable fraction of methane (up to a few percent), with a low molecular hydrogen mole fraction (Table 1). The atmospheric chemistry of methane, coupled to that of nitrogen allows the formation of many organics in the atmosphere, making Titan's environment very rich in organics, in both gas and condensed phases. In addition, the presence of such concentrations of methane in the atmosphere suggests the presence of a methane reservoir on the surface, which could be lakes, small seas or subsurface oceans (Lunine, 1993 ; 1997, and refs. included) of methane and ethane, the main product of methane photochemistry.

Several review papers have already been published on the prebiotic-like chemistry which is going on in Titan's environments (see Coll et al, 1998 ; Coll & Raulin, 1999 ; Raulin & Owen, 2000 ; and refs. included). Very recently the exobiological implications of the internal structure model of Titan has also been discussed (Fortes, 2000). In the present paper, we will mainly consider some new data related to Titan's organic chemistry in its atmospheric environment and will briefly

J. Chela-Flores et al. (eds.), First Steps in the Origin of Life in the Universe, 307–314.

discuss their implications for the future exploration of Titan, through the Cassini-Huygens mission.

TABLE 1. Titan's stratosphere : main composition, detected organics and O-compounds, and comparison with the products of simulation experiments (adapted from Gautier & Raulin, 1997)

Compounds	Stratosphere Mixing Ratio (E=Equ.; N=North Pole)		Production In Simulation Experiments *
Main constituents			
Nitrogen N_2	0.90- 0.99		
Methane CH_4	0.017 - 0.045		
Argon Ar	0 - 0.06 ?		
Hydrogen H_2	0.0006 - 0.0014		
Hydrocarbons			
Ethane C_2H_6	1.3×10^{-5}	E	Maj.
Acetylene C_2H_2	2.2×10^{-6}	E	Maj.
Propane C_3H_8	7.0×10^{-7}	E	++
Ethylene C_2H_4	9.0×10^{-8}	E	++
Propyne C_3H_4	1.7×10^{-8}	N	+
Diacetylene C_4H_2	2.2×10^{-8}	N	+
N-Organics			
Hydrogen cyanide HCN	6.0×10^{-7}	N	Maj.
Cyanoacetylene HC_3N	7.0×10^{-8}	N	++
Cyanogen C_2N_2	4.5×10^{-9}	N	+
Acetonitrile CH_3CN	few 10^{-9}		++
Dicyanoacetylene C_4N_2	Solid Phase	N	+**
O-Compounds			
Carbon dioxide CO_2	1.4×10^{-8}	E	
Carbon monoxide CO	2.0×10^{-5}		
Water H_2O ***	few 10^{-9}		

* Mean relative abundance : Maj.= major product >> ++ >> +

** Coll *et al.*, 1999b; *** Coustenis *et al.*, 1998

2. Organic chemistry in gas phase

From theoretical modeling and experimental laboratory simulation, Titan's main atmospheric molecular composition is one of the most favorable for gas phase organic syntheses. Energy deposition from solar UV and Saturnian magnetospheric electrons should allow the formation of many organic compounds which should be present at the trace level in the atmosphere. Those are mainly hydrocarbons, both saturated and unsaturated, and N-organics, essentially nitriles. Indeed, many organics have already been detected (see for instance Lebreton, 1997), and many other should be present, both in the gas and in the condensed phases.

An efficient way to study such chemistry is the use of simulation experiments in the laboratory. Many have already been carried out to mimic Titan's organic chemistry (see Coll & Raulin, 1999, and refs. included). Until recently, only C_4N_2 remained detected (in condensed phase) in Titan and not in laboratory simulation. A full program of experimental research has been developed at LISA to clarify this problem, and to reassess the quality and representativeness of the simulation. By using silent discharge and flow systems, able to operate at low temperature, at last the formation of C_4N_2 was clearly demonstrated (Coll et al, 1999b), but at a very low mole fraction (below the ppb level). Several other organics were also detected, including other compounds never reported before in such experiments such as triacetylene, C_6H_2, and cyanobutadiyne, HC_5N. The presence of polyunsaturated compounds suggests that pathways of complexification of the chemistry in Titan's atmosphere involve polyynes and cyanopolyynes, and that the end-products of such chemical evolution, Titan's tholins, should also include such chemical families and be made of C, H and N atoms.

Several of these polyynes and cyanopolyynes should be detectable in Titan's atmosphere from observation in the IR and UV ranges. These compounds have not been widely studied in the laboratory because their preparation through the usual ways of organic chemistry is very difficult, and because they are very unstable. However, in spite of those difficulties, a systematic program of spectroscopic studies both experimental and theoretical is in progress at LISA. The UV and IR spectra of diacetylene, triacetylene and tetraacetylene are already available, with band intensities. The same kind of data should be available soon for cyanobutadiyne, HC_5N (Vuitton et al, 2001a & 2001b, and refs. included).

3. Titan's tholins

The word « tholins » was invented by Sagan and Khare (1979) (from the Greek word « toloz », meaning muddy or vault) to name the solid organic product obtained after irradiation of a variety of gas mixtures of cosmically abundant molecules (including methane and ethane) by a variety of energy sources of astrophysical importance (including UV and electrons). Indeed, in most of the experiments simulating in the laboratory the chemical organic evolution of various environments of astrophysical interest, whatever are the initial physical states involved (pure gas phase or pure solid phase, or a multiphasic system involving gas, liquid and solid or a combination of two of those) one of the main product is a complex organic solid, of very poorly known (if not known at all) molecular composition. Thus, today, the generic name « tholins » is frequently used to refer to those solid products (and to mask our ignorance).

Although we have, so far no direct information on the chemical composition of particles constituting Titan's haze layers, it seems very likely that they are made of Titan's tholins. The recent detection of C_4N_2 in simulation experiments also supports the hypothesis that the tholins obtained during these recent experiments are good analogues of Titan's aerosols (Coll et al, 1999a). The study of good analogues is essential for retrieving of observational data and developing haze models. In particular, those requires the availability of the spectra of the tholins in a wide range of wavelengths, from the UV to the IR (Khare et al, 1984), with the values of the index of refraction (real and imaginary parts), together with information on the size distribution and morphology of the particles. Systematic determination of the index of refraction (real and imaginary parts) of these good analogues are in progress at LISA. They are already available for the spectral range 0.1 to 1 micron, for tholins obtained at room temperature (Ramirez et al, 2000). They should be available for low temperature tholins, more representative of Titan's aerosols, soon.

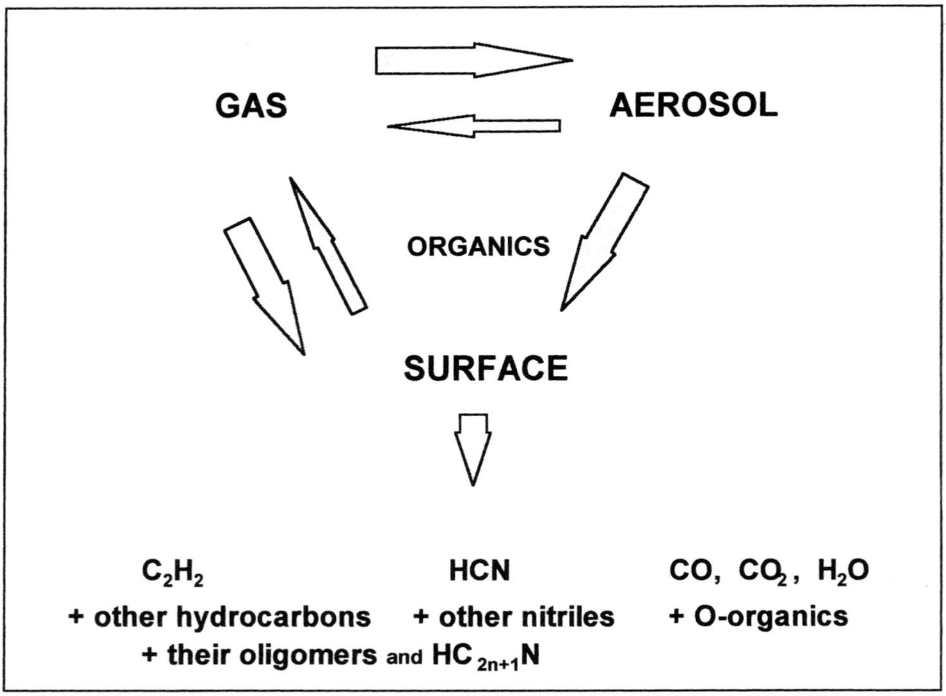

Figure 1. Organic chemistry in the three parts of Titan's geofluid and their couplings

4. The Cassini-Huygens mission.

Water has recently been discovered in gas phase in Titan's atmosphere (Coustenis et al, 1998). Liquid water is not supposed to be present permanently at the surface, although the possibility of the occurrence of liquid water, at least during short period of time during Titan's history has been considered (Thompson & Sagan, 1992). Furthermore, models of the internal structure of Titan suggest the presence in deep layers of liquid mixtures of water and ammonia where the possible presence of life cannot be excluded (Fortes, 2000). But even in the absence of liquid water, Titan can be considered as a full scale prebiotic laboratory with organic processes occurring in the different parts of its « geofluid » (gaseous atmosphere, aerosols and surface), and many chemical and physical couplings (Fig. 1).

Of course, many questions still remain, such as :
What is the origin of the atmosphere ?
What is the source of atmospheric methane ?
What is the origin and distribution of H_2O ?
Are there liquid bodies on Titan's surface ?
What is the chemical composition of the aerosols ?
Is there a prebiotic-like chemistry more complex than expected ?
Does this complex organic chemistry involve oxygen atoms ?
With the presence of homochirality in the organics ?

Several of these questions will be approached in great detail thanks to the Cassini-Huygens mission (Lebreton, 1997). The NASA-ESA Cassini-Huygens mission was successfully launched in October 1997. After a Jupiter fly-by in December 2000, the Cassini spacecraft will reach the Saturn system in 2004. Then it will become an orbiter around Saturn and fly-by Titan, while, simultaneously it will send the Huygens probe into Titan's atmosphere.

One of the main objectives of the Cassini-Huygens mission is precisely to explore in great detail this strange and exotic world which seems to be Titan, with the Cassini orbiter and the Huygens probe (Table 2). Cassini carries twelve scientific experiments which will allow a detailed exploration of the Saturn system. Six interdisciplinary scientists (IDS) will perform complementary studies by using the data derived from these experiments. The Huygens probe carries six scientific instruments. Three IDS programs are associated to Huygens, including the exobiological aspects.

Several instruments of the orbiter and most of the instruments of the probe will provide data of exobiological interest. The optical remote sensing instruments of the orbiter (specially CIRS and UVIS) will determine the chemical composition of different zones of Titan's atmosphere and should be able, in particular, to detect many organics, including new species and allow the determination of their vertical concentration profile. The Cassini Radar will be able to map Titan's surface through the haze layers, and to determine the presence and distribution of liquid bodies, if any.

TABLE 2. Cassini -Huygens Science Instruments and IDS's

Cassini Instruments and InterDisciplinary Programs	P.I., Team Leader or IDS		Exobiological Return
<u>Optical Remote Sensing Instruments</u>			
Composite Infrared Spectrometer (CIRS)	V. Kunde	USA	+++
Imaging Science Subsystem	C. Porco	USA	+
Ultraviolet Imaging Spectrograph (UVIS)	L. Esposito	USA	+++
Visual & I.R. Mapping Spectrometer	R. Brown	USA	++
<u>Fields Particles and Waves Instruments</u>			
Cassini Plasma Spectrometer	D. Young	USA	+
Cosmic Dust Analysis	E. Grün	Germa.	+
Ion & Neutral Mass Spectrometer	H. Waite	USA	+
Magnetometer	D. Southwood	U.K.	
Magnetospheric Imaging Instrument	S. Krimigis	USA	
Radio & Plasma Wave Spectrometer	D. Gurnett	USA	
<u>Microwave Remote Sensing</u>			
Cassini Radar	C. Elachi	USA	+++
Radio Science Subsystem	A. Kliore	USA	+
<u>Interdisciplinary Scientists</u>			
Magnetosphere and Plasma	M. Blanc	France	+
Rings and Dust	J.N. Cuzzi	USA	+
Magnetosphere and Plasma	T.I. Gombosi	USA	+
Atmospheres	T. Owen	USA	++
Satellites and Asteroids	L.A. Soderblom	USA	+
Aeronomy & Solar Wind Interaction	D.F. Strobel	USA	+

Huygens Instruments and InterDisciplinary Programs	P.I. or IDS		Exobiological Return
Gas Chromatograph-Mass Spectrometer	H. Niemann	USA	+++
Aerosol Collector & Pyrolyser	G. Israël	France	+++
Huygens Atmospheric Structure Instrument	M. Fulchignoni	Italy	++
Descent Imager/Spectral Radiometer	M. Tomasko	USA	+++
Doppler Wind Experiment	M. Bird	Germa.	+
Surface Science Package	J. Zarnecki	U.K.	+++
<u>Interdisciplinary Scientists</u>			
Aeronomy	D. Gautier	France	++
Atmosphere/Surface Interactions	J.I. Lunine	USA	++
Chemistry and Exobiology	F. Raulin	France	+++

On Huygens, the GC-MS instrument, a gas chromatograph with three GC capillary columns, coupled to a quadrupole mass spectrometer, will perform a detailed analysis of the atmosphere during the 2.5 hours of descent of the probe, and perhaps of the surface, after landing. The ACP experiment will collect the atmospheric aerosols, heat them at different temperatures and transfer the produced gases to the GC-MS instrument for molecular analysis. This will provide the first direct *in situ* molecular and elemental analysis of Titan's hazes. HASI will determine, in particular, the pressure and temperature vertical profiles. DISR will measure the photon fluxes, determine the cloud structure and take images of the surface. SSP will provide information on the physical state and chemical composition of the surface.

5. Conclusions

Titan does appear as a planet-size prebiotic laboratory, similar in many aspects to the primitive Earth, but also different mainly with the absence of liquid water. Studying Titan's complex organic chemistry, with all the physical and chemical couplings which are involved in the different parts of its geofluid should give us many new insights on the early Earth prebiotic chemistry, and new ideas to approach it.

The occurrence of short episodes with surface liquid water may have allowed some prebiotic chemistry in aqueous solution. Its study would require a dedicated planetary mission to explore more specifically Titan'surface : a long-term forcast (Lorenz, 2000). However, as soon as 2004, the Cassini-Huygens mission will provide many new data on Titan, essential for the field of exobiology. Its scientific return is supposed to be one order of magnitude greater than that of the Voyager mission. As an example, the CIRS experiment has a sensitivity 10 to 100 times higher, a resolution 10 times better and a spectral range much larger than IRIS, the Voyager IR spectrometer. Since 20 years after Titan's encounter by Voyager, IRIS data are still under retrieving, one can easily forcast that it will take more than a century to retrieve all Cassini-Huygens data.

Acknowledgments

This review work has been supported by grants from the French Space Agency (CNES: Centre National d'Etudes Spatiales). Some of the data presented here have been obtained at LISA with the support of CNES grants and grants from the French national program of planetology (PNP). Some have also been obtained in cooperation with UNAM thanks to supports from ECOS-ANUIES French-Mexican programs (Project M97UO1).

References

Clarke D.W. and Ferris J.P. (1997) Chemical evolution on Titan : comparisons to the prebiotic Earth, *Origins of life and Evol. Biosphere*, 27, 225-248.

Coll P., Coscia D., Guez L., Gazeau M.-C. and Raulin F. (1998) Review and latest results of laboratory investigations of Titan's aerosols, *Origins of Life & Evol. Biosphere* **28**, 195-213.

Coll P., Coscia D., Smith N., Gazeau M.-C., Ramirez S.I., Cernogora G., Israel G. and Raulin F. (1999a) Experimental laboratory simulations of Titan's atmosphere: aerosol and gas phase, *Planet. Space Sci.* **47** (10-11), 1331-1340.

Coll P., Guillemin J.-C., Gazeau M.-C. and Raulin F. (1999b) Report and implications of the first observation of C_4N_2 in laboratory simulations of Titan's atmosphere, *Planet. Space Sci* **47** (12), 1433-1440.

Coll P. and Raulin F. (1999) Investigation on planetary atmospheres using laboratory experiments : the example of Saturn's Moon (Titan), *Laboratory Astrophysics And Space Research,* P.Ehrenfreund, C. Krafft, H. Kochan, and V. Pirronello Eds., Kluwer Academic Pub., 183-203, 1999.

Coustenis, A., Salama A., Lellouch E., Encrenaz Th., Bjoraker G.L., Samuelson R.E., De Graauw Th., Feuchtgruber F. and Kessler M.F. (1998) Evidence for water vapor in Titan's atmopshere from ISO/SWS data, *Astron. Astrophys.* **336**, L85-L89.

Fortes A.D. (2000) Exobiological implications of a possible ammonia-water ocean inside Titan, *Icarus* **146**, 444-452.

Khare, B.N., Sagan, C., Arakawa, E.T., Suits, F. Callicott, T.A. and Wilaims, M.W. (1984), Optical constants orf organic tholins produced in a smulated Titanian atmopshere : from soft X-rays to microwave frequencies, *Icarus* 60, 127-137.

Lebreton, J.P., European Space Agency (1997) Huygens : Science, Payload and mission, *ESA SP-* **1177**.

Lorenz, R. (2000) Post-Cassini Exploration of Titan : Science rational and mission concepts, *J. Brit. Planet. Soc.* **53**, 218-2344.

Lunine, J.I.. (1993) Does Titan have an ocean ? A review of current understanding of Titan's surface, *Rev. Geophys.* **31** (2), 133-149.

Lunine, J. I. (1997) Titan surface-atmopshere interactions, *ESA SP-* **1177**, 211-218.

Ramirez S. I., Coll P., Da Silva A, Navarro-Gonzalez R., Lafait J. and Raulin F.(2000) Complex Refractive Index of Titan's Aerosol Analogues in the 200-900 nm domain, submitted.

Raulin F. and Owen T. (2000) Organic chemistry and exobiology on Titan, *Space Science Review.*, in press.

Sagan, C. and Khare, B.N. (1979) Tholins : organic chemistry of interstellar grains and gas, *Nature*, **277,** 102-107.

Thompson R.W. & Sagan C. (1992) Organic chemistry on Titan - Surface interactions, *ESA-SP* **338**, 167-176.

Vuitton V., Gazeau M.-C., Benilan Y.and Raulin F. (2001a) Experimental and theoretical photochemical studies of polyynes : application to Titan's atmosphere , this issue.

Vuitton V., Gazeau M.-C., Benilan Y.and Raulin F. (2001b) A review of gas phase photochemistry of Titan's atmosphere: observations, laboratory simulation experiments and photochemical models, submitted.

EXPERIMENTAL AND THEORETICAL PHOTOCHEMICAL STUDIES OF POLYYNES: APPLICATION TO TITAN'S ATMOSPHERE

V. VUITTON, M.-C. GAZEAU, Y. BENILAN AND F. RAULIN
Laboratoire Interuniversitaire des Systèmes Atmosphériques
61 Avenue du Général de Gaulle, Créteil Cedex, 94010, France

1. Introduction

Titan, the largest moon of Saturn, has a dense atmosphere mainly composed of molecular nitrogen and methane. The dissociation of these two compounds by different energy sources, in particular solar UV photons, initiates complex chemical reactions that form many organic compounds such as hydrocarbons and nitriles [1].

In addition to observations and laboratory experiments, theoretical models have been developed to reach a better understanding of Titan's atmosphere [2-5]. But, the uncertainties in the kinetic and spectroscopic parameters used in these models and the probable omission of critical reaction pathways lead to systematic errors in the calculations [6]. Polyynes are acetylene like linear compounds, with general formula $C_{2n}H_2$. Since they are very unstable in usual laboratory conditions, the lack of data is particularly enhanced for these compounds.

The aim of this paper is to review the current knowledge concerning the chemistry of polyynes. In particular, we point out the parameters (absorption cross sections, quantum yields and rate constants) available in the literature and those that have only been inferred and need to be determined in a near future.

2. Chemistry of polyynes

2.1. PHOTOLYSIS

2.1.1. *Quantum yields*

Vacuum Ultra-Violet photons can penetrate down to Titan's stratosphere [7]. What is the behaviour of polyynes submitted to these long UV photons ? In fact, irradiation products of these compounds have only been determined for diacetylene. It has been shown that, at 147 nm, the main products are C_4H radicals with a quantum yield (Φ) of 0.2, as well as C_2H_2 (Φ=0.1) and a few C_2H (Φ=0.03). With the increasing wavelengths, quantum yields tend to decrease becoming close to zero above 230 nm. Since primary quantum yields are very low, irradiation of diacetylene must lead to the formation of an excited metastable, probably triplet, state (C_4H_2*) [8].

J. Chela-Flores et al. (eds.), First Steps in the Origin of Life in the Universe, 315–318.

2.1.2. *Absorption Cross Sections*

The second parameter of importance when studying the photochemistry of polyynes is absorption cross section. For the two lightest (C_4H_2 and C_6H_2), dissociation coefficients have been determined in gas phase from 150 to 300 nm and at relatively low temperature, more relevant to Titan's atmosphere (≈ 200 K) [9-12]. But, for C_8H_2, data are more sparse since they have been measured only on a short wavelength range (185-215 nm) and at ambient temperature [13]. Finally, for longer carbon chain polyynes, there is no quantitative experimental data.

Nevertheless, semi-empirical calculations have been conducted very recently and they indicate that the dissociation coefficient of polyynes increases very quickly with the number of triple bonds. This result is of particular interest since if polymerisation processes are efficient, we might have here a snowball effect which will lead to the transformation of polyynes to solid material [14].

2.2. POLYMERISATION

2.2.1. *Radical Mechanism*

The kinetic rate constant of the reaction of ethyl radicals with acetylene has been extensively studied [15]. On the contrary, no measurements have been performed on the reaction rates of polyynes with their radicals ($C_{2n}H$). Consequently, modelers have to assume that these reactions have similar rate constants than the one which has been measured and to postulate arbitrary a decrease when the length of the radical chain increases. But, until now, the mixing ratio of diacetylene calculated by photochemical models is at least a factor of five higher than the concentration measured in Titan's atmosphere by Voyager (Table 1). Since mean concentrations appear to be essentially controlled by chemistry, this discrepancy highlights that a loss of diacetylene has probably been forgotten. Thus, it is possible that the role of the excited state of polyynes has been underestimated.

TABLE 1. Comparison of the mixing ratios of C_4H_2 (ppm) at ≈ 100 km measured by Voyager and calculated by theoretical models.

Observations [16]	Model 1 [2]	Model 2 [3]	Model 3 [4]	Model 4 [17]
1.4±0.7	≈ 10	6.8	4.9	≈ 10

2.2.2. *Molecular Mechanism*

The formation of longer carbon chain compounds through reactions of the triplet state of diacetylene with many hydrocarbons has been observed [18-21]. The rate constants of these reactions have been measured relatively to the reaction between excited diacetylene and acetylene that is assumed to be the same as the rate of the reaction between excited and ground state acetylene.

Another crucial parameter is the radiative lifetime of C_4H_2*. Since it has never been measured, it has been set equal, in photochemical models, to the lower-limit value of the lifetime available for C_2H_2* [22]. But, very recently, it has been found that if the lifetime of excited diacetylene was increased by a factor of 5000, the calculated mixing ratio of diacetylene would be in agreement with the observations, without any significant influence on other species [17]. Thus, if the lifetime of C_4H_2* is long enough, it will have the time to react and the molecular mechanism could compete with the free-radical one to form longer carbon chain compounds [22].

3. Conclusion

The chemistry of polyynes is still very badly known. In particular, it is possible that reaction pathways not yet included in the models lead to systematic errors in the calculation. Moreover, only few kinetic parameters have been measured (Table 2) and consequently, many are only inferred from similar reactions. One of the critical processes seems to be the reactivity of excited metastable states. In particular, it seems now that no photochemical model will be representative of the chemistry of polyynes as long as there will be no accurate determination of the lifetimes of their triplet state.

TABLE 2. A review of available experimental parameters relevant to the chemistry of polyynes

Parameter	Reaction	Available data	Improvements
Absorption coefficient	$C_{2n}H_2 + h\nu$	n = 1,2,3 (200 K) n = 4 (300 K)	Low temperature and pressure
Quantum yield	$C_{2n}H_2 + h\nu$	n = 1.2	Dependence on wavelength
Kinetic rate constant	$C_{2n}H + C_{2m}H_2$ $C_{2n}H_2* + C_{2m}H_2$	n = m = 1	Low temperature and pressure
Radiative lifetime	$C_{2n}H_2$*	n = 1 (lower limit)	Need to be done

4. References

[1] Raulin, F., Coll, P., Navarro-Gonzalez, R., Ramirez, S., Bénilan, Y., Shindo, F., Vuitton, V. and Gazeau, M.-C. Titan : Exobiology and the Cassini-Huygens Mission, this issue.
[2] Yung, Y.L., Allen, M. and Pinto, J.P. (1984) Photochemistry of the Atmosphere of Titan : Comparison Between Model and Observations, *Astrophys. J. Suppl. Ser.*, **55**, 465-506.
[3] Toublanc, D., Parisot, J.P., Brillet, J., Gautier, D., Raulin, F. and Mckay, C.P. (1995) Photochemical Modeling of Titan's Atmosphere, *Icarus,* **113**, 2-26.
[4] Lara, L.M., Lellouch, E., Lopez-Moreno, J.J. and Rodrigo, R. (1996) Vertical Distribution of Titan's Atmospheric Neutral Constituents, *J. Geophys. Res.*, **101**, 23261-23283.

[5] Lebonnois, S. (2000) Circulation Générale et Photochimie dans l'Atmosphère de Titan, Thèse de Doctorat, Université Toulouse 3.

[6] Smith, N. (1999) Sensibilité des Modèles Théoriques de l'Atmosphère de Titan aux Incertitudes sur la Photochimie des Hydrocarbures Simples, Thèse de Doctorat, Université Paris 12.

[7] Sagan, C., Thompson, W.R. and Khare, B.N. (1992) Titan: A Laboratory for Prebiological Organic Chemistry, *Acc. Chem. Res.,* **25**, 286-292.

[8] Glicker, S. and Okabe, H. (1987) Photochemistry of Diacetylene, *J. Phys. Chem.,* **91**, 437-440.

[9] Fahr, A. and Nayak, A.K. (1994) Temperature Dependant Ultraviolet Absorption Cross Sections of 1.3-Butadiene and Butadiyne, *Chem. Phys.,* **189**, 725-731.

[10] Smith, N.S., Benilan, Y. and Bruston, P. (1998) The Temperature Dependant Absorption Cross Sections of C_4H_2 at Mid Ultraviolet Wavelengths, *Planet. Space Sci.,* **46**, 1215-1220.

[11] Kloster-Jensen, E., Haink, H.J. and Christen, H. (1974) The Electronic Spectra of Unsubstituted Mono- to Penta- Acetylene in the Gas Phase and in Solution in the Range 1100 to 4000 A, *Helv. Chim. Acta,* **57**, 1731-1744.

[12] Benilan, Y., Bruston, P., Raulin, F., Courtin, R. and Guillemin, J.C. (1995) Absolute Absorption Coefficient of C_6H_2 in the Mid-UV Range at Low Temperature ; Implications for the Interpretation of Titan Atmospheric Spectra, *Planet. Space Sci.,* **43**, 83-89.

[13] Cottin, H., Gazeau, M.-C., Chaquin, P., Raulin, F. and Bénilan, Y. Experimental and Theoretical Studies on the Gas/Solid/Gas Transformation Cycle in Extraterrestrial Environments, submitted.

[14] Vuitton, V., Scemama, A., Gazeau, M.-C., Chaquin, P. and Bénilan, Y. IR and UV spectroscopic data for polyynes : predictions for long carbon chain compounds in Titan's atmosphere, *Adv. Space Res.,* in press.

[15] Pedersen, J.O.P., Opansky, B.J. and Leone, S.R. (1993) Laboratory studies of low-temperature reactions of C2H with C2H2 and implications for atmospheric models of Titan, *J. Phys. Chem.,* **97**, 6822-6829.

[16] Kunde, V.G., Aikin, A.C., Hanel, R.A., Jennings, D.E., Maguire, W.C. and Samuelson, R.E. (1981) C_4H_2, HC_3N and C_2N_2 in Titan's Atmosphere, *Nature,* **292**, 686-688.

[17] Lebonnois, S., Toublanc, D., Hourdin, F. and Rannou, P. Seasonal Variations of Titan's Atmospheric Composition, *Icarus,* in press.

[18] Bandy, R.E., Lakshminarayan, C. and Zwier, T. (1992) Spectroscopy and Photophysics of the 1Du <- 1Eg+ Transition of Jet-Cooled C_4H_2, C_4HD and C_4D_2, *J. Phys. Chem.,* **96**, 5337-5343.

[19] Bandy, R.E., Lakshminarayan, C., Frost, R.K. and Zwier, T.S. (1993) The Ultraviolet Photochemistry of Diacetylene : Direct Detection of Primary Products of the Metastable C_4H_2* + C_4H_2 Reaction, *J. Chem. Phys.,* **96**, 5362-5374.

[20] Frost, R.K., Zavarin, G.S. and Zwier, T.S. (1995) Ultraviolet Photochemistry of Diacetylene : Metastable C_4H_2* + C_2H_2 Reaction in Helium and Nitrogen, *J. Phys. Chem.,* **99**, 9408-9415.

[21] Frost, R.K., Arrington, C.A., Ramos, C. and Zwier, T.S. (1996) Ultraviolet Photochemistry of Diacetylene : the Metastable C_4H_2* Reaction with Ethene, Propene and Propyne, *J. Am. Chem. Soc.,* **118**, 4451-4461.

[22] Zwier, T.S. and Allen, M. (1996) Metastable Diacetylene Reactions as Routes to Large Hydrocarbons in Titan's Atmosphere, *Icarus,* **123**, 578-583.

EXPERIMENTAL SIMULATION OF TITAN'S ATMOSPHERE BY COLD PLASMAS

S. I. RAMIREZ[1, 2], R. NAVARRO-GONZALEZ[1], P. COLL[2] AND F. RAULIN[2]

1. Laboratorio de Química de Plasmas y Estudios Planetarios
Instituto de Ciencias Nucleares, Universidad Nacional Autónoma de México, Circuito Exterior, Ciudad Universitaria, Apartado Postal 70-543, México D. F. 04510 MEXICO

2. Laboratoire Interuniversitaire des Systèmes Atmosphériques
CNRS & Universités Paris 7 et Paris 12, 61, avenue du Général de Gaulle, 94010 Créteil, FRANCE

Abstract. We have used two different types of cold plasmas, corona and glow discharges, to simulate the probable existence of electrical activity in the tropospheric methane clouds and to synthesized an analogue of the stratospheric Titan's aerosols, respectively. A wide variety of hydrocarbons and nitriles were identified after a simulated troposphere was irradiated by positive and negative corona discharges. The energy yields of each one of the identified products were calculated. The energy yields of positive corona are one order of magnitude higher than those of negative coronas. On the other hand, the complex refractive index parameters of the aerosol analogues were calculated within a 200 to 900 nm interval, from transmittance and reflectance graphics. The refractive index (n) varies from 1.53 to 1.63 and the extinction coefficient (k) varies from 2.62×10^{-4} to 2.87×10^{-2}.

1. Introduction

Titan, the largest Saturn's moon and the only one in our Solar System that possesses a dense atmosphere, is besides the comets, Mars, Europe, and Ganymede, one of the planetary bodies of great interesting to astrobiology. The Voyager spacecraft established the existence of hydrocarbons, nitriles, and oxygenated molecules together with layers of aerosols in the satellite's atmosphere. These organics are believed to be the result of photolysis and radiolysis processes that experience the main atmospheric constituents. In the present study we are interested in two particular aspects of Titan's atmosphere: the synthesis of the gas-phase compounds that can be formed in the tropospheric cloud region by a certain type of electrical activity, and in the values of the complex refractive index of a solid aerosol analogue produced under stratospheric conditions by the action of a glow discharge.

During the closest Voyager I encounter with Titan in November 1980, search for emissions corresponding to lightning-like discharges was carried out during 100 minutes covering about the 98% of the satellite's surface (Desh and Kaiser, 1990). The negative results determined that terrestrial-like lightning discharges either were absent or were less energetic than expected. Consequently a different type of electrical activity was thought to develop on Titan. Some of the atmospheric properties that give validity to this argument are the presence of sparse methane clouds (Griffith *et al.*, 1998; Griffith *et al.*,

J. Chela-Flores et al. (eds.), First Steps in the Origin of Life in the Universe, 319–324.

2000), and the registered high levels of atmospherical conductivity (Lindal *et al.*, 1983). Besides, the necessity of an electrical mechanism to adjust the predicted yield of certain detected compounds, like ethene, whose abundance in the photochemical and high energy particle models falls short of the observations by a factor of 10 to 40 (Borucki and Chameides, 1984). Corona discharges, a type of cold plasma, seem to be an appropriate way of electrical activity on Titan because they are a natural electrical phenomenon that can fit the requirements of energy ($\leq 10^6$ Joules) and frequency (≥ 100 flash) expected for the satellite's electrical discharges (Desh and Kaiser, 1990; Navarro-González and Ramírez, 1997).

The aerosols in the stratosphere of Titan play an important role in the absorption, scattering, and reflection of the solar radiation in the atmosphere (McKay *et al.*, 1989). The majority of the models proposed to explain an aspect of Titan's atmosphere like its microphysics (McKay *et al.*, 1989; Toon *et al.*, 1992), the physical and chemical properties of the aerosols (Rannou *et al.*, 1995; Tomasko *et al.*, 1997), or the surface characteristics (Coustènis *et al.*, 1995), need to compute the values of Titan's geometric albedo. Among the parameters used to interpret the geometric albedo is the aerosol complex refractive index defined by $m = n + ik$. We determine, under laboratory conditions that better resemble the Titan's stratospheric environment, a new set of values for the complex refractive index of aerosol analogue films produced by a glow discharge.

2. Experimental

Details of the experimental set up, analysis and identification conditions used for the irradiation of the Titan's simulated troposphere as well as the energy yield calculation details are described in Ramírez and Navarro-González (1999). For the synthesis of the aerosol analogue films, a mixture of 2% CH_4 in N_2 (99.99995% purity from Linde) was allowed to flow inside a "U" type Pyrex reactor at 1 mbar and 293 K. The glow discharge takes place inside the reactor with a potential difference of 4 kV at a current output of 70mA is applied between two iron electrodes. Thin films of the aerosol analogues were deposited on quartz plates after 120 and 240 minutes of irradiation. The samples were recovered in the absence of terrestrial atmospheric contamination conditions. The complex refractive index values were estimated solving the Abelès equation system (Abelès, 1967) provided the reflectance, transmittance and thickness of the films were known. Transmittance and reflectance measurements were performed in the 200 nm to 900 nm interval by using a Cary 5 (Varian) spectrophotometer. The fluorescence of the aerosol analogue films was excited with an Argon laser beam and the films were observed with a confocal microscope (Zeiss LSM 410) in order to determine their thickness.

3. Results and Discussion

The electrical discharges have been better studied for the terrestrial atmosphere. The most frequent are the intracloud and those occurring between the cloud and the surface (Uman and Krider, 1989). A descending or ascending flux of negative or positive corona current

TABLE 1. Energy yields (P) of the gas-phase organics produced by positive and negative corona discharge irradiation of a Titans' simulated troposphere (10% CH_4 in N_2) at 670 mbar and 293 K*.

Compound	Identification technique**	P (molecule J^{-1}) × 10^{14} *Positive polarity*	*Negative polarity*
Ethane	**MS, IR**	**29.8 ± 2.1**	**6.6 ± 0.1**
Ethene	MS, IR	3.8 ± 0.4	1.4 ± 0.4
Ethyne	MS, IR	7.3 ± 0.7	2.2 ± 0.1
Propane	**MS, IR**	**6.2 ± 0.4**	**0.6 ± 0.02**
2-Methylpropane	MS, IR	1.8 ± 0.1	0.3 ± 0.01
2, 2-Dimethylpropane	MS, IR	1.1 ± 0.1	0.4 ± 0.01
Cyclopropane	MS	7.6 ± 0.1	* * *
Propene	MS, IR	0.9 ± 0.01	0.2 ± 0.02
2-Methylpropene	MS	0.6 ± 0.01	* * *
n-Butane	**MS, IR**	**4.3 ± 0.3**	**0.9 ± 0.04**
2-Methylbutane	MS, IR	2.8 ± 0.1	0.7 ± 0.01
2, 2-Dimethylbutane	MS, IR	0.4 ± 0.01	* * *
2, 2, 3-Trimethylbutane	MS, IR	0.4 ± 0.01	* * *
1-Butene	MS	0.2 ± 0.01	0.2 ± 0.02
2-Butene	MS	0.1 ± 0.01	0.1 ± 0.01
2-Methyl-2-butene	MS	0.2 ± 0.01	0.1 ± 0.01
n-Pentane	MS, IR	0.6 ± 0.01	0.2 ± 0.01
2-Methylpentane	MS, IR	1.5 ± 0.1	**0.7 ± 0.01**
3-Methylpentane	MS. IR	0.2 ± 0.01	* * *
2, 2-Dimethylpentane	MS, IR	0.7 ± 0.01	* * *
2, 3-Dimethylpentane	MS, IR	**0.8 ± 0.01**	**0.2 ± 0.01**
3, 3-Dimethylpentane	MS, IR	0.5 ± 0.01	0.5 ± 0.01
n-Hexane	**MS, IR**	**0.4 ± 0.03**	* * *
n-Heptane	**MS, IR**	**0.2 ± 0.02**	* * *
Methanenitrile	**MS**	**1.9 ± 0.2**	**0.6 ± 0.01**
Ethanedinitrile	MS	0.3 ± 0.01	0.1 ± 0.001
Ethanenitrile	**MS, IR**	**1.9 ± 0.02**	**0.9 ± 0.01**
Propanenitrile	MS, IR	0.8 ± 0.01	0.3 ± 0.01
2-Methylpropanenitrile	MS	0.1 ± 0.01	* * *
Butanenitrile	MS, IR	0.2 ± 0.01	0.1 ± 0.01

* Energy yields (P) were calculated from the slope of number of molecules *vs* deposition energy graphics. The main products are boldfaced and the compounds already detected in Titan are in italics.
** MS in electron impact mode at 70 eV. IR at a 4 cm^{-1} resolution. *** Not detected products.

initiates these discharges. The corona discharges postulated for Titan's troposphere are probably initiated in the same manner. Once an electrical dipole is formed inside the methane clouds, the electrical potential across the dipole increases until the local breakdown occurs. The displacement of a negatively charged column towards the surface promotes the development of corona streamers. As the electrical field in Titan's atmosphere is not as intense as the terrestrial one, the development of lightning-like discharges does not occur. Propagation of the ionizing streamers is initiated and when they reach each other, a current flowing is established and a corona discharge is formed. Negative coronas can be initiated in this way, from the cloud base to the surface satellite. Convective currents or some other atmospheric phenomena allowing the horizontal separation of positive charge in the top of the cloud can deform clouds. This can originate positive coronas from the cloud base to the surface. Corona discharges in Titan can also develop between the electric charges accumulated on the base of the cloud and the charges, of opposite polarity, induced on the surface of the satellite in mountain-type elevations. As shown in Table 1, the production of organic compounds obtained when a Titan simulated troposphere is exposed to the action of positive and negative coronas yielded a total of 24 hydrocarbons and 6 nitriles. Positive coronas show higher energy yields and number of identified compounds among which cyclic and saturated hydrocarbons are present. These last ones with methyl and dimethyl substituents. Unsaturated hydrocarbons are detected in both, positive and negative coronas. The products formed exclusively by positive coronas are: cyclopropane, 2-methylpropene, 2,2-dimethylbutane, 2,2,3-trimethylbutane, 3-methylpentane, 2,2-dimethylpentane, n-hexane, n-heptane, and 2-methylpropanenitrile.

Yung *et al.* (1984) proposed that ethane and propane are characteristic compounds of photolysis processes. In this study it is observed that positives coronas produce five times more ethane and butane and ten times more propane than negative coronas. For the rest of the products, energy yields of negative coronas are, in general, one order of magnitude smaller. This means that photoemission is the main driven mechanism in positive coronas, while negative coronas depend basically on ionization by electronic impact (Chang *et al.*, 1991). It seems that there is also a more efficient use of the dissipated energy by positive coronas.

Once the small hydrocarbons and nitriles are synthesized, they participate in secondary processes to form the heavier and more complex compounds that formed the aerosol particles. In the simulation experiments it is confirmed by the formation of a solid phase whose synthesis is promoted by electronic impact and photolysis reactions generated by the glow discharge. Under the experimental conditions of the present study a deposition rate of 1 $\mu m\ hr^{-1}$ of irradiation was determined. The transmittance and reflectance graphics of our aerosol analogue films show a material that highly absorbs in the ultraviolet domain and become more and more transparent towards higher wavelengths, in a similar way as Titan's aerosols do. In Figure 1 are plotted the refractive index (n) and the extinction coefficient (k) calculated from our samples. The error bars resulted from the effect of the sample thickness uncertainty and of the scattered light. Our results are compared with the values of Khare *et al.* (1984), known as the field reference. Some important differences are observed. The variations on our refractive index (n) are much less contrasted and weaker. The resulting graphic lacks the structure between 100 and 400 nm observed in the Khare's et al graphic. A maximum of the extinction coefficient (k) around 300 nm is observed in our values, but is not present in Khare's et al results. These differences can be attributed to a lower methane content (2%) in our simulated stratosphere, to the improvement in the film sample thickness determination, to the absence of terrestrial atmospheric gas contamination, the non-over irradiation of the mixture to obtain the sample films, and in the use of a different energy source. A detailed explanation of these differences can be found in (Ramírez *et al.*, 2001).

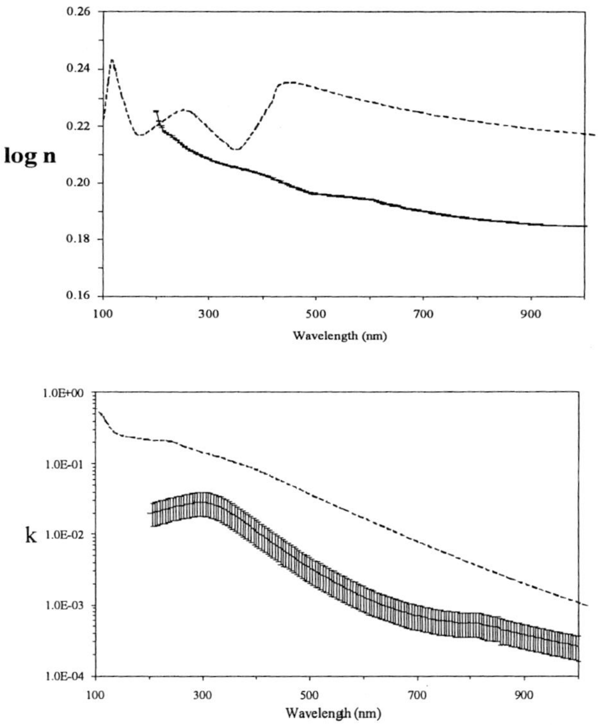

Figure 1. Final determinations of the complex refractive index parameters: n (above) and k (below) from Titan's aerosol analogues. The analogues were obtained after the irradiation of a 2% CH_4 in N_2 mixture with a glow discharge at 2 mbar and 293 K. The error bars correspond to the effect of thickness uncertainty and scattered light. (Khare *et al.*, 1984) values correspond to the dashed curves.

4. Conclusions

A complex gas-phase organic chemistry induced by corona discharges is expected in Titan's troposphere. The detected products include exobiological important molecules like the HCN and also molecules detected in Titan's atmosphere. An important feature in corona discharges is the formation of mostly saturated hydrocarbons and nitriles. In particular the most abundant products are ethane, ethene, propane, n-butane, methanenitrile, and ethanenitrile. Positive coronas report higher energy yields and they seem to preferably produce the ramified and unsaturated identified compounds. The production of ethene (C_2H_4) and its subsequent diffusion into the upper atmosphere set corona processes as an adequate source that may compensate the low abundance reported by photochemical and high energetic-particle simulations. Besides, corona discharges led to nitrogen fixation in the lower atmosphere, demonstrating to be a good abiotic nitrogen fixer. Regarding the solid-phase aerosol analogue films, a new set of values for the refractive index (n) and the extinction coefficient (k) is presented. The films were obtained from experimental conditions that better resemble the Titan's stratosphere conditions, except for the temperature. The optical properties displayed by the synthesized films can contribute to the explanation of Titan's surface bright and dark areas. They will also help in the interpretation of the data collected by the Cassini-Huygens mission that will land on Titan on the 2004 year. This mission means a new opportunity to look for the presence of cloud and electrical activity. The scientific instruments on board the probe and spacecraft will perform also *in situ* analysis of the atmosphere and of the aerosols.

5. References

Abelès F. (1967) Optics of thin films. In *Advanced Optical Techniques* (ed. A. C. S. Van Heel), pp. 143-188. John Wiley and Sons, Amsterdam, North Holland.

Borucki W. J. and Chameides W. L. (1984) Lightning: Estimates of the rates of energy dissipation and nitrogen fixation. *Rev. Geophys. Space Phys.* **22**, 363-372.

Chang J.-S., Lawless P. A. and Yamamoto T. (1991) Corona Discharge Processes. *IEEE Trans. Plasma Sci.* **19**(6), 1152 -.

Coustènis A., Lellouch E., Maillard J.-P. and McKay C. P. (1995) Titan's surface: composition and variability from the near-infrared albedo. *Icarus* **448**, 87-104.

Desh M. D. and Kaiser M. L. (1990) Upper limit set for level of lightning activity on Titan. *Nature* **345**, 442-444.

Griffith C. A., Hall J. L. and Geballe T. R. (2000) Detection of Daily Clouds on Titan. *Science* **290**, 509-513.

Griffith C. A., Owen T., Miller G. A. and Geballe T. (1998) Transient clouds in Titan's lower atmosphere. *Nature* **395**, 575-578.

Khare B. N., Sagan C., Arakawa E. T., Suits F., Callcott T. A. and Williams M. W. (1984) Optical Constants of Organic Tholins Produced in a Simulated Titanian Atmosphere: From Soft X-Ray to Microwave Frequencies. *Icarus* **60**, 127-137.

Lindal G. F., Wood G. E., Hotz H. B., Sweetnam D. N., Eshleman V. R. and Tyler G. L. (1983) The Atmosphere of Titan: an Analysis of the Voyager 1 Radio Occultation Measurements. *Icarus* **53**, 348-363.

McKay C. P., Pollack J. B. and Courtin R. (1989) The Thermal Structure of Titan's Atmosphere. *Icarus* **80**, 23-53.

Navarro-González R. and Ramírez S. I. (1997) Corona Discharge of Titan's Troposphere. *Adv. Space Res.* **19**(7), 1121-1133.

Ramírez S. I., Coll P., Da Silva A., Navarro-González R., Lafait J. and Raulin F. (2001) Complex refractive index of Titan´s aerosols analogues in the 200-900 nm domain. *Icarus* **in press**.

Ramírez S. I. and Navarro-González R. (1999) Quantitative Study of the Effects of Various Energy Sources on a Titan´s Simulated Atmosphere. In *Astrobiology. Origins from the Big-Bang to Civilisation* (eds. J. C. Flores, G. A. Lemarchand and J. Oró), pp. 307-310. Kluwer Academic Publishers, Netherlands.

Rannou P., Cabane M., Chassefière E., Botet R., McKay C. P. and Courtin R. (1995) Titan's Geometric Albedo: Role of the Fractal Structure of the Aerosols. *Icarus* **118**(2), 355-372.

Tomasko M. G., Doose L. R., Smith P. H., West R. A., Soderblom L. A., Combes M., Bézard B., Cousténis A., deBergh C., Lellouch E., Rosenqvist J., Saint-Pé O., Schmitt B., Keller H. U., Thomas N. and Glien F. (1997) The Descent Imager / Spectral Radiometer (DISR) Aboard Huygens. In *Huygens: Science, Payload and Mission* (ed. A. Wilson), pp. 109-138. ESA Publication Division, Noordwijk, The Netherlands.

Toon O. B., McKay C. P., Griffith C. A. and Turco R. P. (1992) A Physical Model of Titan's Aerosols. *Icarus* **95**, 24-53.

Uman M. A. and Krider E. P. (1989) Natural and Artificially Iniciated Lightning. *Science* **246**, 457.

Yung Y. L., Allen M. and Pinto J. P. (1984) Photochemistry of the Atmosphere of Titan: Comparison between Model and Observations. *Astrophys. J. Suppl. Ser.* **55**, 465-506.

Section 8:

New approaches to the detection of extraterrestrial life and intelligent radio signals

WAYS TO DETECT LIFE ON NEW PLANETS

M. OLLIVIER

Max-Planck-Institut für Astronomie
Königstuhl 17
D-69117 Heidelberg - Germany

Abstract: The discovery of new planets around nearby stars sets the question of their potential habitability. This question could appear hopeless taking into account the huge distance between these objects and us that precludes all in-situ observations.

However, adopting a rather generic definition of life, and assuming life developed on the whole planet, we can show that signatures of habitability and life can be found in the chemical composition of its potential atmosphere. A remote spectroscopic observation of extra-solar planets atmospheres allows getting some crucial information.

In this paper, I will discuss several criteria, particularly the "water and oxygen" one and several objections to these criteria. Finally, I will present the DARWIN mission, a space interferometer that should allow us to get spectra of planets around nearby stars.

1. Introduction

The existence of life in the universe is certainly one of the oldest questions of astronomy. After centuries of personal beliefs, we are now able to have a scientific and observational approach.

The number N of "intelligent" civilisations[1] in our galaxy has been theoretically formulated by Drake in his famous equation (Drake 1964) which can be expressed in the form:

$$N = N^* \,.\, f_{pl} \,.\, f_{habit} \,.\, f_{life} \,.\, f_{civil} \,.\, <T_{civil}>/T^* \qquad (1)$$

where
N^* is the number of stars in the Galaxy
f_{pl} is the fraction of them, which developed a planetary system
f_{habit} is the fraction of systems where a planet is habitable
f_{life} is the fraction of habitable planets where life really appeared
f_{civil} is the fraction of planets where life evolved to a civilisation
$<T_{civil}>$ is the civilisation mean lifetime
T^* is the star mean lifetime.

---

[1] "Intelligent" means here "able to communicate". This assumption is of course not a definition and can be discussed for a long time

J. Chela-Flores et al. (eds.), First Steps in the Origin of Life in the Universe, 327–334.

The huge number of stars in the Galaxy $N^* \sim 10^{11}$) is the origin of Fermi's paradox that can be summarised by the question: "where are they?" To answer that question, two approaches can be considered:
- a "global" approach, trying to determine directly N: the SETI approach (trying to detect "noisy" civilisations).

The main drawback of such approach is that a lack of detection has little significance.
- a "step by step" approach, trying to determine individually, each term of the equation. That is this approach I adopt in this paper.

At present, only the first term in equation (1) is well known. Taking into account the distance between near stars and us, an "in situ" observation of these planets can not be considered. The next term of the equation should be known soon by the indirect detection of extra-solar planets (by methods such as radial velocity measurements or transits study). The next ones require direct detection of planets and analysis of them. I will present in this paper a spectroscopic criterion for remote detection of life and discuss it. I will conclude with the presentation of a space mission that should use this criterion to find clues of life in neighbour planetary systems.

2. Detecting life from a distant source

A living system[2] can be defined (Brack, 1993) as a system which:
- has a structure (organisation is quantified by a the "negentropy" analogue to the entropy for disorder),
- is able to replicate,
- is able to evolve by random changes in its information and natural selection.

According to this general definition, detecting life on distant worlds appears hopeless. It is not true if we make the assumption that it developed on a planetary scale. In that case, the presence of life modifies the composition of the planet atmosphere, and a spectroscopic analysis with qualified criteria gives important information.

Several spectroscopic criteria have been proposed:

a) Search for states out of thermodynamic equilibrium

This idea has been proposed by Lovelock (Lovelock, 1965, 1975), who noticed that the large amount of out of equilibrium gas (CH_4 and O_2 for instance) in the terrestrial atmosphere has a biological origin. He concluded this point could be generalised as a criterion for life. It is true that the terrestrial environment is favourable to out of equilibrium processes, but the composition of the atmosphere cannot be obtained without life. This criterion may be used if, at the same time, we have an accurate knowledge of the atmospheric conditions on the exoplanets we are observing, excluding the possibility of abiotic processes. This criterion was applied by Sagan et al. to evidence life on the Earth using Galileo observations of CH_4, O_2 and NO_2 (Sagan, 1993).

---

[2] The number of life definitions is maybe as large as the number of researchers working on that topic. The definition adopted here tries to be as generic as possible.

b) Presence of CH_4 and H_2O

Kasting proposed this criterion, similar in its principle to the previous one. It consists in the observation of the CH_4 concentration in the presence of H_2O (which produces OH that destroys CH_4). The present mixing ratio is $[CH_4] \sim 10^{-2}$, instead of $[CH_4] \sim 10^{-4}$ if CH_4 is not continuously replenished. That means that a source of CH_4 exists, and may be biological Kasting, 1997a).

c) Organic chemistry in solution in liquid water

According to our definition of life, the information requires a support. Among all possible information supports, macromolecules appear to be well suited:
- they can be compared to words written with letters, which are molecular functions,
- a replication process is easy to imagine provided that the chemical functions have or not affinities with different chemical functions.

Of course, we have in mind the case of DNA, that has also catalysis properties, but several other molecules can have the same characteristics (Maurel, 1992).

If linear macromolecules are considered as the support for information, then the carbon chemistry appears to be powerful. Carbon can exist either reduced (CH_4) or oxidised (CO_2), which allows a great variety of chemical species. This property is unique. Silicon is often presented as an analogue to carbon but its reduced form (SiH_4) is unstable and reacts strongly with oxygen. The oxidised form (SiO_2) is very stable (silica). The richness of carbon chemistry is not only limited to laboratory conditions (where only 1% of the 10^7 known molecules (except DNAs) is inorganic). Among the 112 identified species in the interstellar medium, in 1996, 84 contain carbon, and only 8 contain silicon.

As a consequence of carbon based life, the chemical reactions should occur in a solvent. Solid state reactions are too slow, gaseous state reactions require too high temperatures, that would destroy the molecules. Among the possible solvent, liquid water is a special one (Brack, 1993). Its high dielectric constant ($\varepsilon \sim 80$) allows salt ionisation. Thanks to the capability of creating hydrogen-bounds the macromolecules can adopt different conformations with hydrophilic/hydrophobic radicals allowing key-keyhole reactions. In addition, water is chemically active (hydrolysis) but not too much and is indirectly UV resistant by the creation of O_3 that protects the created molecules against photo-dissociation (Owen, 1980).

Ammonia (NH_3) has none of these last two properties. As a conclusion, good candidates for support of life are **organic macromolecules in liquid water as a solvent**.

Consequence 1: definition of the habitable zone

According to the previous criterion, the habitable zone around a star is the region where a planet can sustain liquid water on its surface during part of its local year. This concept has been extended by Kasting et al., defining the continuous habitable zone which supposes that these conditions last enough to allow life to develop and to evolve (Kasting et al., 1993).

Consequence 2: a criterion to detect life

Considering a carbon-water based life, generalised on a planet surface[3], Owen noticed that a good criterion to detect the presence of such a life would be the massive presence of oxygen in the atmosphere (Owen, 1980). A large scale developed and evolving life requires namely a lot of organic carbon (partly reduced C). The abundant raw material is CO_2 (where C is fully oxidised). It is so necessary to reduce C from CO_2 into organic. The most efficient reaction is a photosynthesis-like reaction that can be written as follow:

$$CO_2 + 2\,H_2O + \text{energy} \rightarrow (CH_2O) + O_2 + H_2O$$

where (CH_2O) is an example of simple sugar and energy can be brought by photons (photosynthesis) or by chemicals from the mantel. O_2 is a very reactive gas with rocks or volcanic gases. If it were not continuously replenished, it would disappear from the surface in about 5 Myr ($FeO \rightarrow Fe_2O_3$). As a conclusion, **massive presence of O_2 and H_2O in a planetary atmosphere can be a criterion for carbon chemistry-based life** (Owen, 1980).

3. Qualifying the O_2 - H_2O criterion

The question is now to validate our life criterion, and to identify if they exist, abiotic sources of oxygen. Several hypotheses have been proposed. Some of them lead to situations where abiotic oxygen can be abundant (and the life criterion based on oxygen production fails), but only in special cases that can be easily identified. Let's review these different hypotheses.

a) Photolysis of CO_2

Rosenqvist and Chassefière studied the possibility of forming a large quantity of oxygen by photolysis of CO_2 (Rosenqvist and Chassefière, 1995). Starting from a large scale of CO_2, water and UV conditions, they came to the conclusion that this process could only form a small amount of CO2 (at best 5 to 10 mbar). **The photolysis of CO_2 does not appear to be an efficient process to form a large amount of oxygen**.

b) Photolysis of H_2O on a cool planet

Oxygen can be formed by the action of photons on water, followed by escape of H_2 from the planetary atmosphere. However, this process is strongly reduced, on a rocky and cool planet, by the presence of a cold trap at the tropopause that limits the concentration of water in the upper part of the atmosphere. Water returns on the ground under the form of rain or snow. In addition, the presence of a buffer gas (N_2) in the

[3] The remote detection of a life limited to narrow niches appears to be a hopeless question. In this paper, we consider only life developed at a planetary scale.

atmosphere limits the escape rate of H_2 that diffuses through the buffer gas. In the same time, reducing rocks (FeO) and gases (CO, H_2), and the effect of weathering, linked to the water cycle in the inner part of the atmosphere are strong sinks of O_2. The result is a small amount of O_2 resulting from the dissociation of water under Earth-like conditions (Chassefière, 1996). These results are confirmed by the fact that on the Earth, 99.9999 % of the breathable oxygen is produced by photosynthesis. Only 1 PPM comes from abiotic processes (Walker, 1977).

c) Planet with moist runaway

In the inner part of the habitable zone, estimated to a < 0.95 AU for an solar system analogue to our (Kasting, 1993), the CO_2/carbonates cycle can fail stabilising the temperature, leading to an increase of the surface temperature, a disappearance of the cold trap, and, as a consequence, H_2O can go up into the stratosphere where the photo-dissociation by the UV is strong. The results are a four step-mechanism:
1) H_2O goes up into the stratosphere, CO_2 is present,
2) the atmosphere contains H_2O, O_2 and H_2 resulting from the dissociation of water, and CO_2,
3) no more H_2O (fully dissociated), O_2 decreases (the sinks effects), CO_2 is still present,
4) neither H_2O nor O_2, and only CO_2 remains. It becomes a Venus-like planet.

This process shows that the second and third steps lead to a massive presence of abiotic oxygen. By chance for our criterion, the phase 2 is very short (~ 0.1 to 0.3 Gyr) compared to the planet lifetime. Such a case is so very rare in terms of statistics. The phase 3 is longer, (~ 1 to 5 Gyr), but H_2O is not present during this phase. Taking into account these two cases, we can modify our criterion that requires now **a massive presence of O_2 and H_2O at a > 0.95 AU.**

d) Icy planet without volcanism

A slow accumulation of O_2 produced by photolysis can happen if the planet has no sink of O_2 (Kasting, 1997a). This requires neither volcanism nor plate tectonics, and concerns thus only small planets. This requires also no weathering, and concerns planets without atmosphere, i.e., icy planets. This kind of planet is very easy to identify using the size, surface temperature, and spectrum of the planet (where a low partial pressure of water particularly can be identified). In that case, our criterion can be checked.

e) Oxygen brought by comets

After observation of O_2 and O_3, formed by the action of protons on ice, in Rhea and Dione spectra, Noll et al. (1997) proposed that, if pre-cometary grains had been irradiated in the proto-planetary disk, these gases could be contained into comets and injected in terrestrial planets atmospheres, leading to a false detection of life according to our criterion.

However, we have shown that this objection has neither theoretical nor observational support (Léger et al., 1999). First, recent models of proto-planetary nebula show that the amount of oxygen and ozone formed by irradiation and contained in the grain is not sufficient to form the required quantity of gases. Then, O_2 has not been detected in the Halley comet composition (i.e. $[O_2]/[H_2O] < 0.5$ %). Finally, study of terrestrial geological history shows that O_2 was *not* present on young Earth during the heavy bombardment period.

As a conclusion: apart from life, no process has been identified to produce a huge amount of oxygen during a long period on an Earth-like planet. Except in several well-identified cases, that are easy to recognise, we can trust on the life criterion based on simultaneous detection of water and oxygen. This work is not complete, and studies are still necessary.

4. Remote detection of O_2: The DARWIN mission

a) Ozone as a signature of oxygen

Oxygen has spectral signatures in the near infrared (at 720 and 760 nm), and in the millimetre range (~5 mm). In the first case, the contrast between the planet and the parent star is huge ($5x10^9$), in the second, the signal from the planet is too small (in the N band, where an Earth-like planet at 300 K peaks, the Earth receives only ~1 photon/s/m^2). Detection of oxygen as proposed by Owen is difficult (Owen, 1980).

The signal of a planet at 300 K is maximum (and the contrast with the star minimum, "only" $7x10^6$) in the thermal infrared. Oxygen, which is a homopolar molecule, has no IR signature. However, ozone was proposed to be a signature of oxygen (Angel et al., 1986). It is formed by photo-dissociation of oxygen and has a signature at 9.6 μm. Even if the processes of formation from O_2 and dissociation of ozone are complex, O_3 appears to be a good tracer:

- the amount of ozone does not appear to vary a lot in various stellar environments from sun-like to cool stars (Kasting, 1997b, Ollivier, 1999, Selsis, 2000). This point requires extra-work to evaluate the consequences on the planet spectrum.
- $[O_3]$ has a logarithmic dependence with $[O_2]$. Compared to O_2, it allows the detection of life at an earlier state (Léger et al, 1993).

In addition the thermal infrared spectral range has the advantages to exhibit signature of CO_2 that gives information on the nature of the planet, and H_2O that reveals the presence of liquid water at its surface, and so, its habitability. CH_4 can also be detected and reveal a pre-biotic chemistry in the case of cold planets (Titan-like planets).

b) How to get spectra from planets?

If we want to detect an Earth-like planet around a nearby star, we must try to detect an object $7x10^6$ times fainter than its star, at about q=0.1 arcsec from it (for a sun-like star at 10 parsecs). In other words, it is like trying to catch the light of a lightworm, 30 cm away from a lighthouse at a distance of 600 km. At 10 μm, it requires a cool 25 m space-telescope, coupled with a high-performance coronograph, unreachable in the next decades. Another solution, that conjugates high angular resolution and high dynamics range was proposed by Bracewell and is known as a "nulling interferometer" (Bracewell, 1978). The idea is to use two small telescopes, pointed both in the direction of the star and to combine their beam. Taking into account the diffraction effects, each telescope does not resolve the system star-planet. On the beam combiner the wavefronts from both star and planet create interference. If we introduce a p phase-shift in one arm of the interferometer, the star interference is destructive (the optical path difference is equal to zero). In the direction of the planet, and because of the angular distance to the pointing direction, the path difference is not equal to zero. Adjusting the distance B between the two telescopes, one can obtain that the path difference (B.sin(θ)) introduces a phase-shift that compensates the previous one. One gets so a constructive interference in the direction of the planet. The distinction between the star and planetary photons is so performed by the study of their phase.

c) The DARWIN mission

The DARWIN mission, and its U.S. analogue, the Terrestrial Planet Finder mission, are space interferometers dedicated to the search for and the spectral analysis of extra-solar planets (Angel et al., 1986, Léger et al, 1993b, Beichman, 1999, Fridlund, 2000). Working on a principle derived from Bracewell's interferometer, they will combine their 4 to 6 telescopes to get spectra of planets in the range 6-18 mm with a resolution from 20 to 100. Such a spectral range is only accessible from the space. Present projects are based on several free flying telescope-configurations orbiting around L2. A complex system of optical metrology allows the pointing and the co-phasing of the array. These observatories should send their firsts data around 2015-2020, and give orbital parameters and spectra of the possible planets around about 100 stars. Such instruments are still under definition, and only laboratory demonstrators are available (Wallace et al., 2000, Ollivier et al., 2000)

5. Conclusion

After centuries of suppositions and personal beliefs, we have reached an epoch where the question of life in the universe can be studied under the observational point of view. The consequence of the discovery of an alien life would be huge on the Earth and its population. In that way, we must be able to give information that cannot be contested or misinterpreted. The development of criteria for the detection of life and tools to use them is thus an important step in the quest. The first instruments that are yet under development are the first steps on the road that leads to these new worlds. The road is long, and full of traps. The first one is our own mind itself that tends to see only what he is prepared to see. Even if the law of physics and chemistry are the same in the universe, even if alien elements are common to us, we must be prepared to "expect the unexpected".

6. References

Angel J.R.P., Cheng A.Y., Woolf N.J., 1986, A space telescope for IR spectroscopy of Earth-like planets., *Nature*, **322**, p. 341-343

Beichman C.A., Woolf N.J., Lindensith C.A. eds, 1999, Terrestrial Planet Finder, Origins of Stars, Planets and Life, JPL Publication 99-3

Bracewell R.N., 1978, Detecting nonsolar planets by spinning infrared interferometer, *Nature*, **274**, p. 780

Brack A., 1993, *Origin of Life and Evolution of Biosphere*, **23**, p. 3

Chassefière E., 1996, Hydrodynamic Escape of Oxygen from Primitive Atmospheres: Applications to the Cases of Venus and Mars, *Icarus*, **124**, p. 537

Drake F., 1964, in Manukurian and Buggs (eds), *Current Aspects of Exobiology*, Pergamon press

M. Fridlund ed., 2000, DARWIN The Infrared Space Interferometer, ESA-SCI(2000)-12

Kasting J.F., Whitmire D.P., Reynolds R.T., 1993, Habitable Zone around Main Sequence Stars, *Icarus*, **101**, p. 108-128

Kasting J.F., 1997, *Origins of Life and Evolution of Biospheres*, **23**, p. 3

Kasting J.F., Whittet D.C.B., Sheldon W.R., 1997, Ultraviolet radiation from F and K stars and implications for planetary habitability, *Origins of Life and Evolution of Biospheres*, **27**, p. 413-420

Léger A., Pirre M., Marceau F.J., 1993a, Search for primitive life on a distant planet: relevance of 02 and 03 detections, *Astron. Astrophys.*, **277**, p. 309-313

Léger A., Mariotti J-M., Puget J-L., Rouan D., Schneider J., 1993b, DARWIN: Mission concept for the ESA Horizon 2000+ program.

Léger A., Ollivier M., Altwegg K., Woolf N.J., 1999, Is the presence of H_2O and O_3 in an exoplanet a reliable signature of a biological activity ?, *Astron. Astrophys.*, **341**, p. 304-311

Lovelock J.E., 1965, *Nature*, **207**, p.568

Lovelock J.E., 1975, Thermodynamics and the recognition of alien biospheres, *Proc. Roy. Soc. London*, **B189**, p.167

Maurel M-C., Décout J-L., 1992, Studies of nucleic acid-like polymers as catalysts, *J. Mol. Evol.*, **35**, p. 190-195

Noll K.S., Roush T.L., Cruikshank D.P., Johnson R.E., Pendleton Y.J., 1997, Detection of ozone on Saturn's satellites Rhea and Dione, *Nature*, **388**, p. 45-47

Ollivier M., 1999, Contribution à la recherche d'exoplanètes - Coronographie interférentielle pour la mission DARWIN, PhD Thesis,University of Paris-Sud

Ollivier M., Mariotti J-M., Léger A., Sékulic P., Brunaud J., Michel G., 2000, Interferometric coronography for the DARWIN space mission - Laboratory demonstration experiment, *Submitted to Astron. Astrophys. Sup. Ser.*

Owen T., 1980, The search for early forms of life in other planetary systems: future possibilities afforded by spectroscopic techniques, in ``Strategies for the search for life in the universe", M Papagiannis Ed., Reidel, p. 177

Rosenqvist J., Chassefière E., 1995, Inorganic chemistry of O_2 in a dense atmosphere, *Planet. Spa. Sci.*, **43**, p. 3

Sagan C., Thompson W., Carlson R., Gurnett D., Hord C., 1993, A search for life on Earth from the Galileo spacecraft, *Nature*, **365**, p. 715

Selsis F, 2000, Modèle d'évolution physico-chimique des atmosphères de planètes telluriques. Application à l'atmosphère primitive terrestre et à la détection de la vie sur les exoplanètes, PhD Thesis, University of Bordeaux I

Walker J.C., 1977, Evolution of the Atmosphere, Macmillan, N.Y.

Wallace K., Hardy G., Serabyn E., 2000, Deep and stable interferometric nulling of broadband light with implication for observing planets around nearby stars, *Nature*, **406**, p. 700-702

SEARCHING FOR WATER IN EXOPLANETS

CRISTIANO BATALLI COSMOVICI

Istituto di Fisica dello Spazio Interplanetario, CNR,
100, Fosso del Cavaliere, 00133 Roma, Italy
cosmo@ifsi.rm.cnr.it

Up to now about 50 extraterrestrial planets have been discovered with different optical methods mostly based on the Doppler velocity technique (1, 2). This successful technique allows the detection of giant planets with sufficient mass in order to excite a gravitational wobbling of the central star and it allows the determination of mass, period and distance from the star and eccentricity, but it does not give any chemical information on the atmospheric structure.

In 1994 the impact of 21 nuclei of Comet Shoemaker/Levy with the atmosphere of Jupiter gave a unique opportunity to study the consequences of a catastrophic impact in a planetary atmosphere and the subsequent changes in the chemistry and in the excitation conditions of molecular species.

By using a new fast multichannel spectrometer (up to 128.000 channels), coupled with the 32 m dish of the Medicina Radiotelescope (Bologna, Italy), we were able to detect for the first time a MASER water emission line at 22 GHz (1,35 cm) originated from a comet and due to a catastrophic impact (3).

Our measurements have shown that under particular physical conditions, like a catastrophic impact generated by comets delivering water or infrared pumping from a nearby M-Star, water MASER emission could be observed from exoplanetary atmospheres in a range of 50 light years.

Thus water- (and probably other molecules like SiO-) MASER emission could be used as a powerful diagnostic tool for planetary search outside the Solar system where cometary bombardments occur today as they occurred on our planet four billion years ago.

By using the same multichannel spectrometer we started the search of the 22 GHz water line on 10 selected targets shown in Table 1 (cf., the page after the references).

The multichannel spectrometer is a very high frequency and time resolution Fourier transform spectrometer and its tremendous computation rate allows us to compute the FFT and the square magnitude "on fly"over an input bandwidth of 6 MHz, with an efficiency of 30-35 %.

During the first observations the spectrometer and converter rack were configured to 8192 frequency channels in a 4.5 MHz bandwidth centered at 22.23508 GHz (4).

J. Chela-Flores et al. (eds.), First Steps in the Origin of Life in the Universe, 335–338.

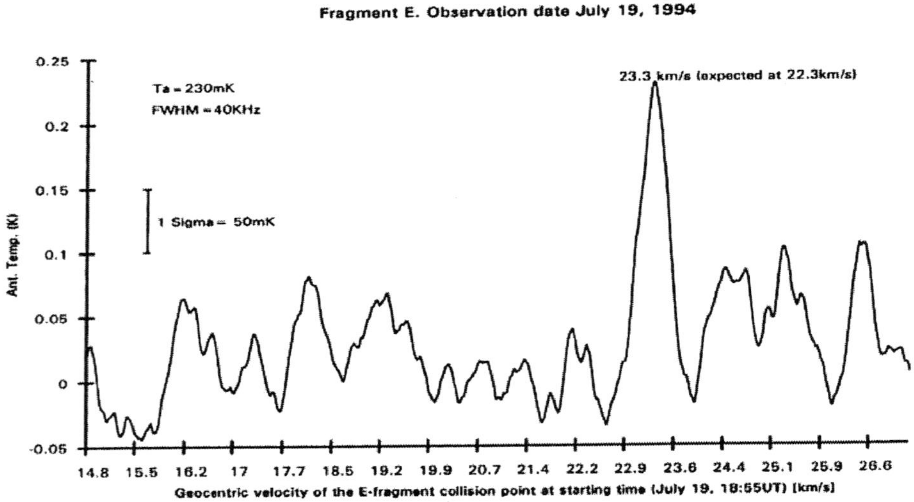

Fig. 1 Detection of the 1.35 cm (22 GHz) MASER water line during the SL/9-Jupiter collision

Since planetary parameters, as rotation and revolution period, are unknown, a very sophisticated software is under development in order to allow addition of integration times over long different observational periods as it is necessary to shift one spectrum with respect to another with a proper velocity.

The targets given in Table 1 include objects like epsilon Eridani where cometary bombardment is supposed to take place at present time, objects where giant planets have been already discovered, or peculiar dwarfs like Lalande 21185 of the M spectral type. With our 32 m dish about 100 hours/object would be necessary for a significant S/N ratio.

With the 64 m dish under construction in Sardinia, which will be ready in 2004, the integration time would be reduced by a factor of 100 and the range of observations could be easily enlarged to about 100 light years.

For very interesting objects in the southern hemisphere, like beta-Pictoris, we are preparing observations with the 64 m radio telescope at Parkes in Australia.

References

1) Mayor M.,Queloz D., Udry S. and Halbwachs J.L. in "Astronomical and Biochemical Origins and the Search for Life in the Universe" Eds. C.B. Cosmovici, S.Bowyer and D.Werthimer, p .313, Editrice Compositori, Bologna, Italy, 1997.

2) Marcy G.W.,Butler P.R., Fischer D.A. ,and Vogt S.S. in "A New Era in Bioastronomy", *ASP Conf. Series,* **213**, 85, 2000

3) Cosmovici C.B., Montebugnoli S., Orfei A., Pogrebenko S. and Colom P., *Planet. Space Sci.*,**44**, 735, 1996

4) Cosmovici C.B.,Pogrebenko S., Montebugnoli S. and Maccaferri G. in "A New Era in Bioastronomy", *ASP Conf.Series*, **213** , 151, 2000

TABLE 1 (on the following page): A selection of candidates for the search for water in the exoplanets.

CANDIDATES FOR 22 GHz WATER SEARCH

		distance (psec)	Sp.Type
DRAKE'S SETI CANDIDATES:			
	EPSILON ERIDANI	**3.3 (10.8 ly)**	**K2V**
	TAU CETI	**3.7**	**G8V**
STARS WHERE PLANETS HAVE BEEN DISCOVERED IN THE OPTICAL			
	GLIESE 876	4.7	M4V
	51 PEG	14.7	G2IV
	TAU BOO	**15.0**	**F7V**
	55N CnC	13.4	G8V
	ups And	**13.5 (44.14 ly)**	**F8V**
OTHER INTERESTING CANDIDATES :			
	LAL 21185 (red dwarf)	**2.0**	**M2**
	GL229 (brown dwarf)	6.7	M1/M2V
BEST CANDIDATE IN THE SOUTHERN EMISPHERE:			
	BETA PICTORIS	**18**	AV5

THE SAHA CRATER RADIOASTRONOMIC AND SETI OBSERVATORY

Set-up of the data link

G. GENTA
Mechanics Department, Politecnico di Torino
Corso Duca degli Abruzzi 24, Torino, Italy

Abstract

The farside of the Moon is the best place for a radioastronomic and SETI observatory and Saha Crater is a very suitable location. A first step is the landing of an automatic radiotelescope on the floor of the crater, remotely controlled from the Earth through a radio station located in the Mare Smythii, connected to the radiotelescope via a data link which does not produce electromagnetic noise in the moon farside environment. Three laser relay stations, landed in suitable places located not too far from the optimal position for transmitting and receiving and then accurately positioned using walking rovers are shown to be the simplest, cheapest and safest way to establish the data link.

1. Introduction

The electromagnetic pollution due to human activities and particularly to telecommunication satellites is increasingly limiting the possibilities of conducting SETI studies from the surface of the Earth. Actually the problems are not limited to SETI, which by definition deals with very weak signals and includes the search on the whole radiowave spectrum, but also affect radioastronomy in general, particularly on some frequencies. The use of a space-based radiotelescope on LEO or even in one of the Lagrange points of the Earth-Moon system does not solve the problem and indeed can make some of the interference issues worse. The far side of the moon is the only site not too far from the Earth which is completely free of electromagnetic noise, at least until exploration and exploitation activities on the Moon may jeopardise also this still untouched environment [1], [2].

Steps must be taken to preserve the interference free environment of the far side of the moon, but it would be meaningless to advocate that the whole farside of the Moon be protected as a sort of electromagnetic waves free conservation area, as this would conflict with the very exploration and exploitation activities from which the present project depends and also because it would be untenable in the long run. Instead of protecting selected frequencies in the whole far side, as it was asked by the International Telecommunication Union Recomendation (ITU 1997), the promoters asked for protecting all frequencies in a selected small area [2]. Such zone should be

J. Chela-Flores et al. (eds.), First Steps in the Origin of Life in the Universe, 339–344.

(1) located inside a strongly walled crater, whose rim protects from lunar based electromagnetic noise, (2) large enough to deploy the required equipment and to show potentialities for growth, (3) in the equatorial region, to have access to the whole sky and (4) as close as possible to the visible side, for easy access and communications.

The Saha Crater is the best candidate, as its position (102°E, 2°S) guarantees that it is never in view of the Earth (by 2° even with 7° libration) while being very close to the equator and to the near side of the Moon. It is surrounded by a circular rim 3,000 m high protecting it from Moon generated interferences and is large enough (100 km diameter). A regolith trail 350 km long can be built from Mare Smithii, where a future robotic or even manned base may be built and where the radio station for communications with the Earth can be conveniently located (Figure 1).

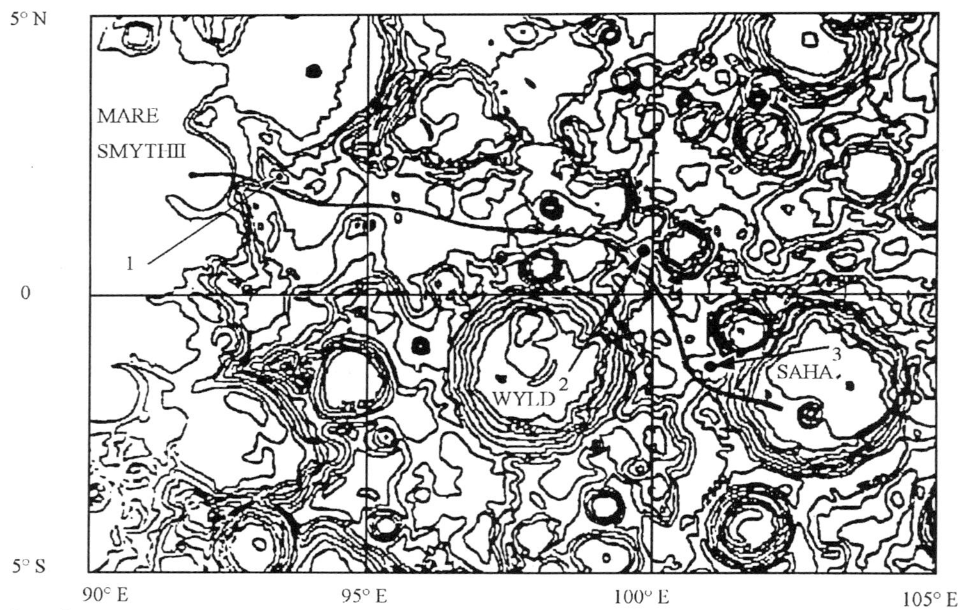

Figure 1.
Map of the zone of the Moon between 95° and 105° E, and 5°S and 5°N, with Saha crater and the regolith trail linking the crater with Mare Smithii. The points labelled 1, 2 and 3 are the possible locations for the relay stations: 1) Eastern hills of Mare Smithii, 2) NE of Wyld Crater, 3) Near the pass in the western rim of Saha Crater (NASA Map).

Once lunar exploration and perhaps exploitation is under way and lunar bases are built, the construction and operation of the Saha Crater observatory will be straightforward, but in this way the whole project is delayed of several decades, with the risk of seeing irreversible contamination of the radio environment in the relevant area. However, it is not impossible to start establishing a small automatic radiotelescope in the crater, operated from the Earth through a radio station located in the Mare Smithii and a suitable data link. This solution can be fairly simple and relatively low cost and, although the performances of the radiotelescope installed in the crater will be modest, can be considered as a starting point.

Two alternatives data link configurations have been proposed: a cable and a set of three laser relay stations [3]. The main problems related to the construction of a cable data link between Saha crater and Mare Smythii are related to the weight of the cable and to the need to deposit it without causing any damage on rough terrain. The slightest damage to the cable will cause a complete failure of the mission, as no repair mission can be conceived, if not at a cost comparable with that of the original one. An adequate redundancy must be planned, e.g. the use of multiple strands of cable running in parallel and joined at intervals, in such a way that even if multiple interruptions occur in the various wires or fibres, the signal can nevertheless go through the link.

An optical fibre cable is lighter than a conventional metal cable, but its mass is still about 10 g/m as a minimum, with a total of 3500 kg for the whole link. If a suitable redundancy is built in, this value may prove to be severely underestimated. Moreover, the attenuation of the cable makes it necessary to amplify the signal in several places along the route. It is impossible to give an accurate evaluation of the number of the amplifying stations and of their power because these issues depend on the bandwidth of the signal and on the performance of the link, in terms of information rate. With an attenuation of about 4 dB/km, an amplifying station must be located every 20-30 km, yielding a total of 12 to 17 of such devices, each one with its own power supply, with an estimated mass of about 150 kg.

Although suggestions of deploying of the cable in space before landing or shooting it using a harpoon from either the radiotelescope or the radio station have been forwarded, the only realistic way to build a cable data link is to use a rover to lay the cable along the 350 km trail from the radio station to the radiotelescope or viceversa. As the radio station is far lighter than the cable itself, it is convenient that the rover carries also the radio station so that a single lander is needed for the whole mission: it can land in the Saha crater, deploy the radio telescope and the rover which will then travel all the way up to Mare Smythii. This is not in itself a very difficult task: using a vehicle similar in principle to the Lunar Rover carried to the Moon by the Apollo astronauts but much larger (10 - 12 tons, including the cable and the laying mechanism) and assuming an average speed of 10 km/h, the whole job could be performed in about 35 hours under remote control from the Earth. During this phase a radio link is needed, but there is no difficulty in leaving one or more satellites in orbit around the Moon to perform provisionally this task.

As for this mission a high speed is not required, a twin-rigid frames hexapod legged vehicle has been considered [3, 4, 5] as it has several advantages, like the very good ratio between the payload and the structural mass, the capability of moving on very rough terrain and walking up very steep gradients, allowing to connect the two places with a shorter route, and the simplification of the cable unreeling process.

Preliminary computations lead to an estimate for the mass of the rover plus payload of about 7500 kg and to an average speed of 320 m/h, with a total of 1100 hours (46 Earth days) needed for the whole cable lying procedure. The power needed for this operation, not exceeding 1 kW, can be supplied by solar panels, if the rover is allowed to stop during the night, with a total duration of the operation of about 3 months. During all this time the machine must be controlled from the Earth.

The alternative to the use of a cable is to locate three laser relay stations, deposited on the ground by a rover or landing directly in the prescribed positions, in the three locations shown in Figure 1. The laser communication link will consist of two laser terminals (both transmitter and receiver for a two-way link) and three optical relay stations located on the eastern hills of Mare Smithii, NE of Crater Wyld and near the pass in the western rim of Saha Crater (Figure 2). A suitable laser source is described in [6]; by assuming to use in the transmitter a laser with a 1 W output power, in order to get an optical power of 1 μW at the receiving station (sufficient to transmit data to a rate of some Gbps) telescopes with 50 mm aperture for the transmission and reception of the beam can be used [7, 8]. To maximise the power transfer between the 50 mm telescopes of the two laser terminals, the transmitted beam must have a divergence of about 6 arcseconds, corresponding to a width of about 10 m at a 370 km distance. This makes unpractical the utilisation of the same beam for the operations of mutual alignment of the transmitter, receiver and of the relay stations, particularly in view of an autonomous implementation of these operations and an auxiliary laser diode is needed to perform the initial alignment of the line of sight of the main laser to an accuracy of about 1 arcsecond. The divergence of the beam of the laser diode, not subject to any telecommunication constraint, can be set to a larger value; the corresponding decrease in the power density at the receiver is compensated by the high sensitivity of the CCD detector (for the alignment operations, the low detection frequency is not an issue). Thanks to the wavelength difference between the main laser and the auxiliary one, the two optical paths can be separated by means of dichroic beamsplitters. Once the alignment has been completed the diode lasers are switched off to avoid any possible interference with the main laser.

A preliminary layout of the stations, including, is described in [6].

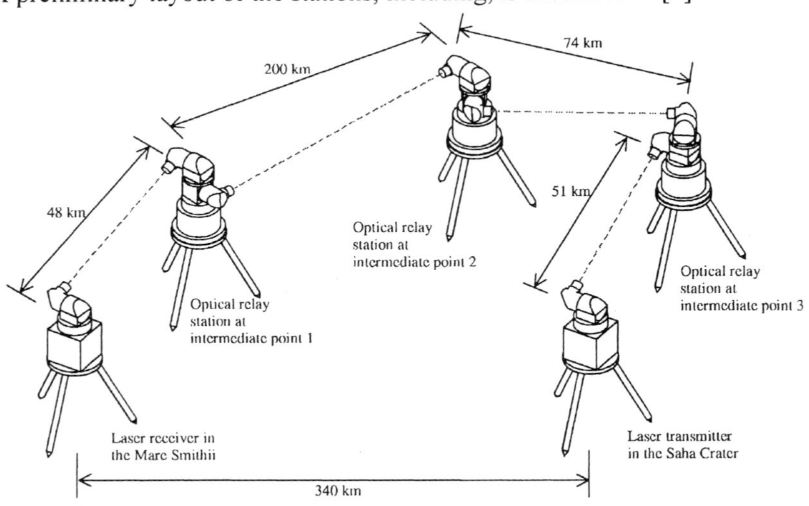

Figure 2. Functional configuration of the laser communication link. The distances are only indicative.

The mass of each terminal and each relay station is estimated of the order of 60 kg and the power consumption of the transmitter is estimated to be of the order of 50 W.

The total mass of the relay stations is of about 180 kg, far smaller than that of the cable. It is possible to use a rover to carry the three laser relay stations into position; however, the relay stations must be positioned in exactly chosen locations, where they are in optical sight of each other, and the rover has to leave the trail to climb on rough terrain. A preliminary study [3] for a twin rigid frames walking machine yields values of 360 kg for the overall mass of the rover and of 500 m/h for its maximum speed. The time needed for the deployment operation is then of about 520 hours (22 Earth days), neglecting the time needed for the actual laying down of the stations. If the 150 W required for propulsion are supplied by solar energy, the total deployment time is increased to about one month and a half, as the rover must stop during the night.

The alternative which appears as the most suitable is that of landing separately the three relay stations in their final positions. As it is impossible to guarantee that the stations are landed in exactly the best position for performing their task, it is wise to provide them of means to move around and to perform fine position adjustments. This allows also to land them in places chosen to maximise the chances of success of this operation, without bothering of doing it in close proximity of the final position of the station. In this case the advantage of mounting each station on its own twin-frames walking rover is noteworthy. Practically the rover is nothing else than a self-propelled stand, taking advantage of the low speed and high precision positioning requirements for this task (Figure 3).

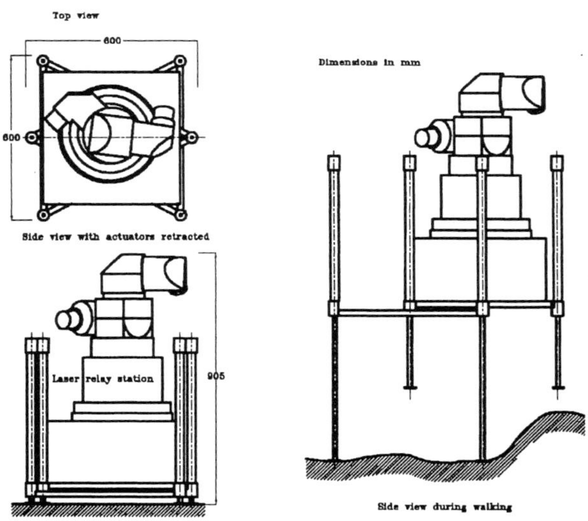

Overall mass (kg)	90
Maximum speed (level ground) (m/h)	190
Maximum speed (30% grade) (m/h)	160
Maximum obstacle height (mm)	400
Maximum longitudinal slope (deg)	30
Power at maximum speed (level ground) (W)	45
Power at maximum speed (30\% grade) (W)	55

Figure 3. Sketch and main characteristics of the self-propelled laser relay station.

Assuming that the landing occurs within a range of 1 km from the final spot, the time needed for the deployment operation is of less than 6 hours. A low value of the speed has been assumed to reduce the power consumption to about 50 W, as in the present case the distance travelled is at any rate short. Also in this case a satellite supplying a radio link between the rover and the Earth is required, but it is not necessary that the coverage is continuous, as the rover can remain in standby when the satellite is below its horizon.

3. Conclusions

A radio observatory located in the Saha crater will be a very valuable asset for both SETI and radioastronomy to fight the problems which will be increasingly caused by the growing electromagnetic pollution due to earth- and space-based human activity. To establish the observatory without the long delays linked with the timeframe of other exploration and exploitation lunar missions, a small automatic radiotelescope, operated from the Earth must be considered. The data and control link with the Earth must be supplied by a radio station located in Mare Smythii, connected to the radiotelescope via a cable or a number of laser relay stations. The latter solution is by far the best choice for what the mass to be carried on the Moon (hence the costs), the reliability and the ease of construction and maintenance are concerned, particularly if the three relay stations are landed near the required position and then carried in place using twin rigid-frames walking rovers. This requires a provisional radio link using a satellite of the Moon, active only for short periods of time for installation and maintenance, and no complex mechanical operations performed by robots or teleoperating devices.

Since it involves a number of identical small devices, some savings can be due to small scale construction. In case of failure of a single relay station, it can be replaced at a cost which is only a fraction of the cost of replacing a long cable. If some changes in the required positions of the relay stations or even in the minimum number of devices are made during the mission planning phase, it is possible to alter the data-link configuration with a minimum effect on the overall mission design.

The power needed to move the rovers is of the order of 50 W. The power source can be a solar panel of about 1 m^2, if operation of the optical link only during daytime is required. If continuous operation is required, a 50 W RTG has the added advantage of acting as a heating system to allow thermal control during the long nights.

References

1. Heidmann J., *Recent Progress of the Lunar Farside Crater SAHA Proposal*, 48th Int. IAF Congress, Torino, Oct. 1997.
2. Heidmann J., *A New IAA Cosmic Study:Establishing a Radio Observatiory on the Moon Farside*, 50th Int. IAF Congress, Amsterdam, Oct. 1999.
3. Genta G., *Twin Rigid-Frames Hexapod Rovers for the Saha Radioastronomic Missions*, Adv. Space Res., Vol. 26, No. 2, 351-357, 2000
4. Bussolino L., Del Corso D., Genta G., Perino M.A.,Somma R., *ALGEN - A Walking Robotic Rover for Planetary Exploration*, 1977 Int. Conf. on Mobile Planetary Robots & Rover Roundup, Santa Monica, Jan. 1997
5. Amati N., Chiaberge M., Genta G., Miranda E., Reyneri L.M., *WALKIE 6 - A Walking Demonstrator for Planetary Exploration*, Space Forum, to be published.
6. Genta G., Perino M.A, Cesare S., *The Laser Data Link for the Saha Radioastronomic Mission*, I.A.A. Cosmic Study on the Saha Radioastronomic Mission, to be published.
7. Lutz H.P., *Optical Communications in space - Twenty Years of ESA Effort*, ESA Bulletin, n. 91, Aug. 1997.
8. Siegman A.E., *Lasers*, University Science Books, Sausalito, 1986.

NATURAL INTELLIGENCE AND CELLULAR COMPLEXITY

M. RIZZOTTI
Department of Biology, University of Padova
via Bassi 58 B, 35131 Padova, Italy; rizzotti@bio.unipd.it

1. Intelligence regards both natural and artificial systems

Everybody considers it as a difficult task to provide a satisfactory definition of intelligence, but providing a succinct objective definition of it is more useful to science than conserving a hazy image of it. After a brilliant discussion, Barlow (1983) defined intelligence as "the capacity to guess right by discovering new order", but this definition still requires too much interpretation in judging living things in general (for the Primates, see Byrne, 1995) as well as automata (e.g., Russell and Norvig, 1995). I would say that *any material system that connects receptors and effectors can be regarded as intelligent if its effectors have their actions regulated in a strongly changing environment so that they usually maximize the system's success with reference to a given goal.*

The regulation that intervenes between receptors and effectors requires a circuitry that can be made to a large extent of the repetition of some kind of functional unit. The artificial intelligent systems that have been built so far by the mankind rely on many kinds of functional units such as mechanical ones, vacuum-tube electronic ones, or solid-state electronic ones. In contrast, natural intelligent systems can only originate in living things (Rizzotti, 1999) and can only rely, *as functional units*, on suitable cells. Actually, on the one hand, multicellular structures are useless as they do not ensure basic properties that differ from those of single cells; on the other hand, subcellular structures are constrained within a single cell and thus limited in their possibility to build complex circuits.

As regards the goal, that of an artificial system is planned by its constructor, whereas that of a living system is a product of natural selection and cannot be anything else than maximizing reproduction. In comparing different intelligent systems, the speed of performance is also important; it is correlated, in general, to the speed of the events that are usually relevant for the goal.

2. There are only two kinds of cellular organizations

In order to discuss cellular complexity, the possible kinds of cells must be preliminarily considered. At the most fundamental level, all cells of the modern biosphere only show two organizations, i.e., the *prokaryotic* and the *eukaryotic* ones. "The distinction between prokaryotic and eukaryotic cell structure [...] is still accepted today as the most fundamental discontinuity in the living world" (Roger, 1999) in spite of different

J. Chela-Flores et al. (eds.), First Steps in the Origin of Life in the Universe, 345–348.

attitudes on the subdivision of all living things into two or three domains (Mayr, 1998; Woese, 1998). This means that no cellular organization exists either intermediate between Prokaryotes and Eukaryotes or alternative to them.

However, in the last few years, examples of challenging cellular features have been sometimes claimed. One such example comes from a Prokaryote called *Gemmata obscuriglobus* that is endowed with a sort of a nucleus (Fuerst and Webb, 1991), i.e., the distinctive feature of Eukaryotes. However, its nucleoid organelle, unlike true nuclei, includes functional ribosomes and lacks nuclear pores, so that this species is classified without any doubt among the Prokaryotes (Holt et al., 1994).

Other challenging features have come from the eukaryotic side: they basically consist in the absence of the mitochondria, the dictyosomes, or even the nucleus. However, the mitochondria are no longer considered as a distinctive feature of Eukaryotes because they are absent (primarily or secondarily) in at least one thousand species that show a normal nucleus. The dictyosomes, also called Golgi apparatus, are perhaps secondarily absent in a few parasitic species (Katz, 1999), but the parasitic life style leads to simplification that simulates primitivity, so that it is not a good guide to look for true primitivity and perhaps intermediate cellular organization.

In some organisms, the nucleus has some putative primitive feature or is even absent. For example, the nucleus of *Giardia lamblia* (a unicell which is often considered as the product of the first dichotomy of eukaryotic phylogeny, but is parasitic) shows openings in the nuclear envelope besides the nuclear pores (Li, 1999). As another example, most Dinomastigotes (also called Dinoflagellates or Dinophyceans) have no histones in their chromosomes, so that these are not arranged into nucleosomes (Rizzo, 1987), but there is strong evidence that their absence is due to secondary loss (Taylor, 1999). Sometimes, even its permanent absence is insufficient to dispute the eukaryotic appurtenance of a cell. In a number of photosynthetic Eukaryotes there are complex plastids in which an outer compartment derived during evolution from a eukaryotic unicell, but the nucleus never develops in it today (Rizzotti, 2000). In conclusion, the presence of the nucleus in an ancestor is the only requirement to belong to the Eukaryotes.

Other Eukaryotes secondarily lacking nuclei had recently been identified in the epixenosomes of the Ciliate *Euplotidium itoi* on the basis of preliminary results, but at a closer inspection they turned out to be Prokaryotes (Petroni et al., 2000).

3. Cellular complexity can be viewed from different standpoints

Different parameters have been proposed to "measure" the complexity of individual cells both in uni- and multi-cellular organisms: they regard both the genome and the membranous compartments. Their list could include *total genome length*, *number of coding genes*, *number of simultaneously expressed coding genes*, *cytoskeletal complexity*, *kinds of cellular compartments*, *number of concentric membranes*.

The longest genomes are present in some amoebas and in some Animals and Plants that include Dipnoans (lungfish), Urodels (salamanders and newts) and Liliaceans (lilies and onions). All these organisms have genomes about one hundred times longer that of humans, which is about three billion bases long. The length of the human genome is equal to those of the other Mammals and roughly ten times longer those of Insects. In contrast, the average genome of Prokaryotes is three million bases long, but there is an overlap with those of the Eukaryotes as the shortest genome of a Microsporidian (parasitic unicells) is only 2.9 million bases long (Li, 1999).

However, a genome can include long stretches that do not code for proteins. This is particularly true in Eukaryotes, although their coding genes are usually many more than in Prokaryotes all the same. To date, the highest number determined in Prokaryotes is 5570 (in *Pseudomonas aeruginosa*, Stover et al., 2000), whereas the lowest number found in Eukaryotes is 5885 (in *Saccharomyces cerevisiae*), but lower numbers are expected among Microsporidians. The highest number in the whole biosphere is perhaps that of Mammals, which is about 100,000. In any case, the coding genes are expressed together all the time neither in uni- nor multi-cellular organisms. In the last few years, increasing evidence is emerging that the neurons, i.e., the nervous cells, hold the record for the number of expressed genes per cell. Probably, these are related to the fine modulation of the cytoskeleton of the projections (dendrites and neurites), the closeness of the cell-cell junctions (synapses), and the vesicle traffic (in synaptic buttons).

As regards the cytoskeleton, it is virtually absent in Prokaryotes, whereas high levels of complexity is very common in the unicellular Eukaryotes. In the Ciliates, in particular, there is a sort of intracellular differentiation of cytoskeletal organules that mimics the differentiation into multicellular organs of multicellular organisms.

The cellular compartments consist at most in isolated vesicles in Prokaryotes, whereas they can be very developed in Eukaryotes. As regards the kinds of compartments, the photosynthetic Eukaryotes have more than heterotrophic ones because of the presence of plastids. In particular, some complex plastids, i.e., those of the unicells Cryptomonads and Chloraracnids, also have a residual nucleus, so that they have perhaps the highest number of different compartments among all cells.

The highest number of concentric membranes is present in the complex plastids of some eukaryotic unicells, in particular, Dinomastigotes, because they seem to have sometimes more than four concentric membranes. In general, the cells of unicellular Eukaryotes are more complex than the individual cells of multicellular organisms for both the protein cytoskeleton and the membranous compartments.

In conclusion, we may say that the neurons are by no means the most complex cells in every respect: they are mainly characterized by a high number of expressed genes.

4. There are very different views on natural intelligent systems

In order to give rise to intelligence, there must be also a complex circuitry, then many neurons, then a nervous system. The Eukaryotes are optional members of a biosphere (Rizzotti, 1999), and the Animals to the more so. However, once Eukaryotes originate, animal-like organisms are likely to appear, sooner or later. Intelligence is only possible in advanced Animals, which are the eukaryotic organisms characterized by multicellular receptors, multicellular quick effectors (the muscles), and a multicellular fast connection and regulation circuitry (the nervous system).

The neurons differentiate during development from epithelial cells, and the same basically happened during evolution. This means that neurons appeared after the epithelial cells, and thus after the pre-animal organisms that, in turn, appeared after the first multicellular Eukaryotes. The late origin of neurons rules out any direct descent of them from precellular aggregates such as the microspheres (Fox et al., 1998).

The centralization of the nervous system, i.e., the formation of a brain, improves intelligence. In any case, the brain is also highly ordered (like a computer, to some extent). This means that chaotic dynamics (Newman, 1999) is completely misleading, as an approach to its operations, because it would better emerge from a chaotic brain. Also the brain size improves intelligence. At the present time, the largest brains are found in Primates and Cetaceans. According to the encephalization quotient, a primate species overtook the best-equipped Cetaceans only two million years ago (Marino, 1998).

Intelligence implies learning capacity, which has also been described as the ability to imitate cultural achievements dubbed "memes" (Dawkins, 1976; Blackmore, 2000). These are still frequently said to be transferred to the next generation according to a Lamarckian model (e.g., Diambrini-Palazzi, 1999). As a counterexample, we may remark that a meme can also be transferred to a previous generation (for instance, when an old person learns a computer game from a child or when an old Japanese macaque learns how to wash the sand off sweet potatoes from a young), so that the right meme transfer model is rather an infective one (see Boyd and Richerson, 1988). As regards the evolution of the memes in the environment made up by the brains, they are supposed to undergo Darwinian selection.

5. References

Barlow, H. B. (1983) *Nature* **304**, 207-209.

Blackmore, S. (2000) *Sci. Am.* **283**, 52-61.

Boyd, R., and Richerson, P. (1988) *Culture and the evolutionary process*, Univ. Chicago Press, Chicago.

Byrne, R. (1995) *The thinking ape*, Oxford Univ. Press, Oxford.

Dawkins, R. (1976) *The selfish gene*, Oxford Univ. Press, Oxford.

Diambrini-Palazzi, G. (1999) in R. Colombo et al. (eds.), *Origin of intelligent life in the universe*, New Press, Como, pp. 215-228.

Fox, S., Pappelis, A., and Grubbs, R. (1998) in J. Chela-Flores and F. Raulin (eds.), *Exobiology*, Kluwer, Dordrecht, pp. 87-92.

Fuerst, J. A., and Webb, R. I. (1991) *Proc. Natl. Acad. Sci. USA* **88**, 8184-8188.

Holt, J. G., Krieg, N. R., Sneath, P. H. A., Staley, J. T., and Williams, S. T., eds. (1994) *Bergey's manual of determinative bacteriology, 9th ed.*, Williams & Willimas, Baltimore.

Katz, L. A. (1999) *Am. Nat.* **154** (suppl.), S137-S145.

Li, J.-Y. (1999) *Endocytobiosis & Cell Res.* **13**, 1-86.

Marino, L. (1998) *Brain Behav. Evol.* **51**, 230-238.

Mayr, E. (1998) *Proc. Natl. Acad. Sci. USA* **95**, 9720-9723.

Newman, D. V. (1999) in R. Colombo et al. (eds.), *Origin of intelligent life in the universe*, New Press, Como, pp. 93-106.

Petroni, G., Spring, S., Schleifer, K.-H., Verni, F., and Rosati, G. (2000) *Proc. Natl. Acad. Sci. USA* **97**, 1813-1817.

Rizzo, P. J. (1987) in F. J. R. Taylor (ed.), *The biology of dinoflagellates*, Blackwell, Oxford, pp. 143-173.

Rizzotti, M. (1999) in R. Colombo et al. (eds.), *Origin of intelligent life in the universe*, New Press, Como, pp. 107-116.

Rizzotti, M. (2000) *Early evolution*, Birkhäuser, Basel.

Roger, A. J. (1999) *Am. Nat.* **154** (suppl.), S146- S163.

Russell, S. J., and Norvig, P. (1995) *Artificial intelligence*, Prentice-Hall, Upper Saddle River.

Stover, C. K., and 30 coworkers (2000) *Nature* **406**, 959-964.

Taylor, F. J. R. (1999) *J. Phycol.* **35**, 1-3.

Woese, C. (1998) *Proc. Natl. Acad. Sci. USA* **95**, 11043-11046.

EXTRATERRESTRIAL LIFE AND COMMUNICATION PERSPECTIVES

VALERIA ASCHERI [1]

Università di Genova
Dipartimento di Filosofia - Sezione Epistemologia
Via Balbi 4, 16126 Genova - Italia
E-mail: ascheri@nous.unige.it

Abstract: The aim of this paper is to propose a stronger connection and collaboration between SETI program and the space science research on extraterrestrial life. Considering the continuos progress in the whole space science research, like in astronomy and bioastronomy, a more interdisciplinary approach in organizing SETI projects may be not only useful but really necessary to improve and to suggest new lines of research. Working side by side with astronomers and bioastronomers and looking at their discoveries, for example about new extrasolar planets and first possible tracks of primitive life, SETI may get useful suggestions to follow in order to detect where extraterrestrial forms of intelligent life might develop in the future or instead, unlikely, are to be born. Thus, the organization of a task force to search for extraterrestrial life, called here *"the Macrocosm project"*, and the drafting of as much as possible *interstellar maps* as exhaustive as possible to use in extraterrestrials' hunting, are suggested in order to renew SETI programs and to look for new chances of success.

1. Introduction: the SETI strategy

As Tough (2000) reminds us, the SETI project is articulated traditionally in three main strategies:

- radio SETI;
- optical SETI;
- *active* SETI.

This strategy of research[2] so far has not given any really interesting result, mainly due to three reasons:

- the immensity of our galaxy, and of the whole universe, to search in,
- the limits of technology,

[1]I would like to thank the organization of the conference that supported me with a full grant to let me participate to the meeting.

[2]Actually the strategy is basically one and it consists in waiting for receiving extraterrestrial signals (a message) by radio frequencies or optical ones. Active SETI is a really secondary strategy because only very few terrestrial messages, both radio and material ones, have been sent towards interstellar area.

J. Chela-Flores et al. (eds.), First Steps in the Origin of Life in the Universe, 349–352.

- last but not least, the doubts on the existence of extraterrestrial life itself.

It is extremely difficult to develop or modify these methods of research considerably because the first point mentioned above is a handicap and the third one is out of the range of humans. On the contrary, the SETI program should aim at progress in the "human" limitations, both in technology and in organization. I will attempt to suggest how it would be possible to obtain a new perspective in the Search for Extra Terrestrial Intelligence.

2. Assumptions in the search of extraterrestrial life in the universe

A rational approach in the search for extraterrestrial intelligence is constituted by the following assumptions:

- our planet, solar system and the origin of life are not rare cases but something that may reasonably be repeated;
- from primitive life a biological evolution of many varied species, up to the intelligent being, may occur under certain physical conditions;
- due to the fact that SETI is led by human beings and with human terrestrial methodology and technology, the extraterrestrial intelligence should be of "terrestrial" type because otherwise, the recognition, contact, and eventually communication, may not be possible.

The "philosophy"[3] of this way of working is based on "analogy making: starting from what we are and know, in this case about life, we look for something comparable in order to find something already known or not completely ignored at all, so as to get a basis from which to begin to make hypothesis or theories to verify afterwards. Thus, in the search for life in the universe, the track scientists are following is based on what terrestrial life is.

The birth of new scientific disciplines and their constant progress, particularly bioastronomy and astrobiology, besides great developments in biochemistry and biophysics, have contributed to the search for extraterrestrial life in the universe, providing a lot of information and making discoveries almost daily. On the other hand, astronautics and engineering have been engaged in several successful missions, even if the costs are extremely high and there have been serious mistakes.

Comparing this impressive field of research with the SETI program, I have the impression that SETI has remained somewhat behind such progress; a proof of this situation is that its program is not generally endorsed by the scientific community at large [4]. This situation has several causes: first the objective difficulty of the search itself, second the emphasis on the problems, third the isolation of its research activity, as well as of some of the research, finally the intrinsic difficulty of the technology and means of communication that have been suggested.

SETI would have a new start and gain much more credit if it were directly involved and settled in the search for extraterrestrial life *tout court*, as its "final phase". Following results for the search for primitive life and astronomical discoveries step by step would be the best guarantee to succeed in finding extraterrestrial intelligence, since,

[3] It is an ancient philosophy explained and used, just to quote two great masters, by Aristotle and Tomas Aquinas.

[4] SETI program current state of search corresponds to what the philosopher of science T. Kuhn (1962) calls "paradigm not progressive but regressive" because so far it has no results, neither can it assure when it may get them. Hence comes the difficulty to gather funds to continue such line of search.

as Harrison and Dick (2000) have stated: "confirmation of extraterrestrial intelligence could come about in any number of ways, *but given our known laws of nature and current technology*[5], some procedures seem more promising than others"(p. 22).

3. The Macrocosm project

Recently ESA has accepted to set up a very important and ambitious mission, called GAIA. Its targets are to get new and detailed information about our galaxy's cosmological history; in particularly the Gaia satellite would detect the composition, formation and evolution of our galaxy in order to create a precise three-dimensional map of more than one billion stars, also beyond the Milky way. It is the latest example of the space exploration's efforts, successor of the Hipparcos satellite, and it should answer some questions such as, where the sun sisters are, how our galaxy and its neighbors originate and evolve.

Looking at GAIA mission as an example to follow, my proposal is to set up what may be called "*the Macrocosm project*": a great task force of scientists, composed by all the people who work in the study of the origin, evolution and distribution of life on Earth and in the universe and by SETI scientists, who should contribute, each according to his own competence, to analyze as many different aspects of the universe as possible.

As early as 1978, J. G. Miller had proposed a model called Living System Theory (LST) to study extraterrestrial life based on terrestrial one. Harrison and Dick (2000) explain that LST is: "a universal theory of behavior and historical epochs, it is an interdisciplinary framework that integrates all the biological and social sciences from cellular biology to international relations" (p. 24). Moreover, they suggest a kind of collaboration similar to the Macrocosm project: "Living System theory and SETI scientist are interested in the same phenomena and apply similar intellectual tools. Researches from both traditions are accustomed to thinking in terms of energy, matter, and information and in terms of interconnections and systems and in terms of microcosm and macrocosm"(p. 24).

Thus, now it would be really possible to join effectively the great and varied efforts in the search for extraterrestrial life, exploiting advanced technologies, such as satellites, probes and the support of the new spatial station. Scientists of all research fields should consider life as a unique complex phenomenon to investigate without disciplinary, or conceptual barriers.

3.1. SETI AS AN "ARCHEOLOGICAL" PROJECT

If we compare SETI search to an archeological mission on an island in Asia or in Africa, *interstellar maps* would be an indispensable instrument to indicate where the target - the treasure – is, that is, the site where extraterrestrial intelligent forms of life may exist.

Researchers should provide a sort of *radiograph* of selected parts of the universe, starting from those produced by GAIA and other satellites; the maps should be completed with all the *parameters* suitable for discovering life: the chemical-biological

[5] I have added the italics.

elements, the physical and geological features of the environment, the astronomical history as well as probable future cosmic events.

The interstellar maps should indicate where to search at the moment and where to focus our attention in the future[6]; meantime they should reveal which areas are to be avoided because they may be less hospitable for life. With this data, the SETI program may organize its search: the radio telescopes - in the next few years more powerful or more numerous - may be pointed to certain areas rather than towards less favorable ones[7].

The Macrocosm project might be a good way to set the SETI program in the core of space exploration and the search for life in the universe. Clearly, the SETI objectives are the most ambitious, but so far with no reproducible signals. It may have new opportunities and, above all, it may be no longer considered an endless quest but as a field of research in real progress.

4. References

Ascheri V. (2000) "A new methodological approach to communicate with extraterrestrials" in Smith, Joe Q. et al. *Bioastronomy '99 - A New Era in Bioastronomy*, Proceedings of a Conference held on the Kohala Coast, Hawaii, 2-6 Aug. 1999, ASP Conference Series 213, p.603

Dick S. J. (1998) *Life on other worlds*, Cambridge University press, USA

Harrison A. A., Dick S. J. (2000), "Contact: Long Term Implications for Humanity", A. Tough (ed.) *When SETI succeeds: The Impact of High-Information Contact,* Foundation for the Future, Bellevue, Washington, USA, pp. 7-31

Kuhn T. (1962) *The Structure of Scientific Revolutions*, The University of Chicago Press, Chicago, USA

Miller J. G. (1978) *Living systems*, McGraw-Hill, New York, USA

Tough A. (2000) "How to Achieve Contact: Five promising strategies" in A. Tough (ed.), *When SETI succeeds: The Impact of High-Information Contact,* Foundation for the Future, Bellevue, Washington, USA, pp. 115-125

[6] It may be very useful to know predictable times of development of forms of life on the grounds of the physical conditions.

[7] A possible collaboration between astronomers and SETI researchers may work like this: examining the radio spectra of a well-determined space area, or better, of an extra solar planet and its atmosphere, perhaps it would be possible to identify its chemical features and physical conditions. Hubble telescope can already send us maps of Titan and other small satellites inside the solar system but in the next years it will be possible to improve such capacities.

ON THE LAST TERMS OF DRAKE EQUATION: THE PROBLEM OF ENERGY SOURCES

PAOLO MUSSO
University of Genova - Dep. of Philosophy (Section of Epistemology)
V. Balbi 4, 16129 Genova (GE) - Italy
musso@nous.unige.it

1. The Drake Equation

One of the main tools for the evaluation of the possible number N of technological civilizations with which we could enter in touch via radio-waves is the famous Drake Equation

$$N = R* \cdot f_p \cdot n_e \cdot f_l \cdot f_i \cdot f_c \cdot L$$

where:
$R*$ = rate of star formation;
f_p = fraction of stars having planets;
n_e = number of suitable planets per ecosphere;
f_l = fraction of suitable planets on which life starts;
f_i = fraction of life starts that evolve into intelligence;
f_c = fraction of intelligent civilizations that develop the technology to attempt communication;
L = mean longevity of a technological civilization.
Only $R*$ has a reasonably certain value (that is, more or less, 10 per year), while we can only make suppositions about all the other terms, because we have neither, at present, a deep enough knowledge of the laws of nature involved in the related processes, nor the possibility to observe them directly (even if the most recent discoveries of extra-solar planets, although only indirect, allow us to suppose that f_p should be very high, maybe near 1). Anyway, progress both theoretical and technological should provide, in the next future, an increasing amount of data about these subjects as far as f_i, so that our assessments about the first five terms of the Equation should be improved more and more in time.

2. The number of technological civilizations: the role of energy sources

On the contrary, this task sounds particularly difficult with reference to the sixth and seventh terms, f_c and L, because in these cases we seem not to have any law of nature at all to base on, in order to carry out our estimations. The development and the longevity of a technological civilization able to communicate with us, indeed (as well as the decision of actually communicating), seems to depend only on the intelligence, the enterprise and the free choices of its members, which, obviously, cannot be possible

J. Chela-Flores et al. (eds.), First Steps in the Origin of Life in the Universe, 353–356.

objects for science.
Anyway, it is possible to find some constraints on this process, which, on the contrary, *can* be scientifically evaluated. One of them (maybe the main one) is, in my opinion, the problem of energy sources.
The starting point is the consideration that any technological civilization needs, for its development, a large amount of free energy. Obviously, we cannot know what kind of energy sources could be used by another, more advanced civilization. However, it is possible to make some reliable considerations basing on what we can deduce about this subject looking at *our own* history, because we *are* able to communicate via radio-waves, consequently our present degree of technological development, as primitive as it could be, is a good touchstone for our purposes.

3. Life and fuels: a cosmic intercourse

If we consider carefully the energy sources we know, we'll find a surprising circumstance: they are all strictly related with life. For example:
1) All the simplest energy sources (that is wood and fossil fuels) are organic and derive directly from living beings.
2) Uranium is much more abundant in the Earth's surface than in the inner because of its transformation into soluble UO_2, due to the oxidant atmosphere, which is produced by plants. On the other side, it plays a crucial role in keeping the Earth warm both with its radioactivity and its taking part to the CO_2 cycle.
3) Deuterium (which is indispensible for nuclear fusion) is abundant and easily available on Earth due to the presence of a large amount of liquid water, which is indispensible also for life and can still exist because the Earth itself has just the right size and temperature, which are also the right ones for life.
4) Renewable energy sources (as solar, aeolian, hydro-electric, hydrothermal and, obviously, biomasses) are all involved in the basic phenomena of life (i.e. photosynthesis, the presence of great masses of water and a not too rarefied atmosphere, the existence of a warm core keeping the planet at a suitable temperature)...
...and we could go on further.
In conclusion, it's very likely that, if somewhere there is life, then there will be also energy sources enough to develop technology, because the same processes seem to be involved in generating both.

4. The "Techno-Anthropic" Principle

What we have said implies also that many of the "cosmic coincidences" related with life in the so-called "Anthropic Principle" (as the "fine-tuning" of many cosmic constants, the "carbon cycle" in the stars, the creation and diffusion of heavy elements in the supernovae, the crucial role of comets in bringing water on planets and so on) hold for the energy sources, too. But there are also others, more specific. For example:
1) Most of fossil fuels became available only about 100 millions years ago (that is last week, making the Earth's age equal to a year), just in time to be at our disposal.

2) Their abundance is *almost exactly* the one we need to learn how to use the other energy sources before they run dry: only a 10 factor less would be probably enough to make it impossible.
3) In a just a little higher-density universe than ours the abundance of deuterium would be exponentially decreasing, so that fusion would be impossible.
4) The total amount of solar energy per year reaching the Earth is about 100.000 times (2.301×10^{24} J vs. 1.833×10^{19} J) the total amount of energy needed by our technological civilization, and this enables it to exist: with a smaller rate, indeed, our energy production would be too great, in proportion, to be absorbed by the environment without causing terrible unbalances...
Once more, we could go on further.
So, we could even speak about a "Techno-Anthropic" Principle, meaning by this that: "Our universe is shaped in such a way that its laws allow the birth of technological civilizations".

5. Fuels and matches: a universal relation

But, from another point of view, just from a very rough observation we can easily see that, *while the total amount of fuel needed to develop a given amount of energy tends to decrease with the development of technology, on the contrary the size of the "match" needed for it tends to increase.* The reason is that at every step we are going deeper and deeper into the inner structure of the matter. We don't need a very deep knowledge, nor a complex technology to get thermal energy from wood or coal, being enough to burn them; but we need something more to build up a steam engine or a ydro-electric central, and something much more to get nuclear fission or, worse, fusion; finally, renewable energies, such as, e.g., solar, in which there is no "fuel" in the proper sense, require (at least at present) even much wider apparatuses to be exploited efficiently.
This relation seems to hold universally, because it is based on the laws of nature. So, it is very unlikely that a technological civilization could ever arise without needing to use, at least for a while, the simplest energy sources, among which a crucial role is played by fossil fuels.
Thus, on one side we should admit that every intelligent species in the universe should have a very high *a priori* probability (very likely equal to 1) to develop technology. On the other, some precise constraints are implied by what we have said, both on the rising and the duration of a technological civilization.

6. Constraints on technological development related with f_c

1) <u>Intelligent life must not evolve too soon</u>, otherwise it could become extinct before a sufficient amount of fossil fuels have formed. Further analyses will be needed to evaluate in a quantitative way, as far as possible, this range of time.

2) <u>Intelligent life must evolve on dry land</u>, because the simplest ways to get energy from nature always imply the need of *burning* something: so, maybe a technological

civilization could *develop* also under the sea, but certainly it could never *begin* down there.

3) Just for the same reason, intelligent life must evolve in beings of a suitable size (that is not too small) and provided with something like hands, otherwise they could never be able to manage fire, as requested at the previous point.

7. Constraints on technological development related with L

4) Scientific progress and the development of technology cannot work under a critical speed, otherwise it would become impossible to reach the degree of "know how" needed to use other energy sources before fossil fuels run dry (or generate a devastating "greenhouse effect"). Further studies will be needed in order to determine the width of this "window", which, however, does not seem likely to be very large (probably no more than some centuries): so, our fastest technological development, about which we are so worried, could turn out to be, in reality, the only possible way for every civilization, with all the related risks we know very well.

5) Technological progress does not necessarily imply a moral one and vice versa. A more "altruistic" attitude than ours, which made energy sources accessible to all people just from the beginning in a more equalitarian way, indeed, would have unavoidably the effect of shortening the "window" we have spoken about at point 4, asking them for a greater (and therefore more difficult) effort in order to develop their scientific knowledge faster. While a more "egoistic" civilization could be even favoured from this point of view, but would run a greater risk of self-destruction.

Acknowledgments

I want to thank very much Dr. Carlo Sozzi (*Institute of Plasma Physics* of *CNR*, Milano), who has suggested me this idea and has given me also a great aid in adjusting some technical points.

References

1. Assessorato alla Cultura del Comune di Bologna (ed.) (1986) *Energia e materia*, Paolini Editore, Bologna.
2. Assessorato alla Cultura del Comune di Bologna (ed.) (1986) *Energia e vita*, Paolini Editore, Bologna.
3. Assessorato alla Cultura del Comune di Bologna (ed.) (1986) *Energia e società*, Paolini Editore, Bologna.
4. Caira, M. et al. (1999) MARS: un reattore per la seconda era nucleare, *Le scienze* **368**.
5. Clò, A. (1989) Il mito delle risorse energetiche rinnovabili, *Le Scienze* **245**.
6. Fells, I. (1998) The need for energy, *Europhysics News* **29**, n. 6.
7. Fermi, E. (1972) *Termodinamica*, Boringhieri, Torino.
8. Garozzo, M. (ed.) (1997) Energie pulite, *Quaderni di Le Scienze* **97**.
9. Hawking, S. and Penrose, R. (1996) The Nature of Space and Time, *Scientific American*, July 1996.
10. National Ignition Facilitiy, *Web Site*, http://lasers.llnl.gov/lasers/nif.html.
11. Nordland, R. (1999) Where is the Next Chernobyl?, *Newsweek* Oct. 18, 1999.
12. Press and Public Relations Group, *JET Joint Undertaking Web Site*, www.jet.uk.
13. Ristinen, R. A. and Kraushar, J. J. (1998) *Energy and the Environment*, Wiley and Sons, New York.
14. Rubbia, C. (1996) The energy amplifier: a description for the non specialists, *CERN/ET/Internal Note* **96-01**, Jan. 5th, www.cern.ch/CERN/Divisions/SL/EET/ea/htmldoc/nonspec/nonspec.html.
15. Williams, P. and Woessner, P. N. (1996) The Real Threat of Nuclear Smuggling, *Scientific American*, January 1996.

Section 9:

Deeper implications of the detection of extraterrestrial life

ORIGINS AND CREATION

GEORGE V. COYNE, S.J.
Specola Vaticana
V-00120 Città del Vaticano

1. Introduction

It is arguably difficult to find a more heated topic of discussion than that concerning the origins of the universe, and especially of life and of intelligence, and whether such origins can be understood without evoking a Creator God. Responses range from the extremes of a Stephen Hawking or a Pope Pius XII to almost all conceivable intermediate positions. Hawking claims that, if his quantum cosmological theory of the origins of the universe without boundary conditions is correct, then we have no need of God. Pius XII attempted to claim that with Big Bang cosmologies scientists were coming to discover what had already been known from the Book of Genesis, namely that the universe had a beginning in God's creative action. In between we have such positions as evolutionary naturalism and episodic divine intervention. Evolutionary naturalists would claim that, although our scientific knowledge of evolution is limited, the best explanation of the universe and all that it contains is through complexification in an expanding, evolving system in which both deterministic and chance processes play out their roles in a universe abundant with opportunities, 15 billion years old and containing 10^{22} stars. Those who profess episodic divine intervention would claim that divine activity is required, at least in some phases of the evolutionary process and, in particular at the occurrence of human life and intelligence, because natural processes alone are not adequate to explain the end result. What is one who is both a believing Christian and a scientist to make of all of this?

2. Scientific Evidence for Universal Evolution

Let us take a sweeping view of a reasonable scientific picture of things. By reasonable I mean that, while we do not have all the answers, the following picture can stand up to any scientific critique. If we look today in infrared light at the center of Orion we see boiling gas and dust. The fact is that stars are being born in this gas. And where the hottest, most massive and, therefore, brightest stars are already born, they are irradiating the gas, and it is giving off hydrogen alpha radiation. In this way we can identify star birth regions. The region of star birth in

J. Chela-Flores et al. (eds.), First Steps in the Origin of Life in the Universe, 359–364.

Orion is just a little part of our Milky Way. Our Milky Way, like most other spiral galaxies, measures 100,000 light years across and it contains about a hundred billion stars. It has several beautiful spiral arms and the sun is located in one of the outer arms, about two-thirds of the distance from the nucleus of our galaxy.

How is a star born? It happens by the laws of physics. A cloud of gas and dust, containing about 100 to 1,000 times the mass of our sun, gets shocked by a supernova explosion or something similar and this causes an interplay between the magnetic and gravity field. The cloud begins to break up and chunks of the cloud begin to collapse. The mass is so great that the internal temperature reaches millions of degrees and thus turns on a thermonuclear furnace. A star is born. Thermonuclear energy is the source whereby a star radiates to the universe. You need a very hot piece of the universe to do this, and so you can only get this thermonuclear furnace by having a cloud collapse and raise the temperature. You can only get it, in other words, in stars, with one exception, namely, in the very hot early universe before galaxies or stars were born.

Stars also die. A star at the end of its life can no longer sustain a thermonuclear furnace and so it can no longer resist against gravity. It collapses for a final time, explodes and expels its outer atmosphere to the universe. This may happen nice and peacefully or it may happen in a violent cataclysmic explosion, called a supernova. The most famous of these is the Crab Nebula which has a pulsar at the middle as its dead star. So stars are born and stars die. And as they die they spew left over star matter out to the universe. The birth and death of stars is very important. If it were not happening, you and I would not be here, and that is a scientific fact. In order to get the chemical elements to make the human body, we had to have three generations of stars.

Obviously this story of star birth and death is very important for us. Out of this whole process around one star, which we call the sun, a group of planets came to be, among them the little grain of sand we call the Earth. An amazing thing happened with that little grain of sand when, in the 16th and 17th centuries with the birth of modern science, we developed the capacity to put the universe in our heads. We do that by using mathematics, physics, chemistry and biology.

Let us now review what we know of the history of the expanding universe. As it aged, distances got larger in the universe. As this happened certain key events took place. Quarks combined to form elementary particles, which in turn formed atoms and then molecules. The universe became transparent and the cosmic background radiation came to be. Galaxies and stars were formed. The first microscopic life forms came to be after twelve billion years in a fifteen billion year old universe. Why did it take so long to make even an amoeba? We have already discussed one reason. We did not have the chemistry to make even an amoeba until we had had three generations of stars.

3. Human Origins

How did we humans come to be in this evolving universe? It is quite clear that we do not know everything about this process. But it would be scientifically absurd to deny that the human brain is a result of a process of chemical complexification in an evolving universe. After the universe became rich in certain basic chemicals, those chemicals got together in successive steps to make ever more complex molecules. Finally in some extraordinary chemical process the human brain came to be, the most complicated machine that we know.

Let us pause for a moment to review the degree of certainty which we can place in the above scenario. We certainly do not have the scientific knowledge to say how each living creature came to be in detail. We do not know precisely how each more complex chemical system came to contribute to the process of self organization which brought about the diversity of life forms as we know them today. Most importantly, we do not know with scientific accuracy the sufficient elements in nature to have brought about the unbroken geneological continuity in evolution that we propose actually happened. There are, in brief, epistemological gaps which prevent natural science from saying that a detailed theory of biotic evolution has been proven. What I have presented is the most adequate account conceivable at this time considering the available empirical data. And that empirical data, with respect to biotic evolution, comes from various independent scientific enterprises, including molecular biology, paleontology and comparative anatomy.

Did we come about by chance or by necessity in this evolving universe? The problem is not formulated correctly. There is a third element that is very important. It is what we may call "opportunity." What this means is that the universe is so prolific in offering the opportunity for the success of both chance and necessary processes that such a character of the universe must be included in the discussion. The universe is 15 billion years old, it contains about 100 billion galaxies each of which contains 100 billion stars of an immense variety. For 15 billion years the universe has been playing at the lottery. What do I mean by the lottery? When we speak about chance we mean that it is very unlikely that a certain event would happen. The "very unlikely" can be calculated in mathematical terms. Such a calculation takes into account how big the universe is, how many stars there are, how many stars would have developed planets, etc. In other words, it is not just guesswork. There is a foundation in fact for making each successive calculation.

By using the mathematics of nonlinear dynamics, one can say that as this process goes on and more complex molecules develop, there is more and more direction to this process. As the complexity increases, the future complexity becomes more and more predetermined. There are also necessary, deterministic processes occurring. But there is a lot of chance as to what the exact conditions are, so that the necessary process may take place. There are definitely both chance and necessity, but they are happening in a universe that is so prolific that the eventual outcome has a kind of predetermined nature.

This predetermined nature may be represented by a tree, the Tree of the Universe. It is a strange tree in that everything that ever happened in the universe, from the making of quarks to the making of toenails, is all here. Even those processes that never succeeded, that failed, every dead leaf and dead branch has been conserved. Every meeting of molecules in inopportune circumstances is there. The tree has never been pruned. But blow a quiet breeze through this tree and what will you see? You will inevitably see something that resembles the bare trunk of a tree with certain branching to various forms of life and ending at the top with the human being. The result is inevitable because with a combination of chance and necessary processes in a very prolific universe with so many opportunities there is a narrowing down of the evolutionary process due to the nature of physics, chemistry, biology and non-linear dynamics.

If we truly accept that there are chance processes involved, then the branching of the tree could be somewhat different. But since complexity proceeds towards an ever more determined direction the trunk of the tree could not be very different. The paleontology, biology and chemistry behind all of this is quite uncertain, but it is clear that something like this would happen. Why is the human being at the top? It is because we are ignorant. We do not know what else to put at the top. The human brain is the most complicated mechanism we know.

4. Implications for Religious Belief

How are we to interpret this scientific picture in terms of religious belief. Do we need God to explain this? Very succinctly my answer is no. In fact, to need God would be a very denial of God. God is not the response to a need. One gets the impression from certain religious believers that they fondly hope for the durability of certain gaps in our scientific knowledge of evolution, so that they can fill them with God. This is the exact opposite of what human intelligence is all about. We should be seeking for the fulness of God in creation. We should not need God; we should accept him when he comes to us. The religious believer is tempted by science to make God "explanation." We bring God in to try to explain things that we cannot otherwise explain. "How did the universe begin?", "How did we come to be?" and all such questions. We sort of latch onto God, especially if we do not feel that we have a good and reasonable scientific explanation. He is brought in as the Great God of the Gaps.

The scientific picture traced above deals with the questions of origins, of how what we observe and experience today came to be? The question of creation, and therefore of a God Creator, responds to the question of why is there anything in existence. Creation is not one of the ways whereby things originated as opposed to other ways that can be thought of, including quantum cosmology and evolutionary biology. The claim that all things are created is a religious claim that all that exists depends for its existence on God. It says nothing scientifically of how things came to be, although beautiful stories are told in the Book of Genesis, to elaborate on the dependence of all things for their existence upon God.

Having opened the Pandora's box of the Bible, let us elaborate a bit upon it. The Bible is a collection of writings by various authors at various epochs using various literary genres. And so it best serves reason if one speaks of a specific book rather than of the Bible in general. It is clear that the overall intention of the authors of Genesis is to evoke religious faith, an adherence to the God of Abraham, Isaac and Jacob, and not to teach. There is simply no scientific teaching in Genesis. In the Judaic/Christian tradition, the roots of religious belief reach to 2,000 years before Christ with the prophet Abraham. But Modern Science cannot be dated before the 16th or 17th century, roughly from the time of Galileo and then through many others to Newton, with the discovery of the universal law of gravity, the differential calculus, etc. The modern science that speaks to religion today is born much later than the religion to which it speaks. It has to be recognized that the religious tradition is historically much longer and to a certain extent has that richness of the past that modern science does not. It is unfortunate that, at least in America, creationism has come to mean some fundamentalistic, literal, scientific interpretation of Genesis. Judaic-Christian faith is radically creationist, but in a totally different sense. It is rooted in a belief that everything depends upon God, or better, all is a gift from God. The universe is not God and it cannot exist independently of God. Neither pantheism nor naturalism is true.

But if we confront what we know of origins scientifically with religious faith in God the Creator, in the senses described above, what results? I would claim that the detailed scientific understanding of origins has no bearing whatsoever on whether God exists or not. It has a great deal to do with my knowledge of God, should I happen to believe he exists. Let me explain. Take two rather extreme scientific views of origins: that of Stephen Gould of an episodic, totally contingent and, therefore, non-repeatable evolutionary process as contrasted to a convergent evolutionary process such as that of Christian de Duve, in which the interplay of chance, necessity and opportunity leads inevitably to life and intelligence. In either case, it is scientifically tenable to maintain an autonomy and self-sufficiency of the natural processes in a natural world, so that recourse to God to explain the origins of all that exists, is not required. It is not a question of chance in nature, excludes God; destiny in nature requires God. In neither case is God required.

If, however, I believe in God then what nature tells me about God in one case is very different from what nature tells me about God in the other. Please note that I am not calling upon faith to adjudicate between contrasting scientific viewpoints. I do think that convergent evolution is more consistent with God's revelation of himself in the Book of Scripture, so that, as Galileo was fond of stating, the Book of Scripture and the Book of Nature speak of the same God.

If we take the results of modern science seriously, it is difficult to believe that God is omnipotent and omniscient in the sense of the scholastic philosophers. Science tells us of a God who must be very different from God as seen by the medieval philosophers and theologians. This is not to place a limitation upon God. Far from it. It reveals a God who made a universe that has within it a certain dynamism and thus participates in the very creativity of God. Such a view of creation can be found in early Christian writings, especially in those of St. Augustine in his comments on Genesis. If they respect the results of modern science, religious believers must move away from the notion of a dictator God, a Newtonian God who made the universe as a watch that ticks along regularly. Perhaps God should be seen more as a parent or as one who speaks encouraging and sustaining words. The universe has a certain vitality of its own like a child does. It has the ability to respond to words of endearment and encouragement. Words which give life are richer than mere commands or information. In such wisdom does God deal with the universe.

These are very weak images, but how else do we talk about God. We can only come to know God by analogy. The universe as we know it today through science is one way to derive analogical knowledge of God. For those who believe modern science does say something to us about God it provides a challenge, an enriching challenge, to traditional beliefs about God. God in his infinite freedom continuously creates a world which reflects that freedom at all levels of the evolutionary process to greater and greater complexity. God lets the world be what it will be in its continuous evolution.

LIFE IN THE UNIVERSE: PHILOSOPHICAL AND THEOLOGICAL ISSUES

ROBERT JOHN RUSSELL
The Center for Theology and the Natural Sciences
The Graduate Theological Union
2400 Ridge Road, Berkeley, California, USA 94709

1. Introduction

Over the past four decades, the interdisciplinary field of 'theology and science' has undergone tremendous growth involving scholars from philosophy of science, philosophy of religion, the natural sciences, theology, ethics, history and related fields.[1] Topics range from comparative methodologies to the relations between theologies of creation, divine action, and redemption in light of Big Bang, inflationary and quantum cosmologies, quantum physics, evolutionary and molecular biology, the neurosciences, anthropology, sociobiology, behavioral genetics, etc. Originally the work drew primarily on Christian theology, but today representatives of many other religions are active contributors. Surprisingly underrepresented in this rapidly growing interaction, however, is a focus on the philosophical and theological issues raised by the possibility of extraterrestrial life (EL) and extraterrestrial intelligence (ETI).[2] This is particularly curious since historians of science have shown that Christian theology contributed in significant ways to the assumption that ETI does in fact exist.[3] It is particularly timely, then, that this Sixth Trieste Conference on Chemical Evolution includes a section on the philosophical and theological implications of extraterrestrial intelligence.

There are, in fact, a wealth of topics which could be addressed here.[4] For the purposes of this short presentation, I will focus on the following three: 1) Will the discovery that life is abundant or rare in the universe influence its value or meaning? 2) Will intelligent life be capable of both rationality and moral reasoning? 3) Will ETI experience moral failure or be entirely benign? First, two caveats: 1) I will speak from the context of Christian theology and welcome others to widen the circle of religious discussions of EL/ETI. 2) I will assume that we will share modes of sensory awareness and rationality with ETI. It may be that some extraterrestrial civilizations are millions or even billions of years older than ours; for such advanced ETI, the usual 'contact' scenarios may simply not apply.[5] For the present purposes, however, I will focus on the possible discovery of ETI for which 'contact' would be a reasonable hypothesis.

J. Chela-Flores et al. (eds.), First Steps in the Origin of Life in the Universe, 365–374.

2. Three issues from the perspectives of philosophy and theology

2.1 WILL THE DISCOVERY THAT LIFE IS ABUNDANT OR RARE IN THE UNIVERSE INFLUENCE ITS VALUE OR MEANING?

The relative abundance of life in the universe is one of the pivotal issues of our conference. Hopefully, we will learn the answer in the reasonably near future, either by projects such as the exploration of Mars and Europa, at least regarding pre-biotic and microbial life, or by projects like SETI and the discovery of extraterrestrial intelligence. For now, attempts at estimating the abundance of life and intelligence remain highly controversial. On the one hand, Stephen Jay Gould has stressed the overwhelming role of chance in the history of life on earth.[6] Thus, if life has evolved elsewhere, we should expect radical diversities in its morphologies reflecting differing evolutionary histories shaped by strongly contingent events. Paul Davies[7] and Christian DeDuve[8], on the other hand, have argued that evolution is tightly governed by the laws of physics and biology. Julian Chela-Flores has taken De Duve's point one step further, arguing that once life originates, the evolution of eukaryotes will lead to procaryotes. From there the 'convergence' of life towards intelligence is assured, although there may well be significant diversities in the morphologies which underlie intelligence.[9]

Underlying these issues, however, is a more fundamental question: Is the meaning and value of life a function of its abundance in the universe, or is it essentially meaningless *regardless* of what we find through the exploration of our solar system or SETI? It is this latter question which has received considerable philosophical and theological reflection.

Some scientists have suggested that biological life *per se* has little significance whether or not we are alone in the universe. They see life as essentially meaningless, a random product of physics and chemistry of no more significance than the wetness of water or the structure of Saturn's rings. Biological processes are just what matter does when really unusual conditions occur, but the universe, "at rock-bottom", is just endless mass-energy and curving spacetime. Such 'cosmic pessimism' is of course a *philosophical* interpretation of nature; it is not science, per se, nor is it one which can be 'proved' by science, but it is one that has been widely propounded by eminent scientists such as Bertrand Russell[10] and Jacques Monod[11]. It is certainly the impression Steven Weinberg gave in his often-quoted conclusion to *The First Three Minutes*: "(H)uman life is ... just a more-or-less farcical outcome of a chain of accidents reaching back to the first three minutes ...The more the universe seems comprehensible, the more it also seems pointless."[12]

Others, though, disagree with this view, arguing instead for a philosophy in which life is a clue to the meaning of the universe. In *Disturbing the Universe*, Freeman Dyson writes: "I do not feel like an alien in this universe. The more I examine the universe and study the details of its architecture, the more evidence I find that the universe in some sense must have known that we were coming." And in his 1985 Gifford Lectures, *Infinite in All Directions*, Dyson explicitly rejects Weinberg's opinion, telling us instead he sees "...a universe growing without limit in richness and complexity, a universe of life surviving forever and making itself known to its neighbors across the unimaginable gulfs of space and time...Twentieth-century science provides a solid foundation for a philosophy of hope."[13] Writing in a similar vein, Paul Davies has depicted life in terms of "teleology without teleology"[14] and William Stoeger has written about the "immanent directionality" of evolution.[15]

The difference in these views may stem in part from a further division between reductionist and non-reductionist philosophies. If one assumes that the processes and properties characteristic of living organisms can be fully explained by physics and chemistry, there may be little if any basis for attributing meaning and value to life. Non-reductionist arguments on the other hand, such as those deployed by Francisco Ayala[16], Ernst Mayr[17], and Charles Birch[18], offer a basis within natural processes for attributing varying degrees of meaning and value to organisms with differing levels of complexity and organization. Non-reductionist epistemologies, in turn, play a crucial theological role in a variety of views often referred to collectively as 'theistic evolution.' This perspective includes two central themes: *creatio ex nihilo* and *creatio continua*. 1) God as transcendent creates the universe out of nothing (*creatio ex nihilo)*, holding it in existence at each moment and maintaining its law-abiding character which we express scientifically as the laws of nature.[19] 2) God as immanent creates the universe continuously in time (*creatio continua*), working "in, with, under and through" the processes of nature[20], as Arthur Peacocke nicely phrases it. Scientists and theologians have developed these themes in light of physical cosmology, quantum physics, chaos and complexity theory, evolutionary and molecular biology, anthropology, the neuro and cognitive sciences, etc.[21] Most hold that the multi-leveled complexity of living organisms points to the intrinsic value of life. Arguably the most remarkable construction in the galaxy is the primate central nervous system. The number of connections between the neurons of the human brain is greater than the number of stars in the Milky Way. This staggering complexity makes possible the almost unimaginable feat of self-consciousness, of knowing oneself as a free, rational and moral agent in the world. Thus on our planet, at least, we are privileged to discover a hint of what God's intentions might have been in creating a universe like ours, with its particular laws of physics. For when the evolutionary conditions are right as they have been on Earth, and as they may be elsewhere in our universe, God, the continuous, immanent, ongoing creator of all that is, working with and through nature, creates a species gifted with the "image of God" (the *imago dei)* including the capacities for reason, language, imagination, tool-making, social organization, and self-conscious moral choice, a species capable of entering into covenant with God and in turn with all of life.

Thus if it took the precise characteristics of this universe to allow for the possibility of the evolution of life, even if life is scarce in the universe[22], then it is life as such that gives significance to our universe --- and even if ours is only one of a countless series of universes, as some inflationary and quantum cosmologies depict[23]. In short, I see life as the enfleshing of God's intentions amidst biological evolution which, in turn, is the ongoing expression of God's purposes in creating all that is. God thus offers to nature nature's conscious experience of the God who acts within nature.

2.2 WILL INTELLIGENT LIFE BE CAPABLE OF BOTH RATIONALITY AND MORAL REASONING?

We turn here to the question of the biological origins of ethics.[24] If our human experience of moral capacity, like our capacity for rational thought and relationality, is a gift of God, the *imago dei*, rooted in our biological nature and bequeathed us by God acting in and through evolution, does this suggest that wherever evolution results in creatures capable of rationality they would also be equipped by God with a capacity for relationship and moral reasoning?

Sociobiologists and, more recently, behavior geneticists, have explored the biological basis of human social behavior in order to determine the relation between evolutionary and genetic constraints on the one hand and their cultural expression on the other. Many of them, such as E. O. Wilson, Richard Dawkins and Michael Ruse, are unabashedly reductionistic, interpreting their scientific research in strictly deterministic and functionalist accounts of human behavior. Ruse, for example, has argued extensively that both the capacity and the content of human morality are entirely the products of evolution.[25] He defines biological altruism as any cooperative behavior between organisms that increases 'evolutionary gain', while moral altruism refers to our considered choices to help others because it is seen as 'right' to do so. For Ruse, moral altruism is a product of (nonmoral) biological altruism with "... no objective foundation. It is just an illusion, fobbed off on us to promote 'altruism'".[26]

Geneticist Francisco J. Ayala takes a very different position. For nearly three decades he has argued against reductionism in biology[27], disagreeing dramatically with Ruse over the evolutionary origins of human moral capacity. According to Ayala, evolution selected for intelligence in our ancestral hominid line; one of the many byproducts of intelligence is ethics. "Ethical behavior came about in evolution not because it is adaptive in itself, but as a necessary consequence of man's (sic) eminent intellectual abilities, which are an attribute directly promoted by natural selection."[28] In short, ethics is open to cultural, philosophical and religious views. We can chose to act against natural predispositions, such as selfishness, if they are judged morally unacceptable. Conversely, some moral norms, such as justice and benevolence, may be inconsistent with behaviors favored by natural selection. Similar arguments against reductionism have been developed by a wide range of scientists and philosophers, including Theodosius Dobzhansky and Ayala, Ernst Mayr, Arthur Peacocke[29], Ian Barbour[30], Nancey Murphy[31] and George Ellis.[32]

How does this bear on the question of ETI? If what we have found about life on earth can be generalized, we may expect that moral capacity will be found wherever life has evolved to the point of intelligence in the universe, and that evolution will play a relatively strong / weak role in determining its contents depending on where one stands on the reductionist / non-reductionist debate above.

What sorts of responses might Christian theologians offer to the discovery of ETI with rational and moral capacities? Physicist Paul Davies predicts it would "(shatter) completely the traditional perspective on God's relationship with man (sic)."[33] But theologian Ted Peters finds "little or no credible evidence" for Davies' view.[34] Instead there is rich evidence in the history of Christian thought in support of a "plurality of worlds" and extraterrestrial life in the universe.[35] Contemporary theology, too, has been genuinely open to the possibility of rational and moral ETI. Peters cites Roman Catholics, such as George van Noort, Theodore Hesburgh, Hans Küng, Karl Rahner, and Francis J. Connett, conservatives such as Billy Graham, and Protestants such as Krister Stendahl, A. Durwood Foster, and Paul Tillich.[36]

I would find the discovery of ETI endowed with rational and moral capacities to be a wondrous exemplification of the intentions of God in creating a universe like ours, namely in order to achieve the evolution of creatures capable of bearing the *imago dei* and entering into genuine community and covenant with God and all of life.[37] I am also persuaded by the philosophical arguments of non-reductionists that ethical choice and the contents of our moral codes remains at least partially a 'free variable'. I would therefore expect that ETI will be 'accountable' for its choices in some way even as humanity is. This leads directly to the challenge of moral failure, our final issue.

2.3 WILL ET EXPERIENCE MORAL FAILURE OR BE ENTIRELY BENIGN?

This question embodies a tragic reality at the heart of human existence. Why do we act with a level of violence against our own kind and other species which far exceeds the needs of survival and the level of violence of all other forms of life on Earth? Why do we lust after unlimited power and indulge in travesties like racism, sexism and specism? Put theologically, why do we sin? Why do we fail to love and serve God, one another, and all of God's creation, and instead indulge ourselves in unbridled pride and inordinate sensuality? One form of Christian response to this foundational question has been to assert the paradox that sin is not an intrinsic part of human nature, yet we all sin inevitably.[38] Making it intrinsic would rob us of our individual and corporate responsibility on which our legal and political systems are built; failing to recognize its inevitability would lead to the false hope that we can free ourselves of it without depending on the grace of God. In sum, each of us inherits both the *imago dei*, the 'image of God', and the inevitability of sin; both are unique to our species. [39]

This traditional response conforms nicely to our contemporary scientific understanding of biological evolution, particularly with the philosophical theme of 'novelty within continuity. Thus we as a species inherit diverse propensities from our pre-hominid past, but in *homo sapiens* something strikingly new emerges. This 'newness' is manifest both in the *imago dei*, including our capacity for relationship, abstract thought, formal language, complex technologies, art, ethics and science, and in the reality of human sin, including ruthless violence and our insatiable appetites for power and control. It is only through the grace of a loving God that our lives can be transformed into the fullness of what it truly means to be alive. Conversely, the formation of authentic human personhood requires a lifetime of genuine wrestling with tough moral choices and the repentance of moral failure.

What then about ETI and the 'domain' of moral failure: is it truly universal or is it limited to terrestrial history? I have suggested for scientific, philosophical and theological reasons that the essential characteristics of human life are a genuine clue to the nature of life in the universe and not just an evolutionary fluke of the evolutionary processes on Earth. I therefore expect that ETI will experience much the same kind of moral dilemma that characterizes human experience, though obviously differing in its 'moral morphology' --- the personal and social form of ethics. Jill Tarter has argued that extraterrestrial civilizations will be far older than ours, and to achieve such longevity, they will have had to overcome the temptation to warfare.[40] I tend to agree with her observation, but it still leaves open the question of how such temptation is overcome?

My assumption is that ETI will experience an empowering for their struggle by a source which transcends ETI's natural capacities. Put into theological language, I believe --- perhaps I should say I predict! --- that God will be present to the struggles of life everywhere, and that God's grace will redeem and sanctify every species in which reason and moral conscience are kindled.

As a Christian theologian and scientist, this leads me to a final question: should Christians expect that a single Incarnation of Christ in the person of Jesus is sufficient for the redemption of all life in the universe, or should we expect there to be an Incarnation of Christ in each species of ETI? An adequate discussion would take us far beyond the limits of this essay, but I do want to note in closing that modest support for both options can be found among both Protestants and Roman Catholics: a single, universally efficacious Incarnation is suggested by Protestants Ted Peters and Wolfhart Pannenberg and Roman Catholics L. C. McHugh and J. Edgar Bruns, while multiple Incarnations have been considered by Protestants Paul Tillich and Lewis Ford and the Roman Catholics Karl Rahner, E. L. Mascall and Ernan McMullin.[41] All agree, however, that wherever ETI exists, it will be the creation of a loving and redeeming God.

3. Conclusion

Clearly the empowerment by God of the full flowering of authentic personhood is at the heart of the Biblical witness. Taking this to entail such flowering throughout the diverse species of ETI throughout the universe leads to a profound reformulating and creative transformation of Christian thought and action, whose consequences would affect our ongoing search for deeper religious pluralism and a fuller understanding of the relation between humanity and the plenum of species in our terrestrial environment. Thus, regardless of what lies ahead as we await 'first contact' with ETI, pursuing these kinds of questions and reflections will be immensely valuable.

ENDNOTES

[1]For a scholarly introduction, see Ian G. Barbour, *Religion in an Age of Science* (San Francisco: Harper & Row, 1990); Nancey Murphy, *Theology in the Age of Scientific Reasoning* (Ithaca: Cornell University Press, 1990); Arthur Peacocke, *Theology for a Scientific Age: Being and Becoming* (Minneapolis: Fortress Press, 1993); John

C. Polkinghorne, *The Faith of a Physicist: Reflections of a Bottom-up Thinker* (Minneapolis, Minn.: Fortress, 1994); W. Mark Richardson and Wesley J. Wildman, eds., *Religion and Science: History, Method, Dialogue* (New York: Routledge, 1996). For a less technical introduction see John F. Haught, *Science & Religion: From Conflict to Conversion* (New York: Paulist Press, 1995); Ted Peters, ed., *Science & Theology: The New Consonance* (Boulder, Colorado: Westview Press, 1998); Christopher Southgate et. al., eds., *God, Humanity and the Cosmos: A Textbook in Science and Religion* (Harrisburg: Trinity Press International, 1999). For a recent survey article with extensive references see my "Theology and Science: Current Issues and Future Directions" at www.ctns.org.

[2]See Ted Peters, "Exo-Theology: Speculations on Extra-Terrestrial Life," *CTNS Bulletin* 14.3(Summer 1994) (Berkeley: Center for Theology and the Natural Sciences). For a recent survey of Roman Catholic views, see Douglas A. Vakoch, "Roman Catholic Views of Extraterrestrial Intelligence: Anticipating the Future by Examining the Past," in *When SETI Succeeds: The Impact of High-Information Contact*, ed. Allen Tough (Bellevue, Washington: Foundation for the Future, 2000). See also Ernan McMullin, "Life and Intelligence Far from Earth: Formulating Theological Issues," in *Many Worlds: The New Universe, Extraterrestrial Life & the Theological Implications*, ed. Steven Dick (Philadelphia: Templeton Foundation Press, 2000); Coyne, S. J., George V., "The Evolution of Intelligent Life on Earth and Possibly Elsewhere: Reflections from a Religious Tradition," in *Many Worlds* ed. Steven Dick. For a recent survey of ethical implications see Richard O. Randolph, Margaret S. Race and Christopher P. McKay, "Reconsidering the Theological and Ethical Implications of Extraterrestrial Life," *CTNS Bulletin* 17.3(Summer 1997) (Berkeley: CTNS).

[3] Steven J. Dick, *Plurality of Worlds* (Cambridge: Cambridge University Press, 1982), esp. Ch. 2.

[4]In my opinion the discovery of EL/ETI would not undermine the central claims of 'established religion', particularly Christianity, as many authors assume, although it would challenge a literal and inerrant reading of scripture in *any* religion. For Christians who take scripture as normative but not as literal and inerrant, there is plenty of room for a creative interaction between faith and science, including the discovery of EL/ETI.

[5]See for example Freeman Dyson, *Disturbing the Universe* (New York: Harper and Row, 1979), Ch. 19; Steven J. Dick, "Extraterrestrials and Objective Knowledge," in *When SETI Succeeds* ed. Allen Tough.

[6] Stephen Jay Gould, *The Flamingo's Smile: Reflections in Natural History* (New York: W. W. Norton, 1985).

[7] Paul Davies, "Teleology Without Teleology: Purpose Through Emergent Complexity," in *Evolutionary and Molecular Biology: Scientific Perspectives on Divine Action*, ed. Robert John Russell, William R. Stoeger and Francisco J. Ayala (Vatican City State; Berkeley, Calif.: Vatican Observatory Publications; CTNS, 1998).

[8] Christian De Duve, *Vital Dust: Life as a Cosmic Imperative* (New York: Basic Books, 1995).

[9] Julian Chela-Flores, "The Phenomenon of the Eukaryotic Cell," in *Evolutionary and Molecular Biology* ed. Robert John Russell, et. al. Also see his paper in this volume.

[10] Bertrand Russell, "A Free Man's Worship," in *Mysticism and Logic* (London: Allen & Unwin, 1903).

[11] Jacques Monod, *Chance and Necessity*, trans. Austryn Wainhouse (New York: Vintage Books, 1972).

[12] Steven Weinberg, *The First Three Minutes* (New York: Basic Books, Inc., 1977), 154.

[13]Freeman Dyson, *Disturbing the Universe*, 250. Dyson, *Infinite in All Directions* (New York: Harper and Row, 1988), 117-18.

[14] Paul Davies, "Is the Universe Absurd?" in *Science and Theology* ed. Ted Peters, esp. 72-76; Davies, *The Cosmic Blueprint* (New York: Touchstone, 1989); Davies, "Teleology Without Teleology"; see his paper in this volume.

[15] William R. Stoeger, "The Immanent Directionality of the Evolutionary Process, and Its Relationship to Theology," in *Evolutionary and Molecular Biology* ed. by Robert John Russell et. al..

[16] Francisco J. Ayala, "Introduction," in *Studies in the Philosophy of Biology: Reduction and Related Problems*, ed. Francisco J. Ayala and Theodosius Dobzhansky (Berkeley: University of California Press, 1974); Francisco J. Ayala, "Reduction in Biology: A Recent Challenge," in *Evolution at a Crossroads,* ed. David J. Depew and Bruce H. Weber (Cambridge: MIT Press, 1985), 67-78; Francisco J. Ayala, "Can 'Progress' Be Defined as a Biological Concept?" in *Evolutionary Progress*, ed. Matthew H. Nitecki (Chicago: University of Chicago Press, 1988), 75-96.

[17] Ernst Mayr, "How Biology Differs from the Physical Sciences," in *Evolution at a Crossroads* ed. by David J. Depew and Bruce H. Weber, 67-78.

[18] Charles Birch, *A Purpose for Everything* (Mystic, Connecticut: Twenty-Third Publications, 1990); Charles Birch, *Feelings* (Sydney: University of New South Wales Press, 1995); Charles Birch, "Neo-Darwinism, Self-Organization, and Divine Action in Evolution," in *Evolutionary and Molecular Biology* ed. by Robert John Russell et. al..

[19]Arguments from "t=0" in Big Bang cosmology are often used to support this view, even though the theological claim is primarily about existence *per se* and not temporal origins. For recent references, see Robert John Russell, "Finite Creation Without a Beginning," in *Quantum Cosmology and the Laws of Nature: Scientific Perspectives on Divine Action*, ed. Robert J. Russell, Nancey C. Murphy and Chris J. Isham (Vatican City State; Berkeley, Calif.: Vatican Observatory Publications; Center for Theology and the Natural Sciences, 1993).

[20]Note: I am assuming a non-interventionist view of God's immanent action here, one consistent with science. For details, see the CTNS/Vatican Observatory conference publications referenced in these Notes.

[21]See references to the writings of Barbour, Birch, Ellis, Haught, Murphy, Peacocke, Peters, Polkinghorne and others.

[22]Scarcity can indicate value: The discovery of a single palm tree in a vast desert does not mean that the tree is insignificant. Instead its scarcity makes it a tremendous discovery, for a hidden spring of life-giving water lies at its roots. I feel this way about Earth whether or not life is found elsewhere in the universe. Our planet is like the palm tree in what might in fact be a vast interstellar desert. If it takes a thousand million stars to produce the conditions for the possibility of a sea urchin, if it takes a billion years of tinkering with genetic dice to produce a hummingbird, and if it takes a million years of scratching on bark and vocalizing intentions to produce a child who can reach out through human artifacts and chalkboard calculations and touch the edge of the visible universe, then life per se is surely "the pearl of great price" (Matt. 13:45-46) and a clue to the theological meaning of the universe as a whole.

[23]See the extensive discussion of "the Anthropic Principle" (AP) in the literature, including *Quantum Cosmology and the Laws of Nature* ed. by Robert John Russell, et. al.. 'Many worlds' scenarios via inflationary Big Bang and quantum cosmologies are often deployed to undermine the 'fine-tuning' argument supporting divine design in the AP.

[24]This is *not* an attempt to derive ethics from biology, the so-called 'naturalistic fallacy.'

[25]See for example Michael Ruse, *Sociobiology: Sense or Nonsense?* (Dordrecht: Reidel, 1979); Michael Ruse, *Darwinism Defended* (Reading: Addison-Wesley, 1982); Michael Ruse, *Taking Darwin Seriously* (Oxford:

Blackwell, 1985); Michael Ruse, "Evolutionary Theory and Christian Ethics: Are They in Harmony?" *Zygon* 29.1(March 1994).

[26] Michael Ruse, "Evolutionary Ethics," in *Biology, Ethics and the Origins of Life*, ed. Holmes Rolston III (Boston: Jones and Bartlett Publishers, 1995), 95, 100. As Rolston summarizes Ruse: "Ethics is not true, though it is functional. (But) ethics cannot be functional unless it is believed to be true in an objective sense, a false belief." Holmes Rolston, ed., *Biology,* 8. Rolston is sharply critical of Ruse in *Genes, Genesis and God* (Cambridge: CUP, 1999).

[27]See for example Ayala, "Introduction"; Ayala, "Reduction in Biology"; Ayala, "Can 'Progress' Be Defined?"

[28] Francisco J. Ayala, "Ethical Behavior as an Evolutionary Byproduct," in *Biology*, ed. Holmes Rolston III, 118.

[29] Arthur Peacocke, "Reductionism," *Zygon* 11.4(December 1976); Peacocke, *Creation and the World of Science* (Oxford: Clarendon Press, 1979); Peacocke, *God and the New Biology* (San Francisco: Harper & Row, 1986); Peacocke, *Theology for a Scientific Age.*

[30] Ian G. Barbour, *Issues in Science and Religion* (New York: Harper & Row, 1971); Ian G. Barbour, "Ways of Relating Science and Theology," in *Physics, Philosophy, and Theology* ed. by Robert J. Russell et. al., 21-48; Barbour, *Religion in an age of science*; Ian G. Barbour, "Five Models of God and Evolution," in *Evolutionary and Molecular Biology* ed. by Robert John Russell et. al...

[31] Murphy, "Theology in the Age of Scientific Reasoning."; Nancey Murphy, *Beyond Liberalism and Fundamentalism* (Valley Forge, Pa.: Trinity Press International, 1996); Nancey Murphy, "Supervenience and the Nonreducibility of Ethics to Biology," in *Evolutionary and Molecular Biology* ed. by Robert John Russell et. al.; Nancey Murphy, "Nonreductive Physicalism: Philosophical Issues," in *Whatever Happened to the Soul?* ed. Warren S. Brown, Nancey Murphy and and H. Newton Malony (Minneapolis: Fortress Press, 1998); Nancey Murphy, "Supervenience and the Downward Efficacy of the Mental: A Nonreductive Physicalist Account of Human Action," in *Neuroscience and the Person: Scientific Perspectives on Divine Action*, ed. by Robert John Russell et. al. (Vatican City State; Berkeley, California: Vatican Observatory Publications; Center for Theology and the Natural Sciences, 1999).

[32] George F. Ellis, *Before the Beginning: Cosmology Explained* (New York: Boyars/Bowerdean, 1993); George F. Ellis, "The Theology of the Anthropic Principle," in *Quantum Cosmology and the Laws of Nature* ed. by Robert J. Russell, et. al., 367-406; George F. Ellis, "Ordinary and Extraordinary Divine Action: The Nexus of Interaction," in *Chaos and Complexity: Scientific Perspectives on Divine Action*, ed. Robert J. Russell, Nancey C. Murphy and Arthur R. Peacocke, Scientific Perspectives on Divine Action Series (Vatican City State; Berkeley, Calif.: Vatican Observatory Publications; Center for Theology and the Natural Sciences, 1995), 359-96; Nancey Murphy and George F. Ellis, *On the Moral Nature of the Universe: Theology, Cosmology, and Ethics*, (Minneapolis, Minn.: Fortress Press, 1996); George F. R. Ellis, "The Thinking Underlying the New 'Scientific' World-Views," in *Evolutionary and Molecular Biology* ed. by Robert John Russell et. al.; George Ellis, "Reflections on Quantum Theory and the Macroscopic World," in *Quantum Physics: Scientific Perspectives on Divine Action*, ed. Robert John Russell, Philip Clayton, et al. (2001).

[33] Paul C. Davies, *God and the New Physics* (New York: Simon & Schuster, 1983), 71.; Paul Davies, "Transformations in Spirituality and Religion," in *When SETI Succeeds* ed. by Allen Tough.

[34]Peters, "Exo-Theology: Speculations on Extra-Terrestrial Life," 1. I agree with Peters (see Note #3 above). As Peters points out (p. 2; 7), it may have been the Protestant Fundamentalist reaction to UFOs in the 1970s that gave rise to the view of Christianity that Davies critiques. See also McMullin, "Life and Intelligence Far from Earth," 164-67.

[35]As Peters shows, Greek atomists such as Democritus and Epicurus argued for many worlds, but Aristotle rejected their views. Thomas Aquinas sought to reconcile Aristotle with Christian tradition, defending a "one

world" view for philosophical reasons. Others, such as John Buridan, Nicole Oresme, Albertus Magnus, the Jewish scholar Hasdai Crescas, and Nicholas of Cusa, favored the "many worlds" position. See Dick, *Plurality of Worlds*.

[36] Peters, "Exo-Theology: Speculations on Extra-Terrestrial Life," 4-5.

[37]For God as continuous creator and a theological response to sociobiology see Arthur Peacocke, "Sociobiology and Its Theological Implications," *Zygon* 19 (1984): 171-84; Arthur R. Peacocke, "The Incarnation of the Informing Self-Expressive Word of God," in *Religion and Science* ed. by W. Mark Richardson and Wesley J. Wildman, 321-42; Arthur Peacocke, "Biology and a Theology of Evolution," *Zygon* 33.1(March 1998): 695-712; Philip Hefner, *The Human Factor* (Minneapolis: Fortress Press, 1993); Philip Hefner, "Theological Perspectives on Morality and Human Evolution," in *Religion and Science*, 401-24; Philip Hefner, "Biocultural Evolution: A Clue to the Meaning of Nature," in *Evolutionary and Molecular Biology* ed. by Robert John Russell et. al..

[38] Reinhold Niebuhr, *The Nature and Destiny of Man: I* (New York: Charles Scribner's Sons, 1941 (1964)), VI - X.

[39]It is interesting to compare the Augustinian view with contemporary cultural positions. For example, consider our two questions about sin: 1) is it in our nature? 2) is it avoidably by human effort alone? The Augustinian response is: 1), no; 2), no. A secular humanist would probably answer: 1), no; 2), yes. Interestingly, Carl Sagan seemed to take the position: 1), yes (the tripartite brain and its conflicts); 2), yes (through science).

[40]Tarter claims that ET's longevity is inconsistent with them having "organized monotheistic religions" since they are the cause of warfare on earth. Clearly sustained warfare is inconsistent with cultural longevity, but her suggestion of a causal relation between monotheism and warfare would be hard to defend critically. Of course religion is often used as an excuse for violence, but the real question is why humans are capable of such violence. It also seems odd that she speculates favorably about a "universal religion" having a "highly established code of ethics" centered on "the perpetuation of individuals" when these tenets are found in the monotheisms which she rejects. (Jill Tarter, "Implications of Contact with ETI Far Older Than Humankind," in *When SETI Succeeds* ed. by Allen Touch.)

[41]For references, see Peters, "Exo-Theology: Speculations on Extra-Terrestrial Life," 5-6. See also Eric L. Mascall, *Christian Theology and Natural Science* (New York: Longmans, Green & Co., 1956), 36-45; McMullin, "Life and Intelligence," 171-73. Coyne, in "The Evolution of Intelligent Life," leaves the question open. In order to pursue this systematically, one would need to consider a variety of theological issues. Though I agree with Davies that the discovery of EL/ETI will bring a profound transformation in "spirituality and religion", Christological suggestions in the literature do not seem to have what Davies caricatures as "an air of absurd theatricality" or constitute a Catholic "heresy". (See Davies, "Transformations") McMullin views the arguments by Davies seem "simplistic" (p. 172).

IMPLICATIONS OF POSSIBLE BIOLOGICAL EVOLUTION OUTSIDE HABITABLE ZONES IN SOLAR SYSTEMS

JULIAN CHELA-FLORES

The Abdus Salam International Centre for Theoretical Physics
P.O.Box 586; Strada Costiera 11; 34136 Trieste, Italy and
Instituto de Estudios Avanzados Caracas 1015A, Venezuela.

Abstract: We discuss the interface between astrochemical to astrobiological studies. In particular, we concentrate on the discussion of a conjectured universal evolutionary pathway to eukaryogenesis, multicellularity, neurogenesis, synaptogenesis, cerebral ganglions and eventually intelligence. This point of view advocates in favour of the soundness of the search for extraterrestrial intelligence (SETI) by the standard optical and radio techniques. On the other hand, comparative analyses enable some questions regarding limited aspects of brain evolution to be answered. Differences in the neocortex consistently take the form of module formation. For this reason changes in, for instance, internal organisation as well as changes in connection patterns suggest that critical alterations in brain evolution are generated by similar mechanisms. Such mechanisms mentioned above were probably present very early in mammalian evolution. Since mammals represent a direct descent of primitive multicellular animals on Earth, these remarks have to be evaluated seriously, due to their relevance in the appearance of intelligent life elsewhere. Indeed, it is conceivable that the common mechanisms are so fundamental that future changes in human brain evolution will be governed by similar mechanisms, as has been conjectured already by evolutionary neuroscientists. Similarly, within the context of astrobiology, we take one step further and conjecture that evolution elsewhere in the universe is governed by the same (universal) mechanisms characteristic of brain evolution on Earth. However, it is important to rationalise the exceedingly variable cognitive abilities observed across species, for they can be understood, in spite of the existence of the conjectured mechanisms that are underlying brain evolution; for instance, small contingent changes that occur in the genome may be the source of the diversity in the evolution of the neocortex in mammals. Our conjectured evolutionary pathway offers the basis for introducing some rationalisation in the planning of experiments that are worthwhile, and feasible, in the forthcoming campaign of Europa and Mars missions. Within this context, in the present work we review one class of possible experiments. We restrict the consequences of the conjecture of the universal pathway to intelligence to two aspects: firstly, its possible implications regarding biological evolution outside habitable zones in solar systems and, the likelihood of the onset of other intelligence; secondly, the present discussion gives us an opportunity for rationalising the dialogue that should be maintained with philosophers and theologians with respect to the incorporation into our cultural background of a new view on our position in the universe.

J. Chela-Flores et al. (eds.), First Steps in the Origin of Life in the Universe, 375–380.

1. Introduction

By now, most of us agree on what is precisely the subject matter of exobiology; but in my talk we shall use the synonym "astrobiology". In other words, today I will define the subject matter as follows; astrobiology denotes the science of:

1. origin,
2. evolution,
3. distribution and
4. destiny of life in the universe.

I only have a few words to say on the first two topics. These are well established research areas. After the pioneering works of Alexander Oparin and Stanley Miller, together with other researchers, the origin of life was, indeed, turned into a scientific activity.

On the other hand, Darwin had done so earlier for the evolution of life. In my talk, I will focus on the last two topics of our discipline:

3. distribution and
4. destiny of life in the universe

For the distribution of life, I will attempt to review how to arrive at the basis of a well-established science. Instead, for the second part of my talk, that is for the destiny of life in the universe, we shall find ourselves in the company of elder brothers: philosophers and theologians. They have thought about these questions, from a different angle, for a much longer period than ourselves. Although our activities are intrinsically different, a dialogue will be of mutual benefit. Returning to the distribution problem, we ask:

2. Which assumptions do we have to begin with?

First of all, the only intelligent life that we know about in the universe is life on our own planet. We have no evidence of how representative it might be of life elsewhere, although we have conjectured that, given suitable environmental conditions, if life started somewhere else, it would also be constrained to take an evolutionary pathway to eukaryogenesis and multicellularity.

This is my main thesis for the first part of this work. I will just remind you that eukaryogenesis means the origin of cells with an internal nucleus, the main morphological characteristic of advanced multicellular organisms, such as ourselves.

We shall now argue that terrestrial life teaches us that multicellularity is inexorably linked with neurogenesis. This, in turn, is linked to the process of synaptogenesis and, finally, to the formation of cerebral ganglions at low levels of a given tree of life, and their evolution into large brains typical of intelligent beings.

Two of the basic principles of terrestrial biology are natural selection and the existence of a common ancestor for all life on Earth. This can be interpreted as evidence in favour of the fact that to a certain extent, at a cosmic level evolution is predictable, and not entirely dependent on chance. The underlying question concerns the relative roles of adaptation, chance and history, a topic which is, strictly speaking, subject to experimental tests. Bacterial populations, for example, have been shown to propagate in identical environments, achieving similar fitness, independent of prior history, or even of subsequent chance events.

Further support from independent teams suggests that natural selection overrides the randomness of genetic drift. In other words, natural selection seems to be powerful enough to shape terrestrial organisms to similar ends, independent of historical contingency.

On the other hand, some arguments suggest that a human-level of intelligence is a consequence of a conjectured universal constrained evolutionary pathway towards intelligence as it occurred on Earth.

Certainly, in an extraterrestrial environment the evolutionary steps that led to human beings would probably never repeat themselves; but that is hardly the relevant point: the role of chance in evolution has little bearing on the emergence of a particular biological property.

The inevitability of the emergence of particular biological properties is a phenomenon that has been well recognised by students of evolution for a long time. This powerful process is known as 'convergent evolution', which may be illustrated with examples taken from ornithology and neuroscience. I will go through these two examples in turn:

• First of all, an example of convergent evolution is provided by swallows, a group which is often confused with swifts, but are not related to them. In fact, the taxons to which these birds belong are orders, rather than families.
Members of these two orders differ widely in anatomy. Their similarities are the result of convergent evolution, on different species that have become adapted to the same ways of living, in ecosystems that are similar.

• Our second example concerns brain structures that look the same, but have arisen independently, in different lineages. They are considered to have arisen by convergent evolution. In particular, neurones can be classified as to whether they are driven by one eye or the other. This property is called 'ocular dominance'. From a sketch of the organisation of the visual pathway, we can appreciate that when an electrode penetrates the cortex, all neurones encountered in the column defined by the penetration, have the same property of being driven by one eye or the other. This has been observed in cats and monkeys. Hence, this property is likely to have arisen form convergent evolution, since cats and monkeys are very distantly related species, and other intervening lineages do not posses this feature.

In the light of these 2 examples it is not an obvious conclusion that our intelligence is unrepeatable: Its presumed uniqueness depends on a combination biological, geological and possibly other factors. In simpler terms, the metaphor that "if the history of evolution were to be repeated, it would not lead to man again" is, in my opinion, very misleading. The relevant question is rather the emergence of intelligence elsewhere, not the emergence of man. In order to ask the pertinent question, we must appeal to the space sciences through, for instance, the SETI project.

To sum up, there is selective advantage for multicellularity of eukaryotic cells that have already become neurones; such an event occurred very early on the pathway to multicellular evolution on Earth.

The coelenterates, such as the medusae, form the first taxon, or group, in the evolution of multicellular animals. In these organisms neurones and true nervous systems appear for the first time in terrestrial evolution. There are individual interconnected neurones transmitting, chemically, or electrically, in both directions, forming a sort of nerve net (Villegas *et al.,* 2000). Hence, for life on Earth, neurones appear at almost the lowest level of multicellularity of the eukaryotes.

These arguments strongly advocate in favour of the existence of other human-level of intelligence, elsewhere in the cosmos. Essentially, the argument hinges on the basic transition form prokaryotes to eukaryotes.

3. Is eukaryogenesis itself bound to take place?

The main reason why we believe that eukaryogenesis is universal is that not only physico-chemical processes regulate the structure, and evolution of the universe; but besides, we assume that all principles for understanding nature are of general validity, irrespective of the location of the observer in a given galaxy. In particular, this "strong" principle of universality of the natural sciences, forces upon us the consideration that the scientific bases of biology, as known to us on Earth, also apply elsewhere.

Hence, according to such universality of the natural laws, evolution of biological complexity in our cosmic environment is constrained by the two basic principles of biology: natural selection and common descent. But there are compelling reasons to raise the question of biology in our solar system, especially in the Jovian system. Amongst other factors, there is some evidence in favour of Europan life, mainly the very likely presence of the Europan ice-covered ocean, which has been gradually forced upon us by measurements from the Galileo orbital mission, as shown to us by Cynthia Phillips. In fact, the possibility of Europan life is independent of the likely existence of hydrothermal vents at the bottom of the ocean; besides that first factor, the ices of water and carbon dioxide covering Europa, might also support a radiation-based ecosystem of micro-organisms. (This was originally suggested by Christopher Chyba.)

It is conceivable, therefore, that independent of whether there are hydrothermal sources at the bottom of the ocean, a thriving community of micro-organisms might have evolved in that environment anyway. This possibility does not require the technological difficulties of research from a submersible that would search out hydrothermal vents. A lander would be sufficient to process melt water that may already contain micro-organisms due to the rapid observed resurfacing.

4. How do we test the proposed universality of eukaryogenesis?

There are at least two cases of some interest:

- Europa may be a typical biofriendly body amongst extrasolar planets, a question discussed by Tobias Owen. The main problem is how to select appropriate experiments in situ, after surface landers are able to filter melt water from Europa's frozen surface.
- Evolutionary tests are also suitable in potential water sources under the Martian surface, such as in the location 71° S on a wall of a depression. These environments may contain micro-organisms, an exciting possibility raised by David McKay's team with the Allan Hill meteorite, an important contribution that stimulated discussion in the possibility of life elsewhere in our solar system.

In these cases, testing the degree of evolution of possible solar system micro-organisms is possible.

I will not be entering into too much detail; the work was already published two years ago in the journal of ISSOL, our society (Chela-Flores, 1998, 2000).

We only need to recall that the ultimate aim of a biology experiment is to develop robotic tests that are compatible with the necessarily reduced dimension of a Europa lander. A Martian mission to subterranean pockets of liquid water is another possibility. In either case miniaturisation is essential. Many difficulties though are inherent in the eventual design of a test that would intend to identify eukaryotes, robotically, in any extraterrestrial environment.

This question begins to be important, in view of the decisions that have to be made in the selection of biological experiments that should be performed in situ on Europa and on Mars. Two examples may serve to illustrate the underlying difficulties:

- First of all, the occurrence of an internal membrane-bounded cell compartment, wholly containing the genomic nucleic acid, is no guarantee that the cell in question is a eukaryote. In fact, the existence of a genome-bounded non-eukaryotic micro-organism may be illustrated with two different genera of the bacterium planctomycetes: *Pirellula* and *Gemmata.*
- Secondly, some microtubular structures are present in external symbionts in some marine ciliates that were discovered near a beach in the Sea of Japan (Euplotidium itoi). There is a resemblance between ectosymbionts and eukaryotic cells, in spite of the lack of an internal membrane, a point emphasised by Giovanna Rosatti.

These are two good illustrations of the extraordinary difficulties involved in preparing a suitable robotic test. To sum up, a robotic test on a Europan, or Martian lander, has to distinguish between a prokaryote and a eukaryote. Yet, we have illustrated that some micro-organisms, at first appearance, do not completely fit in either category. Further tests in the laboratory are still needed to settle their identity.

On the strength of these two examples, we are suggesting that the clearest test for eukaryogenesis is searching for a highly packed genome. In this case the genome responds in an unambiguous manner to well-defined dyes. The result of such an experiment could be recorded with video equipment for later analysis in an Earth-bound laboratory. If these experiments on evolutionary trends were to be successful, they would bridge a remaining gap in astrobiology.

Recalling the two subtopics of astrobiology that we are discussing, namely, distribution and destiny of life in the universe, what we have argued is that if the experiments on evolution were to be successful, the science of the distribution of life in the universe would lie on solid scientific bases. This concludes the first of the two topics I wished to comment upon. Given such bases for the distribution of life in the universe, it does not seem premature to include in our discussions other sectors of our society. Keeping this in mind, we move on to the second topic of the present work, the destiny of life in the universe.

5. What changes can we expect in our culture if there is a prevalence of the living state throughout the cosmos?

Science and religion are both concerned with the common understanding of the destiny of life in the universe. Since these two intrinsically different cultural activities largely address the same questions, they should, at some point, establish a dialogue. Some of the barriers to the dialogue can be traced back, to the fundamental question of how to read the Holy Books of the three monotheistic religions. The dialogue with other religions may be faced with similar barriers. But already in the fifth century, Saint Augustine had touched on this question in his book in *"The City of God"*. In fact, Augustine discussed a clear separation of two alternative ways of interpreting the Bible: firstly, as a source of spiritual reflection and ethical values, and secondly, as a substitute for scientific matters. Like most thinkers today, already in the fifth century of our era, Augustine accepted the first interpretation: For Augustine, scientific matters should not be inferred form a literal reading of the Holy Book. In cases of conflict, Augustine advocated in favour of an allegorical interpretation of the Bible, rather than as a substitute for science.

A pioneer in this inter-cultural dialogue was Giordano Bruno, who wrote his first three Italian dialogues during a visit to England in 1584. These writings were stimulated by controversial debates at the University of Oxford. I would only like to recall his third dialogue: *"On the infinite universe and worlds"*. Here, Bruno introduced concepts which are still at the centre of astrobiology: He suggested the removal of the Copernican 'sphere of fixed stars'. Bruno spoke of an infinite universe and the plurality of inhabited worlds. This topic is discussed in the Closing Lecture by Francesco Bertola. Such a cosmological vision matured in Bruno's writings long before the science of astrometry allowed this concept to be brought within the scientific domain. Bruno's remarkable achievement is underlined by the 50 or so new extrasolar planets that were commented upon earlier by Marc Ollivier. The position of Augustine was shared by Bruno and soon after by Galileo in his writings, such as the *"Starry Messenger"*. This position is now shared by most contemporary thinkers. This progress may open the way to further fruitful dialogue on the question of the destiny of life in the universe.

In spite of all the progress in astrobiology during the last century, some of the answers to key questions of common interest are escaping us. This is a concept which is of interest to all forms of culture. From the *Starry Messenger* we may take, as an illustration, Galileo's sketch of the constellation of Orion. His simplicity is matched by our still-too-simple view of life on Earth, which is summarised in the three domains of the phylogenetic tree of life on Earth. I have argued that at the microscopic level our tree of life may be of universal validity.

6. What are the implications of positive results of the evolutionary experiments elsewhere in the solar system?

Such experiments would undoubtedly have a significant impact in our culture, not just in our scientific outlook. The influence of the new knowledge will also be felt while discussing the deep questions raised by George Coyne and Robert Russell, such as the concepts of creation, morality and redemption. These questions may be discussed not just against a background of our particular evolutionary line, which has been followed up by life on Earth. Such questions ought to be discussed already in terms of the many evolutionary lines that are hinted at by astrobiology. In fact, the intercultural dialogue is urgent for various reasons, which are strongly suggested by astrobiology:

- A human-level type of intelligence may be widespread in the cosmos,
- Within the context of astrobiology it is clear that our human descent does go back all the way to micro-organisms and,
- ultimately, our origins go back to star dust.

Yet, these three items lie outside our cultural patrimony. But it is not an easy task to integrate our knowledge. In this work, my main overall thesis has been that such an integration is not only possible, but it is also timely, and necessary.

7. References

Chela-Flores, J. (1998) *A Search for Extraterrestrial Eukaryotes: Physical and Bio*chemical Aspects of Exobiology, *Origins Life Evol. Biosphere* **28,** 583-596.

Chela-Flores, J. (2000) Testing the Drake Equation in the solar system, in *A New Era in Astronomy,* Lemarchand G.A. and Meech K. (eds.), ASP Conference Series, San Francisco, **213,** 402-410.

Villegas, R, Castillo, C. and Villegas, G.M. (2000) The origin of the neuron: The first neuron in the phylogenetic tree of life, in J. Chela-Flores, G. A. Lemarchand and J. Oro (eds.), *Astrobiology from the Big Bang to Civilisation,* Kluwer Academic Publishers, Dordrecht, The Netherlands. pp. 195-211.

Section 10:

Recent improvements in the infrastructure for the study of the first steps in the origin of life in the universe

COSMOPARK – A NATURAL ASTRONOMICAL OBSERVATORY

JOAN ORÓ AND XAVIER PALAU

Fundació Joan Oró,
Lleida, Spain.

1. The Cosmopark project

The COSMOPARK project involves the construction of an astronomical observatory in a zone of great environmental interest. The place is the mountain of Montsec geographycally located in the North-East of Spain (Latitude 42N, Longitude 1E) south of the Pyrenees and within the Autonomous Region of Catalonia.
The Montsec mountain (altitude 1.600m) has been selected as one of the best sites for astronomical observation because its absence of luminic pollution, great atmospheric transparency, low humidity and an annual pluviometry of less than 500 mm. It is a very dry area, above the fog level that usually settles during the winter in the neigboring city of Lleida. Furthermore, it is quite accessible by road being situated approximately 40 km north of Lleida, and about 150 km west of Barcelona.
In the past the site has been of tremendous interest to geologists because its Triassic sediments which crop out on the surface are well dated and make possible to follow their tectonic and stratigraphic history in great detail for millions of years. From this geological point of view this regional zone has been one of the most visited sites in Europe.
The Montsec mountain is also of significant paleontological interest. In its soil and in the neighboring areas many footprints and fossils of dinosaurs have been found. One year ago the first international meeting on fossilized eggs of dinosaur was held in the town of Tremp, close to the Montsec mountain. The flora and fauna of the region which includes some unique species are also of particular interest.
Recognizing the exceptional qualities of the Montsec mountain and neighboring areas, a Consortium of local institutions (involving the counties of Pallars Jussà, and Noguera and supported by the "Diputació de Lleida", the Government of Catalonia and the European Community) has been established under the name of COSMOPARK which is managed by the Joan Oró Foundation.

2.Objectives

As suggested by the title and the preceding lines the objectives of the project are twofold. To create an Astronomy Observatory and a Natural History Center. The first is to be constructed on the top of the Montsec mountain near the small town of Sant Esteve de la Sarga, and the second center is to be build, at the base of the mountain, near the city of Ager.

J. Chela-Flores et al. (eds.), First Steps in the Origin of Life in the Universe, 383–385.

Both centers would be dedicated to research, training and educational activities in their respective areas of knowledge fulfilling the needs of the University of Barcelona, and the Institute of Space Studies (IEEC) and other educational institutions. They would take advantage of their priviledged environment for the observation of the cosmos, and for the study of the geological, paleontological and natural history of the Montsec mountain and its neighboring areas. It is hoped that such a project will also contribute to the economic development of this pre-Pyrenees area of Catalonia.

3.Astronomy observatory

The Montsec Astronomic Observatory will develop its own Programs of Research following the tradition of the Fabra observatory in Barcelona. From the Fabra Observatory J. Comas i Sola discovered several comets and asteroids (e.g.Titania, Barcelona). In 1907 he discovered for the first time the presence of an atmosphere in Titan, the largest satellite of Saturn. The new observatory at Montsec will join the European Astronomical network making telescopic observations following the orbits of known asteroids and periodic comets, and will also search for possible unknown minor bodies orbiting the Sun. Other programs of research will concentrate on the outer planets of the Solar System, and also on certain variable stars.
The infrastructure of the Observatory should include the following equipment and facilities.

A 2m telescope for professional use, with camera CCD for direct imaging and photometry, spectrograph with appropiate filters, to be used on site, or robotically guided from the Department of Astronomy of the University of Barcelona, the Institute Space Studies of Catalonia (IEEC) and from other centers.

A 0.8m telescope for professional use as well as for amateur astronomers, with similar accesory equipment as indicated above.

A 0.4m telescope, primarly for the use of amateur astronomers and visiting persons. It should be equipped with comercial CCD camera, photometer, photographic camera, filters, as well as other regular equipment for visual observations.

Technical services for operation and maintenance of the Observatory equipment.

Appropiate Housing for the professional observers and for the maintenance servicemen.

Meteorological observatory, and receiving terminal for the reception of meteorological data from satellites.

4.The Natural History Centre

The Natural History Center has to be prepared for scientific lectures and public education in all the areas of knowledge related to the Montsec mountain and its vicinity. Very briefly this should include:

An interactive museum in the areas of geology, paleontology, flora, fauna environmental protection and astronomy as well as research related to the activities of the centers.

A planetarium where in addition to its normal use, lectures and courses can be offered in the area of astronomy.

A laboratory where one can make practical work on equipment, or make experiments in the pertinent areas of research.

An auditorium, or congress hall, where conferences, symposium or educational lectures can be presented.

A computer laboratory with several computers and printers for internet and informational multimedia activities.

Four retractable domes, relatively small, for amateur telescopic observation of the sky. The atmospheric transparency is so good, even in the city of Ager that one can see in clear detail all of our galaxy, the Milky way.

An open site (COSMOPARK OPEN MUSEUM) for the exposition of dinosaurs fossils and our arboretum with different species of plants typical of the area.

SPANISH PHOTOGRAPHIC METEOR NETWORK: An interdisciplinary crosslink.

JOSEP Mª TRIGO-RODRIGUEZ [1,2], JUAN FABREGAT[1], JORDI LLORCA[3,4] & ALBERTO CASTRO-TIRADO[5]

[1]*Departament Astronomia i Astrofísica, Universitat de València.*
[2]*Depto. Ciències Experimentals, Universitat Jaume I.*
[3]*Institut d'Estudis Espacials de Catalunya.*
[4]*Departament Química Inorgànica, Universitat de Barcelona.*
[5]*Laboratorio de Astrofísica Espacial y Física Fundamental (LAEFF-INTA) and Instituto Astrofísica de Andalucía (IAA-CSIC).*

This paper reports the birth of the Spanish Photographic Meteor Network (SPMN), a project that began their activities in 1997 through professional and amateur astronomers, coinciding with the high activity associated to several stream's outbursts. One of the objectives of the network is to develop a continuous fireball monitoring in Spain and to obtain orbital and chemical information on meteoroids. Today the SPMN is a solid project under the auspices of three Universities (University Jaume I, University of Valencia and University of Barcelona) and one Space Institute (Catalonian Institute for Space Studies). The National Institute for Aerospace Research (INTA) and the Scientific Research Council (CSIC) also collaborate in the project.

1. Introduction: Why fireball networks?

The systematic observation of meteors using photographic, video and CCD techniques has become one of the rare fields in astronomy in which amateurs can work together with professionals and provide important contributions to science. The Earth interacts continuously with hundreds of particles of interplanetary matter, produced from the surface degradation of minor bodies: comets and asteroids. The atmospheric entries of large meteoroids produce spectacular fireballs which, except the maximum of major meteor showers, are rare celestial events. The record of these events using all-possible techniques (photographic, video or CCD) provides excellent means to examine physical properties, chemical abundances (using spectroscopic tools such diffraction gratings or prisms) and dynamic evolution using orbital data. On the other hand, meteor studies use the atmosphere as a giant detector to estimate the flux of extraterrestrial matter on Earth.

J. Chela-Flores et al. (eds.), First Steps in the Origin of Life in the Universe, 387–391.

From two or more records of the same meteor obtained from different stations the trajectory in the atmosphere can be inferred and, in the case of big events, the most probable meteorite impact area on the Earth surface can be determined. Using the trajectory data entry and calculating the meteoroid geocentric velocity (using a rotating shutter) the heliocentric orbit can be obtained, which allows to decipher the particle's origin in the Solar System. However, very large atmospheric coverage and long observation times are required for photographic recording techniques. Our first photographic network was developed in the Teruel province during the 1991 Perseids maximum, and the orbits of three members of this stream were obtained (Trigo-Rodríguez & Artés, 1993). Later on, new stations were organised during the magnificent 1993 Perseid display, and a -9 fireball was registered from two of them (Trigo-Rodríguez, 1997). Nowadays, we develop successive campaigns with multiple stations around the year (Trigo-Rodríguez, 2000). The procedure consists in sending the different photographic centres using electronic mail to the different participants, which will receive a stellar chart with the centre of field to point the camera according to their station coordinates.

2. Current activities and recent SPMN fireball orbits.

The activities carried out by the SPMN are mainly based on the study of meteors, fireballs and meteorites. Different types of studies are being performed, such as the calculation of meteoroid orbits from conventional photography, video and CCD techniques, stream spatial fluxes, elucidation of meteor parent bodies, meteor spectroscopy, meteor atmospheric modelling, identification of meteorites, meteorite recovery and meteorite analysis, and development of meteor software. Some trajectories and orbits obtained from our photographic monitoring system appear in tables 1 and 2.

Table 1.- Trajectory data of recent SPMN fireballs. The table show the photographic magnitude (M_f), beginning and ending heights (in km.) of the meteor trail on the Earth's surface, coordinates of the geocentric radiant (Right Ascension and Declination) and velocity in km/s (at infinite, geocentric and heliocentric).

Code	Parent Body	M_f	H_{beg}	H_{end}	RA_g (°)	DEC_g (°)	V_{inf}	V_g	V_h
SPMN981201	Asteroid Phaeton	-5	95.1	62.9	115.56 ±0.05	30.41 ±0.05	37.5 ±0.1	35.8	33.0
990801	109 P/Swift-Tuttle	-3	118.9	93.7	45.89 ± 0.02	58.32 ± 0.02	59.5 ±0.3	58.4	40.2
990802	109 P/Swift-Tuttle	-4	122.8	81.4	47.3 ±0.2	57.07 ±0.02	59.9 ±0.3	58.8	40.6
991101	1P/Encke	-4	90.2	75.4	63.31 ±0.02	23.22 ±0.02	31.40 ±0.08	29.3	37.4
991103	55P/Tempel-Tuttle	-4	112.9	97.6	153.0±0.6	21.84±0.02	71.4±0.2	70.5	41.2
991105	55P/Tempel-Tuttle	-3	110.7	98.7	152.71±0.02	22.03 ±0.02	71.8 ±0.2	70.9	41.6
991118	55P/Tempel-Tuttle	-2	119.8	105.2	153.73±0.02	21.71 ±0.02	71.7 ±0.2	70.8	41.6
991120	55P/Tempel-Tuttle	-3	106.2	90.7	153.39±0.02	22.15 ±0.02	71.4 ±0.2	70.5	41.3
991124	55P/Tempel-Tuttle	-2	111.7	101.6	153.66±0.02	22.57 ±0.02	71.5 ±0.2	70.6	41.5

Meteor spectroscopy can be used to better know the mechanisms of meteor radiation and degradation processes of interplanetary matter during its entry into the Earth's atmosphere. From meteor spectra the chemical composition of these particles usually coming from comets or asteroids can be determined.

Table 2.- Orbital elements of recent SPMN fireballs. Equinox (2000.00).

Code	Date	q	a	i	ω	Ω
981201	13.9416 Dec 98	0.1045 ±0.0008	1.24 ± 0.01	24.2 ±0.2	330.47 ±0.09	261.74997 ±0.00002
990801	13.0993 Aug 99	0.9608 ± 0.0005	10 ± 1	112.1 ± 0.1	153.0 ± 0.2	139.94546 ± 0.00002
990802	13.1118 Aug 99	0.953 ±0.002	8 ±2	114.1 ±0.2	150.9 ±0.6	139.957 ±0.008
991101	17.03951 Nov 99	0.3555 ±0.0008	2.24±0.02	2.42 ±0.02	293.70±0.04	234.24657 ±0.00002
991103	18.1099 Nov 99	0.985±0.002	9 ±2	162.7 ±0.4	172 ±2	235.323±0.003
991105	18.11439 Nov 99	0.9857±0.0001	14.4 ±2.5	162.62±0.04	173.7±0.1	235.32703±0.00002
991118	18.10359 Nov 99	0.9815 ±0.0001	13.5 ± 2.8	162.51 ± 0.06	170.1 ± 0.2	235.31612 ± 0.00002
991120	18.05145 Nov 99	0.9833 ± 0.0001	10.1 ± 0.7	161.96 ± 0.03	171.4 ± 0.1	235.26356 ± 0.00002
991124	18.08686 Nov 99	0.9832 ± 0.0001	12.1± 1.2	161.17 ± 0.03	171.4 ± 0.1	235.29929 ± 0.00002

3. Meteor spectroscopy and atmospheric interaction.

Using diffraction gratings or prisms meteor light can be decomposed, and the different spectral lines enables the identification of the chemical elements present in the ionised column. In addition, using calibrated spectra the chemical abundances of the incident particle can also be obtained from line intensities. In general, however, we found several limitations because this analysis is restricted to elements which are detectable in meteor spectra and it also depends of the detector sensitivity range (usually photographic plates

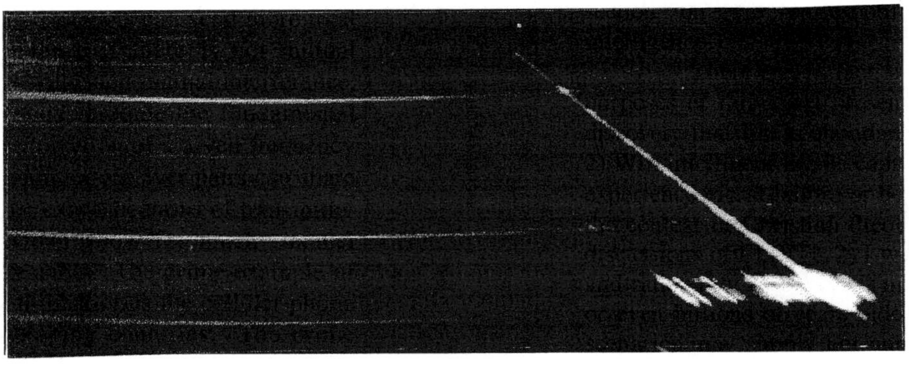

Figure 1.- Fireball spectra produced by a fragment of comet 109 P/Swift-Tuttle. The ending flare show a lot of lines from the different components: Ca II, Fe I and Fe II, Mg I and Mg II, Na and Si II. The brightest line at right belongs to Ca II lines in λ= 3933.7 and 3968.5 Å. We need identify and fit the main lines to calibrate the spectra. This spectrogram has been analysed in the Ph.D. thesis of first author (JMT). The spectra was obtained in the Czech Republic, 13th August 1967 (Courtesy: Jiri Borovicka, Ondrejov Observatory).

or video films). Consequently, the precision of the chemical analysis made in this way is much lower than laboratory studies of meteorites. Despite these restrictions, meteor spectroscopy represents an excellent tool for deepening into the meteoroid-atmospheric interaction and associated processes. In fact, deposition of extraterrestrial atoms and molecules in the terrestrial atmosphere probably had an important role into the Earth crust enrichment and even into the origin of life.

In the last decade some authors have obtained significative progress in meteor spectroscopy. For example, Borovicka (1993) identified the presence of two spectral components with temperatures of about 4,500 and 10,000 K assuming chemical equilibrium in the ionised column. As the gas is usually heated to 4,000-5,000 K only the ablated material radiates making the atmospheric radiation negligible. Atomic and molecular emission are dominant even in the brightest part of the trajectory or in flares thermal emission is also detected. Probably, the high-temperature component is produced in the meteoroid head where the number of atomic impacts are higher. On the other hand, the analysis of the fragmentation processes enables to know the pyrolitic effect on particles during their short entry in the terrestrial atmosphere. Fragmentation of cometary meteoroids produces in some occasions luminous clouds which persist long after the meteor's disappearance. Borovicka et al. (1996) identified the surprising presence of prominent emissions of forbidden lines of ionized atoms such as OII, SII and OIII. The physical conditions that cause this luminescence in the train remain, however, unknown. Probably it could be caused by the interaction between ions, free electrons and atmospheric components in the upper atmosphere. Meteoroid orbital data together with high-resolution spectra can yield new valuable information on important subjects. Several atmospheric processes related with the entry of interplanetary bodies in the Earth's atmosphere are still unknown. Moreover, from orbital data we can obtain interesting information on the population, composition and links between Solar System small bodies. We can use these results for deepening into the role that these bodies had in life's origin. Fireball networks can provide an important source of data for modelling the entry of big bodies in the atmosphere (Borovicka & Spurn_, 1996). At present we plan to install automatic all-sky cameras to cover big fireball events continuously. More information can be seen in the SPMN project homepage: **http://www.spmn.uji.es/**

4. Acknowledgements. Many thanks to the help, data and suggestions received from Drs. Z. Ceplecha, J. Borovicka and Pavel Spurn_ of the Ondrejov Observatory (Czech Republic). J. Castellano Roig realised an excellent work developing part of our software. We also appreciate the decisive collaboration of our Dutch colleagues B. van Gemeren and H. Betlem (Dutch Meteor Society). A lot of amateur astronomers collaborate with the Spanish Photographic Meteor Network, allowing a continuous scan of meteor events from several stations. Our thanks to P. Arranz, P. Alcantara, T. Bernedo, J. M. Bullón, J. Mª Bosch F. Campos, A. del Castillo, E. Coll, E. Fraile, C. Pineda, A. Sánchez, R. Castillo, J. Mª Castro, G. Domínguez, R. Ferrando, J. Gómez, J. González, A. Gutierrez, S. Hernández, J. Izquierdo, C. Labordena, L. Lahuerta, S. Lahuerta, T. Mateo J. L. Mezquita, J. Pastor, J. Patiño, F. Peña, R. Ramírez, J. Ruiz-Garrido, A. del Solar, A. de Ugarte, H. Valero, M. A. Villalonga and F. Villares.

5. References

Borovicka J. (1993); "A fireball spectrum analysis", A&A 279, pp.627-645.

Borovicka J. & P. Spurn_ (1996); "Radiation study of two bright terrestrial bodies and an application to the comet S-L 9 collision with Jupiter", Icarus 121, pp.484-510.

Borovicka J. et al (1996); "The identification of nebular lines in the spectra of meteor trains", Astronomy & Astrophysics 306, pp.995-998.

Trigo-Rodriguez J.M. & Artés J., (1994) "Orbital Elements of Three Photographic 1991 Perseids", WGN 24-1/2, pp.32-34.

Trigo-Rodriguez J.M., (1997) "Impressive Perseid fireball over Spain", WGN 25-4, pp.187-189.

Trigo-Rodriguez J.M., (2000) "The 1997 Leonids outburst", A&A 355, pp.1160-1163.

Pre-dinner Speaker:

Frank Drake

NEW PARADIGMS FOR SETI

FRANK DRAKE

SETI Institute
Mountain View, California, USA

Truly scientific SETI searches have now existed for more than forty years. In the tradition of good science, the search strategies selected to guide most of these searches have recognized that we are unable to predict with any confidence the behavior and technologies of other intelligent civilizations, especially those much advanced than we. In view of this quite profound ignorance, the best we can do is to follow Occam's Razor, and choose the simplest plausible paradigm: THE EXTRATERRESTRIALS WILL BEHAVE LIKE US. This paradigm is still the best we have, but, as we will see below, its implications are calling for major changes in our search strategies as well as humility when it comes to our confidence that we are doing the right thing.

This paradigm has led to the choice of the most common SETI strategy: a search for radio signals in the UHF and microwave regions. Our own civilization has exploited such signals for a variety of purposes: television broadcasts, communication with spacecraft, and radar, for example. Transmissions at these frequencies travel freely through clouds, experience the minimum interference from cosmic radio noise, and are easy to generate and focus into beams and collect with relatively inexpensive antennas. Our history has been one of ever increasing numbers of transmitters for such frequencies, with ever increasing power levels. If this is the undeviating path followed by all civilizations, then our strategy has been the correct one.

While this has been the accepted search paradigm, alternatives have been suggested. These include searches for neutrino beams, gamma rays, gravity waves, infrared transmissions, and, especially, optical signals. Both neutrino beams and gravity waves have been rejected on the basis of the great technical difficulty and cost, relative to the low cost of radio receiving systems, of systems to both transmit and receive detectable transmissions. Similarly, gamma rays have been rejected because of ignorance of technology to generate, efficiently, gamma ray transmissions, and the difficulty in building high sensitivity receptors and the need to place receptors above the atmosphere. Searches for optical signals have been repeatedly suggested for more than thirty years, particularly by Charles Townes, and others. This idea has not been embraced because of the lack of development of powerful coherent light transmitters such as lasers. Furthermore, it has been accepted that plausible detection systems, even those using large optical telescopes, would be burdened with the need to distinguish the

J. Chela-Flores et al. (eds.), First Steps in the Origin of Life in the Universe, 395–398.

laser light signal from the intense scattered light of the star around which the laser orbits.

However, recent developments in both radio and laser technology are rapidly altering our own exploitation of these technologies, and as a consequence may seriously alter choices of search strategy.

First, with regard to the beneficial use of radio technology, a major change rooted in social change is taking place. Heretofore, the very common goal in radio technology was to transmit a limited amount of information to many or distant receivers by creating very powerful, sometimes near omnidirectional radio transmitters. The prime example has been the many thousands of broadcast television transmitters, often transmitting powers of the order of a megawatt to TV receivers with small antennas. This has been efficient, and effective. The ensemble of such signals is probably the most easily detected sign of our existence across interstellar space.

However, and this is a vital point, such high power transmitters cause a certain piece of the radio spectrum to be monopolized by one transmission over a large area. In the very recent past there has been a significant change in the human demand for use of the radio spectrum, and that is to transmit vast amounts of information, much of it intended for only one or a few recipients. This is, of course, the "Information Age", and is simply a consequence of the growth of a more educated and empowered population. As a result, there is a shifting radio technology paradigm from an emphasis on sending a little information to many recipients simultaneously, to an emphasis on sending a great deal of individual information channels to a large ensemble of individual users. Where a wired interconnection is possible, the internet, for example, fulfills this new need. Where such is not possible, the technology is seriously, irrevocably limited by the fact that the maximum information rate in a channel is limited by its frequency bandwidth. This limitation does not grow out of some lack of genius in inventing technology, but comes from basic physics, indeed, the Heisenberg Uncertainty Principle. Thus the free space radio spectrum can only accommodate a certain information rate at any point on Earth.

The upshot of this is that only sharing frequency channels can send more total information world-wide. Such sharing can only take place if there is not mutual interference. There are methods to deal in a limited way with such mutual interference, such as spread spectrum transmissions, but these can not overcome the fundamental physics-based limitations. The best way to increase the utilization of a given frequency channel is to reduce transmitter powers so that more transmitter-receiver pairs can share a channel without mutual interference. It is true that if the exact locations of transmitter and receiver are known, highly beamed transmissions can allow for much channel sharing, but this is an expensive solution and often not feasible. The prime example of the use of low-powered transmitters to increase channel utilization is the cellular phone systems now so ubiquitous in the world, even in developing countries. The prime example of a combination of low power transmissions and crudely beamed signals is the direct-to-home television broadcasts from satellites in synchronous orbit. It is very noteworthy that, in contrast to the megawatt radiated powers of ordinary earth-based TV transmitters, the satellite transmitter power levels are only of the order of twenty watts, and most of this is directed towards the Earth where it is absorbed, leaving a negligible, undetectable signal to escape into interstellar space.

It would appear very likely that a combination of fiber-optic cable systems and direct-to-home satellite TV broadcasts will become the sole source of TV transmissions in a very short time on the cosmic time scale. This example simply illustrates the key point: the societal demand for ever increasing rates of information transfer can cause a planet to become less detectable rather than more detectable. The more civilized and advanced, the LESS detectable may be the radio signals of a civilization.

Thus the basic paradigm still is in place: The extraterrestrials will behave like us. Except our behavior has changed. It would seem that this change will not be reversed: the need for ever more information transfer may very well persist indefinitely in civilizations. There are major implications for SETI radio search strategies. We should expect the detection of signals civilizations use for their own purposes to be more difficult than we have imagined, and therefore the number of detectable civilizations to be less than we have imagined. The most detectable civilizations will be far. This argues for searching in directions where there are as many stars within the telescope reception beam as possible. For us, this means in directions close to the plane of the Milky Way. It argues that "beacon" signals may be the more favored signals for SETI detections. Beacon signals are hypothesized powerful signals transmitted for the benefit of other distant civilizations. Again, since the detectable beacon signals may be rare signals from very distant stars, this argues for concentrating searches in directions where many stars are observed. At present very few, if any, SETI programs follow this strategy. Thus a major change in search strategy appears desirable.

With regard to searches for optical signals, some watershed developments have changed the outlook for such searches. Among these has been the development of inexpensive, high quantum efficiency optical detectors, and very large telescopes such as the 10-meter Keck telescopes and the even larger ones now in the planning stage. However, the most important advance has been the development of extremely powerful short-pulse lasers. These have been developed as part of the US program seeking economically productive sources of energy utilizing nuclear fusion of hydrogen isotopes. In one approach lasers are used to heat pellets containing hydrogen isotopes very quickly to the temperatures required for high efficiency nuclear fusion. A present leading example of such a laser is the "Helios" laser of the Lawrence Livermore Laboratory, which releases a megajoule of optical energy within a nanosecond, so achieving a peak power of about a petawatt.

Such a laser pulse, focused into a narrow beam by a telescope the size of a Keck telescope, creates a power level within the beam which exceeds all the stellar power within the beam by a large margin of a factor of a hundred or more. Furthermore, the number of photons captured by even a small telescope, say of one-meter size, from a solar-type star even many light-years away, may be numbered in the hundreds, and so will be statistically far more numerous than the stellar photons in all of the optical spectrum captured in the time interval of the pulse duration. Thus, sampling of the unfiltered light from the direction of the star at time intervals of the order of nanoseconds can lead to very clear statistical detection of the laser pulse, despite the presence of the starlight. No frequency selection or, equivalently, optical spectral filtering, is required. Such a system solves the old problem of the masking of an optical laser signal by the scattered starlight. It is to be noted, though, that a serious objection to the appeal of this means of detection is the requirement that the signaling civilization be very accurately pointing its transmitter system at the Earth.

Thus is now makes sense to search for such signals, since they could exist, and would be very effective ways to reveal the presence of a technical civilization accompanying a star. Even with modest telescopes, and laser transmitters like the aforementioned, the signals could be detected from perhaps a million or so nearby stars. Utilizing statistics to detect the signal, rather than raw signal-to-noise ratios, gives a great advantage. When it comes to traditional signal-to-noise power ratio, the clear and detectable deviation from Poisson statistics in photon counts which occurs when occasional laser pulses are received can be equivalent to a signal-to-noise ratio of one in million or so.

This new form of search is now being pursued in at least four places. Already important advances in equipment design have been achieved, especially through the suggestion of Daniel Werthimer that the photons captured by a telescope be divided into several streams and sent to several photodetectors, with laser pulses to be distinguished by a timing coincidence between photon bursts detected by several detectors. This approach satisfies the need to establish as spurious the large events which occasionally occur in all photodetectors due to such phenomena as cosmic ray impacts, radioactive decay in the detector casing, corona discharge, and other sources. Only when an event is seen simultaneously in all detectors should the event be considered to be of possible intelligent origin. In practice, this approach is working very well. Thus there is readily available a rather inexpensive way to search for the hypothesized laser pulses, if they exist. Indeed, the optical detection systems are far less expensive than the radio detection systems. So, despite the daunting requirement that the extraterrestrials actively attempt to signal other civilizations, we can expect many optical search programs to exist.

The two changes in paradigm described above should cause us to be cautious and humble when it comes to confidence that we are doing the right thing. There is something profound here. If two such changes can occur in just forty years, what changes might occur over reasonable time scales for civilization evolution? What changes will occur in our understanding over thousands, perhaps millions of years? The best we can do to respond to this quandary is to keep a very open mind.

Closing Lecture

Speaker:

Francesco Bertola

THE PLURALITY OF WORLDS

FRANCESCO BERTOLA

Dipartimento di Astronomia,
Universita' di Padova, Vicolo dell'Osservatorio 5,
I-35122 Padova, Italy
Tel: +39-049-827 8225
Fax: +39-049-875 9840
E-mail: bertola@pd.astro.it

The idea of the plurality of worlds goes back to the Greek atomists. The following sentence is attributed to Democritus by Hippolitus:

> *"There are countless worlds of different sizes. In some of them, neither the sun nor the moon are present; in others, they are larger then ours, still others have more than one of them. These worlds are to be found at irregular distances, more numerous in one direction and less in another; some prosper, other are in decline. In one place they came into being, in another they die and are destroyed as they collide with each other. Some worlds have no animal or vegetable life, nor any water at all."*

The ideas of Democritus were followed up by the Epicurean school. According to Simplicius, Metodorus of Kios, a disciple of Epicurus, stated that:

> *"As it is highly improbable that a large piece of land produces only one ear of wheat, so it is for a world in an empty infinite space".*

It is clear from the above quotations that the term "worlds" does not mean bodies similar to the earth, but rather systems of worlds, that more properly can be called universes. Their nature is not testable and they are mainly the result of our imagination. The ideas of the Greek atomists were sustained in the Latin world by Lucretius. In his work *"De Rerum Natura"* he states that the uniqueness of the sky and the earth is highly improbable.

In complete disagreement with the above philosophers is the conception of the world by Aristotle. According to him, the universe is divided in two parts, the sublunar

J. Chela-Flores et al. (eds.), First Steps in the Origin of Life in the Universe, 401–407.

region, with the earth at the centre, and the surrounding crystalline spheres each one carrying a planet, while the most external one carries the fixed stars. The way the universe of Aristotle is conceived, it is unique and there is no room for others. The centre of the earth is the natural place toward which tend all the fragments of earth, even if they are pushed towards the exterior. The existence of several universes would imply several natural places and this is considered to be nonsense. At the boundary of the spheres there is the Prime Mover, responsible of their evolution. The existence of several universes would imply the presence of several Prime Movers, and once again this is unacceptable.

The debate on the plurality of worlds is transferred from antiquity to Christianity. In this context, the discussion acquires mainly a theological meaning. Hippolitus, Eusebius and Augustine held positions against the existence of more then one world. With the discovery of the ancient text in the 13th century and with the strong influence of Aristotle's thought, the debate became very lively. The questions posed by theologians and philosophers deal with the possibility of the simultaneous existence of several worlds, or rather their existence in successive epochs. In the latter case are included the universes created by God with the same degree of perfection, or is a universe better than the previous one? While in principle the creation power of God was considered not limited, the general tendency was to consider the existence of only one universe, created by God, as the most logic alternative. This opinion was shared by Albert, Thomas, Bacon, Scotus and many others.

An extremely significant episode involving the debate on the plurality of worlds took place in 1277. On request of Pope John XXI, anxious to stop the expansion of the doctrinal errors at the University of Paris, the bishop of Paris Etienne Tempier, on March 7, 1277 published 219 sentences, the content of which had to be accepted otherwise excommunication would follow. Of these sentences that bearing the number 34 is directly related the problem of the plurality of worlds. It says: *Quod prima causa non posset plures mundos facere*, namely: 'Those who believe that God is unable to create many worlds, are heretical and will be excommunicated. The existence of many worlds is not asserted, but only its possibility. Article 34 represents, in any case, a strong attack on Aristotelian philosophy and philosophers. This proposition not only had consequences on theology itself, but also on the history of Science. Such strong rejection of aristotelism can be interpreted as a premonitory sign of the scientific revolution ahead. According to Pierre Duhen, the famous historian of science, the root of modern science goes back to proposition 34. This is his comment:

> *''There has been an effort at the University of Paris to build a new physics not based on the aristotelism, and with this effort the basis of modern science was laid down: we can say that modern science was born on March 7, 1277 from the decree of the bishop of Paris.''*

While in antiquity and during the middle ages the plurality of worlds is intended as an ensemble of universes not directly observable and, therefore of abstract nature, with Copernicus this term receives a different meaning. The exchange of the position of the Sun with that of the Earth has constituted a deep revolution more of ideological nature rather than cosmological, since it has deprived man of its central position. As a consequence, the Earth is not unique in its role, but it is a planet like many others. In this manner, the problem of the plurality of the worlds becomes that of the plurality of planets.

Apart from this, the world system of Copernicus is still medieval: planets move on circular orbits, since the circle is considered to be a symbol of perfection and they are inserted into crystalline spheres. Stars are also bound to the last sphere, embracing the universe, which is finite. However, it is in the work of Copernicus that we can find the roots leading to the demolition of the sphere of the fixed stars and to the concept of an infinite space, infinitely populated of worlds. In chapter VII of Book I of his *"De revolutionibus orbium caelestium"* (1543) we can read: *magnitudo quoque caeli abibit in infinitum* (the size of the sky will be infinite) and further on *at si caelum fuerit infinitum* (if the sky were infinite). As we can realise we are in the presence of the first symptoms leading to the disappearance of the sphere of fixed stars.

The person who for the first time has eliminated this sphere, considering stars as non equidistant objects has been Thomas Digges (1546-1595), an English mathematician and astronomer. He learned these disciplines from his father Leonard. Already in his first works Thomas proposed to use the most recent observations in order to revise and update the Copernican cosmology. In 1576 he added to a new edition of the book of his father *"A prognostication everlasting"* a supplement titled *"A Perfit Description of Coelestial Orbes.* This work contained a faithful paraphrase of the first book of the *"De Revolutionibus"*. This account of the Copernican work became one of the most popular scientific books in England at the time. We know at least seven editions of it from 1576 to 1605. More than any writing it is a figure in the book of Digges which illustrates his cosmological revolution. The figure describes the Copernican system, the Sun at the centre and the planets rotating around it, but the sphere of the fixed stars is removed and the stars are distributed covering the page completely up to its four corners. This was meant to indicate that the stars are filling an infinite space. Only thirty years after the death of Copernicus a new revolution starts opening the way to the modern concept of the universe. However, the thought of Digges is filled with theological arguments. He is unable to give a purely astronomical description of the system of the world, since he is eager to find a space in which to insert the divinity. So, while in the previous Ptolemaic system everything which is of theological nature is placed beyond the sphere of fixed stars, in the Empyrean, according to Digges, the same space filled with stars constitutes the theological space, which is occupied by the angels and God.

Digges does not specify the nature of the stars. Instead Giordano Bruno (1548–1600) had the deep intuition about their physical nature and in this way he is able to build up a true cosmological conception. During his continuous wandering throughout Europe, Bruno spent two years in England from 1583 to 1585 and certainly he acquired knowledge of the work of Thomas Digges. However, the vastness of his thought is such that he has to be considered the first who, overcoming the limitation of the Copernican system, starts modern cosmological thought. He took inspiration from the Latin poet Lucretius (97-55 BC.) who, in his work *"De Rerum Natura"* gave support to the idea of the plurality of inhabited worlds. The intuition of Bruno about the physical nature of the stars, is that they are bodies like our Sun. The scientific demonstration of this assumption came only in 1838 with the measurements of the first parallaxes of the stars and therefore of their distances. Once the latter are known, then is possible to derive the intrinsic luminosity of the stars and to verify that the luminosity of the Sun is similar to that of the stars. Quoting from the Third Dialogue of the work *"De l'infinito universo e mondi"* as Elpino, one the interlocutors, states:

> *"the celestial bodies laying beyond the orbit of Saturn are innumerable suns, more or less visible to us and around them nearby earths rotate, which are invisible to us";*

and again:

> *"There are innumerable suns, there is an infinite number of earths rotating around the suns in the same way we see these seven planets rotating around our nearby sun."*

With these sentences Giordano Bruno not only defends the Copernican system, but puts forward the idea that the solar system is not the only planetary system, that the universe is constituted by an infinite number of planetary systems uniformly distributed in the infinite space. Bruno denies not only geocentricism, but also heliocentricism. In addition, he rules out any anthropocentric concept imagining that the infinite worlds are populated by living beings. From the doubts of Copernicus on the nature of the sphere of fixed stars, to the explanation conceived by Digges and to the interpretation by Bruno, one arrives at the concept of the plurality of worlds, which is intended as a plurality of planetary systems spread in an infinite space and potentially subject to astronomical observations. The problem touched by Elpino concerning "the nearby satellites invisible to us" constitute today one of the most advanced fields of research aiming at the indirect detection of planets around stars. It is a news item of last week that the prestigious Balzan Prize has been awarded to Michael Mayor of Geneva Observatory for his pioneering investigation in this field.

The influence of Bruno was very strong and the pluralism was largely accepted by the 17th century, with special emphasis to the possibility of life in the other planets, either around the sun or around the stars.

Two famous works witness the wide acceptance of the pluralism. The first is authored by Bernard le Bovier de Fontenelle *"Entretien sur la pluralitè des mondes"*, which appeared in 1686 in Paris and successively translated into 9 different languages. It has been reprinted several times during the next 150 years. The success was due also to the elegant prose and easy reading. The second work by the Dutch astronomer Christian Huygens (1629-1695) titles *"Cosmotheoros"* appeared in 1698 and was reprinted in several languages until the end of the 18th century.

Huygens, after discussing the plurality of inhabited systems, takes us back to the concept of the plurality of universes. He doubts whether there are an infinite number of stars and introduces the concept of unreachable entities with properties completely different from those of our universe. This concept is present also in Leibnitz (1647-1716), who says that God has created our universe by choosing the best among the infinity of possible universes.

While the debate on the plurality of planetary systems, possibly inhabited, has continued until the present day, let me recall the 19th century work of Richard Proctor, Camille Flammarion, Giovanni Schiapparelli, who were in favour of pluralism, and, on the side of the anti pluralism the work of Alfred R. Wallace *"Man's Place in the Universe. A study of the result of scientific research in relation to the unity or plurality of worlds"* (1903).

In the same way Giordano Bruno expanded the vision from the uniqueness of the solar system to the plurality of planetary systems, recent cosmological developments consider the universe we live in as only one element of an infinite ensemble.

The Einstein theory of general relativity provides a view of the universe which is quite different from that of the preceding centuries. The main characteristic of the relativistic universe is the fact that it is not static, but in a dynamical state. The universe evolves and continuously changes his aspect. Astronomical observations show that galaxies are receding from us with a speed which is proportional to their distances. From this phenomenon we deduce that the universe, more precisely, that space is expanding and that the motion of galaxies is due to the dragging of space. According to the mean density of matter, whose action consists in decelerating the expansion of the universe, three types of universes are possible. If the density is below the so called critical value, the expansion is endless; on the contrary, the expansion will be halted and followed by collapse. The third case is an intermediate one. The idea of a universe ending with a collapse has attracted in the past many cosmologists and among them one of the founders of modern cosmology, George Lemaître, who in 1933 proposed the idea of an oscillating universe or a phoenix universe, characterised by a series of big bangs followed by big crunches. In this way the universe is cyclically repeating itself. In other words, a series of universes develop with time. This idea has by now only historical interest.

The most recent observations have dramatically changed our view of the universe producing a real revolution in cosmology. The study of supernovae at large distances have revealed that the universe is not decelerating under the action of the gravitational force of its matter content, but it is accelerating, implying an overwhelming presence of a repulsive energy, whose nature is not at all known. This means that our universe is endless with respect to its future.

During the last twenty years a new line of thought has developed, which considers the possibility of multiple or infinite universes, each one ruled by its own physical laws, each one with particular values of the physical constants, each one with particular initial conditions. There are two reasons leading to the development of this new trend in cosmology: first, the realisation that our universe is ruled by the so called "fine tuning", second the introduction of the theory of inflation, which provides a good background for the presence of infinite universes.

The main characteristic of our universe is its continuous evolution, that we are able to describe from the instants immediately after the big bang, until the present and even beyond. This is due to the fact that at the origin of time the conditions of our universe were simple, as it is revealed to us by the uniformity of the background radiation. The way our universe evolves is determined by a series of parameters such as the initial conditions, the form of the physical laws and the value of the fundamental constants. By studying the way the universe transforms itself, one has realised that, in order to get as an end product the universe we live in, the constants of nature and the form of the physical laws have to be fine tuned in the sense that even a minimal displacement from their values and their form, would lead to a universe completely different from ours.

Dozens of examples of fine tuning can be given, so that at the end one arrives at the conclusion that our universe is the product of a unique, precise, well-determined combination. But our universe is a particular one. On Earth life is present, intelligent life, giving rise to beings able to observe the cosmos and to ask questions about it. A universe slightly different, without stars and planets, for instance, does not allow our presence. For life, as we know it, to appear, there is the need of a large universe, extended and expanding just like ours.

These considerations can give rise to questions such as "who did the fine tuning?", which takes us outside the scientific domain. In order to avoid going beyond the frontiers of science, in order to remain bound to the scientific method, the idea has been developed that our universe is not unique, but that all types of universes can exist, characterised by different physical laws and different values of the natural constants. In this way our universe is not a very special one, but merely an accident.

If we were satisfied with only a philosophical approach, we can stop here by admitting the abstract presence of an infinite number of universes. But since the above consideration were originated in the scientific domain, there is an attempt to indicate the physical reason of the existence of an infinite universe. A possible approach to the problem is offered by the inflationary model of the universe. In the 1970s the classical model of the Big Bang had considerable development in the sense that it was giving a description of the universe in agreement with astronomical observations. However, several problems were still open. It was not clear why the background radiation is isotropic, why the mean density of the universe is close to the critical one, the mechanism leading to the formation of galaxies were not understood. In 1979 Alan Guth solved these problems by admitting that in the very first instants of its evolution, the universe underwent an accelerated expansion, called "inflation". But inflation not only furnishes the missing explanations. It becomes a wider theory within which the Big Bang that originated our universe is only one of the aspects. Inflation gives a much wider and comprehensive vision of reality.

At the beginning of the 1980s, Andrei Linde suggests the scenario of what is called the chaotic inflation. According to his scheme the phenomenon of inflation does not concern only our universe, but there are regions of what we can call the metauniverse or the multiverse, where there are favourable conditions for further inflationary phenomena, namely for the formation of other universes. In particular, these universes are supposed to be different from ours in the number of dimensions and in the nature of physical laws ruling them. This situation is called "space-time foam", where some bubbles can inflate more than others and give rise to multiple universes. Inside the universes further inflationary phenomena can take place so that there is the case of universes generating other universes. In this sense the theory of chaotic inflation of Linde goes back to the steady state cosmology, very popular in the fifties. In this case, what is stationary, unchanging, is the multiverse, and within it the process of self reproduction of universes takes place continuously.

The idea of the multiverse is more a speculation that a real theory. Nevertheless, it is proposed not by philosophers but by scientists, so there is hope that efforts are made to indicate a possibility for testing it. The astronomer Martin Rees, who is strongly in favour of the idea of the multiverse said:

> *"I think that the multiverse is really within the territory of science, even if today is not more than an attempt of hypothesis."*

Let me now mention some of the arguments used by those who do not like the idea of fine tuning. They say:

> *"Do we really need a planet like ours to foster life? Is it not possible for life to exist in a huge range of possible environments? ".*

Another way to eliminate fine tuning is to develop a theory, for instance a super theory, in which the constants of physics can be derived from a simple fundamental parameter. According to those in favour of this argument, it is only matter of waiting for the correct formulation to be developed. In the frame of the multiverse it is not yet fully clear why the physical constants should vary from one universe to another. So if the tuning is not so fine and if the constants remain the same there is the chance for life to develop in some other universe different from ours.

AUTHOR INDEX

SUBJECT INDEX

List of participants

Argentina

CIMATO, Pilar
Instituto Nacional Superior del Profesorado
Dr. Joaquin V. Gonzalez. Av. Rivadavia 3577 -
Buenos Aires, Argentina.
E-mail: odonato@yupimail.com

LEMARCHAND, Guillermo A.
Instituto Argentino de Radioastronomía (CONICET),
and Centro de Estudios Avanzados,
Universidad de Buenos Aires, Argentina
Tel: +54-114-774-0667
FAX: +54-114-508-3628
E-mail: lemar@cea.uba.ar

Bangladesh

ISLAM, Jamal
Research Centre for Mathematical and Physical
Sciences, University of Chittagong,
Chittagong, Bengaladesh
Tel: +880(31) 616 780
FAX: +880(31) 610 249
E-mail: nargisnaaz@hotmail.com

Belgium

WHITEHEAD, Edward P.
European Commission
Science Research and Development
- Joint Research Centre,
Rue de La Loi 200
B-1049 Brussels,
E-mail: edward.whitehead@cec.eu.int

Brazil

COSTA Jr., Antonio T.
Dpto. de Ciencias Exatas
Grupo de Física - UFLA ,
Caixa Postal 37 - 37200-000,
Lavras-MG - Brazil
E-mail: antc@stout.ufla.br
Tel. +55 35 829 1379
Fax +55 35 829 1650
antc@stout.ufla.br

GUIMARÃES', Romeu C
Departamento. Biologia General,
ICB - UFMG,
Belo Horizonte, Brazil.
Tel: +55-31-499 2608
Fax: +55-31-274 4988
E-mail: romeucg@mono.icb.ufmg.br

Chile

VICUÑA, RAFAEL
Departamento de Genetica Molecular
y Microbiologia
Pontificia Universidad Catolica de Chile
and
Millenium Institute for
Fundamental and Applied Biology.
Casilla 114-D. Santiago, Chile.
Phone: (562) 686 2663
Fax: (562) 222 5515
rvicuna@bio.puc.cl

Cuba

PEREZ ROJAS, Hugo Celso (Observer)
Grupo de Fisica Teorica,
Instituto de Cibernetica, Matematica y Fisica,
Calle E No. 309, esq. a 15,
Vedado, La Habana 4, Cuba
Tel. 53-7-31-2806,
fax 53-7-33-3373
e-mail: hugo@cidet.icmf.inf.cu,

Czech Republic

KLETETSCHKA, Gunther
NASA Goddard Space Flight Center
Greenbelt, MD 20771
United States of America
gunther@denali.gsfc.nasa.gov

France

BRACK, Andre
Centre de Biophysique Moleculaire,
CNRS, Rue Charles Sadron,
45071 Orleans cedex 2, France
Tel. 00 33 2 38 25 55 76
Fax 00 33 2 38 63 15 17
brack@cnrs-orleans.fr

CORADINI, Marcello
Coordinator, Solar System Missions,
Directorate of Scientific Programme
ESA,
8-10 Rue Mario Nikis
F-75738 Paris Cedex 15,
France.
Tel +33-1- 536 97555
Fax +33-1- 536 97751
mcoradin@hq.esa.fr
Fax +33-1- 536 97751

DOBRIJEVIC, Michel
Observatoire de Bordeaux
BP 89, Floirac 33270
France.
Tel.: 05 57 77 61 24
Fax : 05 57 77 61 10
E-mail: michel@observ.u-bordeaux.fr

PRIEUR, Benoit
ENS, Paris, France
36 rue Monge
F-75005 PARIS,
France
Tél: 33 1 46 33 14 01
E-mail: prieur@technologist.com

PRIEUR, Daniel
Université de Bretagne Occidentale
Institut Universitaire Européen de la Mer
Place Nicolas Copernic
Technopôle Brest-Iroise
29280 PLOUZANE,
France
Tél : 33 (0)2 98 49 87 04
Fax : 33 (0)2 98 49 87 05
e-mail : Daniel.Prieur@univ-brest.fr

RAULIN, Francois
(Co-director)
LISA,
UMR CNRS 7583
Universites Paris 12 & Paris 7
Faculte des Sciences et Technologie
61, Avenue du General de Gaulle
F-94010 Creteil Cedex
France
Tel : 33 - (0) 1 45 17 15 60 / 15 50
Fax: 33 - (0) 1 45 17 15 64
E-mail: raulin@lisa.univ-paris12.fr

RAULIN CERCEAU, Florence
Grande Galerie de l'Evolution
Museum national d'Histoire naturelle
& Centre Alexandre Koyre
36 rue Geoffroy Saint Hilaire
F- 75005 Paris

VANDENABEELE-TRAMBOUZE, Odile
Université des Science Montpellier II
Laboratoire Organisation Moléculaire,
Evolution et Matériaux Fluorés
UPRESA
5073, Bat. 17,
2 place E. Bataillon,
34095 Montpellier cedex 5,
France
Tel. (+33) (0) 4-67-14-39-20
(3851 ou secretariat :3806)
Fax : (+33) (0) 4-67-63-10-46
E-mail : odile_trambouze@post.club-internet.fr

VUITTON, Veronique
LISA, UMR CNRS 7583
Universites Paris 12 & Paris 7
Faculte des Sciences et Technologie
61, Avenue du General de Gaulle
F-94010 Creteil Cedex,
France
Tel : 33 - (0) 1 45 17 15 60 / 15 50
Fax: 33 - (0) 1 45 17 15 64
E-mail : vuitton@lisa.univ-paris12.fr

Germany

HORNECK, Gerda
Institut f. Luft-und Raumfahrtmedizin
Abt. Strhlenbiologie, DLR,
German Aerospace Research Establishment
LinderHohe
51147 Koln,
Germany.
Fax: +49-2203 61 970
E-mail: Gerda.Horneck@dlr.de

KROKER, Holger
Bornheimer Strasse 1 B
D-50969 Koeln,
Germany.
E-Mail: roehrlich.kroker@technik.k.shuttle.de
Tel.: +49-221-3601810
+49-221-3601919
Fax : +49-221-3685399

KUPPERS, Bernd-Olaf
Institute for Philosophy
University of Jena
D-07740 Jena
Germany
Fax +49-3641-944137
E-mail: bernd_kuppers@uni-jena.de

OLLIVIER, Marc
Max Planck Institut fur Astronomie
Konigstuhl 17
D-69117 HEIDELBERG,
Germany
Tel : (+49) 6221 528 223 (Tom Herbst's office)
Fax : (+49) 6221 528 246
E-mail: ollivier@mpia-hd.mpg.de

ROEHRLICH, Dagmar
Bornheimer Strasse 1 B
D-50969 Koeln,
Germany.
E-Mail: roehrlich.kroker@technik.k.shuttle.de
Tel.: +49-221-3601810
+49-221-3601919
Fax : +49-221-3685399

Hungary

CORLISS, John B.
Complex Adaptive Systems Laboratory
Environmental Sciences and Policy
Central European University
Kerepesi ut 87
1106 Budapest,
Hungary
tel. 36-1-327-3887 ,
Tel: +36-1-327-3886
Fax: +36-1-327-3888
E-mail: corliss@syslab.ceu.hu

India

CHADHA Mohindra S.
Bhabha Atomic Research Centre
Bio-Organic Division
400 085 Bombay-Mumbai,
India
Fax: 0091-22-556 0750
E-mail: chadha@soochak.ncst.ernet.in

CHAKRABARTY, S.K.
SN Bose National Centre for Basic Sciences
JD Block, Sector III, Salt Lake,
Calcutta 700091, India
FAX +91-33-335 5706
E-mail: chakraba@astro.bose.res.in

CHANDRA, Suresh
School of physical Sciences
SRTM University,
Nanded 431606, India.
Tel. +91-2462-29242 Ext 217
Fax: +91-2462-29245
sch@iucaa.ernet.in

GATTA, Riccardo S.
Malaria Research Group
International Centre for
Genetic Engineering and Biotechnology
Aruna Asaf Ali Marg
110067 New Delhi,
India.
tel.: +91.11.6175009 / 6189358
Ext.30 Malaria Lab
Fax: +91.11.6162316
E-mail: gatta@icgeb.res.in
gatta@medscape.com

KAMALUDDIN
Chemistry Department,
University of Roorkee, Roorkee 247 667,
India
Fax 91-1332-73560
E-mail: kamalfcy@rurkiu.ernet.in

MUKERJEE, K.
Department of Astronomy and Astrophysics,
Tata Institute of Fundamental Research
Homi Bhabha Road, Colaba, Mumbai 400005, India.
Fax : +91-22-2152110 or 2152181.
phone : +91-22-2152971 extn. 2484.
kmukerji@tifr.res.in

TEWARI, Vinod C.
Wadia Institute of Himalayan Geology
Dehradun- 248001, India, and
The Abdus Salam ICTP, Trieste, Italy.
E-mail: wihg@giasdl01.vsnl.net.in

Israel

SECKBACH, Joseph
Hebrew University of Jerusalem
Home:Mevo Hadas 20, P.O.Box 1132
The City of Efrat, 90435, Israel
Phone/Fax 972-2-993-1832
E-mail: seckbach@cc.huji.ac.il

Italy

ASCHERI, Valeria
University of Genova, Department of Philosophy
Section of Epistemology
V. Puggia 3, 16131 Genova (GE), Italy
Phone +39/010/313396
Fax: +39/010/312042
E-mail: ascheri@nous.unige.it

BARBIERI, Cesare (Observer)
Dept. of Astronomy, University of Padova
Vicolo Osservatorio 5, I-35122 Padova (Italy)
Phone: +39-049-8293434 Fax: +39-049-8293507
e-mail: barbieri@pd.astro.it
http://planet.pd.astro.it/planets/barbieri/

BERTOLA, Francesco
Dipartimento di Astronomia,
Universita' di Padova, Vicolo dell'Osservatorio 5,
I-35122 Padova, Italy
Tel: 049 8293436
Fax: +39- 049 8759840
E-mail: bertola@pd.astro.it

BONACCORSI , Rosalba
University of Trieste.
Dipartimento di scienze Geologiche, Ambientali e Marine,
Via E. Weiss, 2 (c/o Comprensorio di S. Giovanni);
34127 Trieste (Italy).
Phone: 39 (040)6762046 (secretary)
Fax: 39 (040)6762048
e-mail: bonaccor@univ.trieste.it

CARERI, Giorgio
Universita' La Sapienza,
Dipartimento di Fisica
P.le. Aldo Moro,
2, I 00185 Rome,
Italy
Fax: 39 (06) 4957 697
Fernanda.Lupinacci@roma1.infn.it

COLIOLO, Fiorella
IRSPS, Universita' d'Annunzio
Viale Pindaro 42
65127 Pescara,
Italy
Fax. +39-085-4537545
E-Mail: coliolo@tesre.bo.cnr.it

COLLI, Mariaisabella
Politecnico di Torino
c/o Via Vagnono 9
10144 Torino, Italy
Fax: 011 48 2101
sacchi@polito.it

COLLIS, William
Strada Sottopiazzo 18,
14056 Boglietto(AT), Italy.
Tel. 0141-968602
Fax. 0141-968907
Email. mr.collis@physics.org

CORSO, Ezio
Universita' di Trieste,
DEEI, Via Valerio 10,
34100 Trieste, Italy
Fax: +39-040-6763460
shah@spin.it (CORSO, Ezio)

COSMOVICI Cristiano
Consiglio Nazionale Delle Ricerche
Istituto di Fisica
Dello Spazio Interplanetario
IFSI/CNR, Fosso del Cavaliere 100
00133 Rome, Italy.
Tel.+39-06-4993-4388
Fax +39-06-4993-4383
E-Mail: cosmo@ifsi.rm.cnr.it

COYNE George Vincent
Specola Vaticana, Citta' del Vaticano
00120 Roma, Italy.
Fax: (06) 6988 4671
E-Mail: gcoyne@specola.va

FERRARA, Giorgio (Observer)
Istituto di Geocronologia
CNR Pisa, Italy
Tel 0503152266
fax 0503152360
G.Ferrara@iggi.pi.cnr.it

FRANCHI, Marco
Department of Animal Biology and Geneticcs
University of Florence, Via Romana 17/19
50125 Florence, Italy
tel: 055 2288248;
Fax: 055 2288250;
E-Mail: marco.franchi@dbag.unifi.it

GALLORI, Enzo
Department of Animal Biology and Geneticcs
University of Florence,
Via Romana 17/19, 50125 Florence, Italy
Tel: +39-55-228-8241
Fax: +39-55-228-8250

GENTA, Giancarlo
Mechanics Dept., Politecnico di Torino
C. Duca degli Abruzzi 24,
10129 TORINO - Italy
Phone: +39-011 5646922
Fax: +39-011 5646999
E-mail: genta@polito.it

HACK, Margherita
Osservatorio Astronomico di Trieste
Via G.B. Tiepolo 11
34131 Trieste, Italy

MINIUSSI, Emiliano
Universita' di Trieste, DEEI,
Via Valerio 10, 34100 Trieste, Italy
Fax: +39-040-6763460
shah@spin.it (MINIUSSI Emiliano)

MUSSO, Paolo
University of Genova,
Department of Philosophy
Section of Epistemology
V. Montevideo 2A/5, 16129 Genova (GE), Italy
Phone +39/ 0335/ 820 5220
Fax +39/ 010/ 312 042
E-mai: musso@nous.unige.it

PAGAN Fabio
The Abdus Salam
International Centre For Theoretical Physics
Press Office, P.O. Box 586
34100 Trieste,
Italy
pagan@ictp.trieste.it

RIZZOTTI Martino
Universita' degli Studi di Padova
Dipartimento di Biologia,
Via Trieste 75Bassi 58/B
35131 Padova, Italy
Phone: +39-049-827-6208
Fax: +39-049-827-6209;
E-mail: rizzotti@bio.unipd.it

SHAH, Tahir K.
Universita' di Trieste,
DEEI, Via Valerio 10,
34100 Trieste, Italy
Fax: +39-040-6763460
E-mail: shah@spin.it

VANDINI, Claudia
Department of Animal Biology and Geneticcs
University of Florence,
Via Romana 17/19,
50125 Florence, Italy
Tel 39-55-228-8241
fax 39-55-228-8250
gallori@DBAG.UNIFI.IT (VANDINI Claudia),

Japan

MATSUNO, Koichiro
Department of BioEngineering
Nagoka University of Technology
Nagaoka 940-2188,
Japan
http://bio.nagaokaut.ac.jp/~matsuno
Voice & Fax: +81 258 47 9420
kmatsuno@vos.nagaokaut.ac.jp

Mexico

CRUZ-KURI, Luis
Universidad Veracruzana, Instituto de Ciencias Basicas
Av. Luis Castelan S/N (Antes Dos Vistas),
Carretera: Xalapa Las Trancas,
XALAPA, Veracruz, Mexico
Telephone (28) 12-57-45
Fax (28).12-99-63
E-Mails. lckuri@bugs.invest.uv.mx;
luis4321@hotmail.com

LAZCANO-ARAUJO REYES, Antonio
Universidad Nacional Autonoma de Mexico,
Facultad de Ciencias, Apdo. Postal 70 407
Ciudad Universitaria, D.F.,
04510 Mexico City,
Mexico
FAX: 525/616-0451 or 525/622-4823
E-mail: alar@hp.fciencias.unam.mx

NAVARRO-GONZALEZ, Rafael
Universidad Nacional Autonoma
de Mexico, Instituto de Ciencias Nucleares
Lab. de Quimica Do Plasmas y Estudios Planetarios.
A. Postal 70-543, D.F.,
04510 Mexico City, Mexico
Tel.: +52-5-622-4674
FAX: +52-5-616 2233
E-mail: navarro@nuclecu.unam.mx

NEGRON-MENDOZA, Alicia
Universidad Nacional Autonoma
de Mexico, Instituto de Ciencias Nucleares
Apdo. Postal 70-543,
Circuito Exterior. C.U.
Df 04510 Mexico City, Mexico
Tel.: +52-5-622 4674
FAX: +52-5-659 4581
E-mail: negron@nuclecu.unam.mx

RAMIREZ JIMENEZ, Sandra Ignacia
Universidad Nacional Autónoma de México,
Instituto de Ciencias Nucleares, Circuito Exterior,
Zona de Institutos, Ciudad Universitaria,
04510, México City, Mexico.
sramirez@nuclecu.unam.mx

RAMOS-BERNAL, Sergio
Universidad Nacional Autónoma de México,
Instituto de Ciencias Nucleares,
Apartado Postal 70-543, Circuito Exterior,
Ciudad Universitaria. D.F. 04510, Mexico City,
Mexico
ramos@nuclecu.unam.mx

Netherlands

GREENBERG J. Mayo
Leiden Observatory,
Sterrewacht Postbus 9513
NL-2300 Leiden, The Netherlands.
Tel +31-078-527 5804
Fax +31-078-527 5819
E-mail: greenber@strw.leidenuniv.nl

RUITERKAMP, Richard
Raymond and Beverly Sackler
Laboratory for Astrophysics
at Leiden Observatory,
P O Box 9513; 2300 RA Leiden,
The Netherlands
Tel: (31) 715275812
Fax: (31) 715275819
E-Mail: ruiterka@strw.leidenuniv.nl

Nigeria

ASUZU, Isaac Uzoma
Department of Veterinary
Physiology & Pharmacology
University of Nigeria, Nsukka, Nigeria.
EPSEELON@aol.com

LAOGUN, A.A.
University of Benin
P.O Box 2083,
Benin City, Nigeria.

Portugal

DE MATOS, Clovis (Observer)
Scientific Programme Communication Service
ESA-ESTEC
Noordwisk, The Netherlands.

P.R. China

WANG, Wenqing
Beijing University {University of Peking}
Dept. of Technical Physics
100871 Beijing,
People's Republic of China.
E-Mail: wangwq@sun.ihep.ac.cn

Puerto Rico

MÉNDEZ, Abel
Planetary Environment Biophysicist
Department of Physics and Chemistry
University of Puerto Rico at Arecibo
Call Box 4010; Arecibo,
PR 00613, Puerto Rico
a_mendez@cuta.upr.clu.edu

Russia

KRITSKIY, Mikhail,
Head of Laboratory of
Evolutionary Biochemistry
A.N. Bach Institute of Biochemistry
Russian Academy of Sciences
Leninsky Prospekt 33, Moscow 117071, Russia
Phones (+7-095) 9523431 (lab),
(+7-095) 9590773 (home)
Fax (+7-095) 9542732
E-mail: inbio@glas.apc.org

SIMAKOV, Michael B.
Group of Exobiology,
Institute of Cytology
Russian Academy of Sciences
Tikhoretsky Av.,4,
St.Petersburg, 194064,
Russia
fax +7 812 2470341
E-mail: exobio@mail.ru; exobio@usa.net

Spain

GUTIERREZ, Jordi L.
Interdepartamental Group of Astronomy
U P C, Departament de Fisica Aplicada
Universitat Politecnica de Catalunya
Pla del Palau, 18; 08003 Barcelona, Spain.
Phone: (+34) 93-401.79.14
Fax: (+34) 93-401.60.90
E-mail: jordi@rigel.upc.es

ORÓ, John
Fundacio Joan Oro,
Avda. Alcalde Rovira Roure 11,
Alt. Esq. 25006 Lleida, Spain.
Tel.: 34 973 239505,
FAX: 34 973 238609
E-mail: fjo@fjo.es

TRIGO-RODRIGUEZ, Josep Ma.
Red de Investigacion sobre bolidos
y meteoritos (SPMN),
Dept. Astronomia i Astrofísica,
Universitat de València, Dept.
Ciències Experimentals,
Universitat Jaume I,
Spain.
Postal address: c/ Manuel de Falla 26,
12.560 Benicassim (Castello) SPAIN
Phones: (+Spain Code 34) 964 - 282584 / 282968
(office)
(964) 395064 (part.)
Fax: 964 - 285161
E-mail: jmtrigo@ctv.es, trigo@exp.uji.es

Sweden

BALTSCHEFFSKY Herrick
University of Stockholm
Arrhenius Laboratory
Department of Biochemistry
S-106 91 Stockholm,
Sweden
FAX: +46-8-153 679.
herrick@biokemi.su.se

BALTSCHEFFSKY, M.
University of Stockholm,
Arrhenius Laboratory
Department of Biochemistry
S-106 91 Stockholm, Sweden
FAX: +46-8-153 679.
herrick@biokemi.su.se

Sudan

ABDEL-RAHMAN, A.-M.M. (Observer)
Vice-Chancellor,
University of Khartoum,
Khartoum, Sudan
Tel. 772 601
Fax 780 295
abdekmalik@yahoo.com

HABBANI, F.I. (Observer)
Physics Department,
University of Khartoum
Khartoum, Sudan

Switzerland

PUY, Denis
Institute of Theoretical Physics
University of Zurich,
Winterhurerstrasse, 190
CH-8057 Zurich,
Switzerland
fax +41-1-635-5704
puy@physik.unizh.ch

TURIAN, Gilbert
Laboratoire de Microbiologie Generale
Universite' de Geneve, Pavillion des Isotopes
20, boulevard d'Yvoy
CH-1211 Geneve 4, Switzerland
fax +41-22-781 5193
gturian@vtx.ch

United Kingdom

CAIRNS SMITH, Graham
Department of Chemistry,
University of Glasgow
Thorn Cottage, Uplawmoor G 78 4 AF
Glasgow, United Kingdom
E-Mail: g.cairns-smith@chem.gla.ac.uk

DAVIES, Paul
Imperial College, London, United Kingdom
E-Mail: pcwd@camtech.net.au

MOORBATH Stephen
University of Oxford
Department of Earth Sciences
Parks Road, OX1 3PR Oxford
United Kingdom
E-Mail: Stephen.Moorbath@earth.ox.ac.uk

PATEL, Manish
Unit for Space Sciences and Astrophysics
University of Kent,
Canterbury CT2 7NR,
United Kingdom
Tel : +44(0)1227 764 000 Ext. 3266
Fax : +44(0)1227 762 616
mrp4@ukc.ac.uk

United States of America

BAHN, Peter
RR2 Box 239A,
Mount Vernon, Illinois 62864
United States of America
E-Mail: jbahn@midwest.net

DRAKE, Frank
Seti Institute, 2035 Landings Drive
94043 Mountain View (California)
United States of America
Tel.: +1-650-961 6633
FAX: +1(650) 961 7099
E-Mail: drake@seti.or

MCKAY Christopher P.
Nasa Ames Research Center, Ms 245-3,
94035 Moffett Field (California)
United States of America.
Tel: 650-604-6864,
FAX-6779"
E-Mail: MCKAY@galileo.arc.nasa.gov

MCKAY, David
SN-NASA-Johnson Space Center,
Houston TX 77058,
United States of America
Tel: +1-281-483-35048,
Fax::+1-281-244-8892,
E-Mail: david.s.mckay1@jsc.nasa.gov

MILLER, Stanley
Dept of Chemistry and Biochemistry, 0506
University of California, San Diego
9500 Gilman Drive, La Jolla,
CA 92093-0506
United States of America
Tel: (858) 534-3365
Fax: (858)534-6128
E-Mail: smiller@ucsd.ed

MORRIS, Maria
c/o Ms. Brenda Leake,
Dept of Chemistry and Biochemistry, 0506
University of California,
San Diego. 9500 Gilman Drive, La Jolla,
CA 92093-0506
United States of America

OWEN Tobias (Co-director)
Institute For Astronomy,
2680 Woodlawn Drive,
96822 Honolulu (Hawaii)
United States of America
Fax: +1 (808) 956-9580
E-Mail: owen@ifa.hawaii.edu

PAPPELIS Aristotel
Southern Illinois University at Carbondale
Department of Plant Biology
62901-6509 Carbondale (Illinois)
United States of America
Fax: +1-618-453-3441
pappelis@plant.siu.edu, aristote@siu.edu

PHILLIPS, Cynthia
Center for the Study of Life in the Universe
Seti Institute, 2035 Landings Drive
94043 Mountain View (California)
United States of America
tel (650)-960-4528
FAX: +1(650) 961 7099
email: phillips@seti.org

RUSSELL, Robert John
Centre for Theology and the Natural Sciences
2400 Ridge Road, Berkeley California 94709
United States of America
Tel. (510) 848-8152
Fax. (510) 848 2535
E-Mail: rrussell@ctns.org

SCHOPF, J. William
Department of Earth and Space Sciences,
Institute of Geophysics and Planetary Physics
(Center for the Study of Evolution and the Origin of Life),
and Molecular Biology Institute,
University of California, Los Angeles, CA 90095-1567
United States of America.
PHONE: (310) 825-1170
FAX: (310) 825-0097
E-MAIL: schopf@ess.ucla.edu

SHOCK, Everett
Department of Earth and Planetary Sciences
and the McDonnell Center for Space Sciences
Washington University, St. Louis,
Missouri 63130, United States of America.
Fax: +1-314-935 7361
E-Mail: shock@zonvark.wustl.edu

Venezuela

CHELA-FLORES, Julián (Co-director)
The Abdus Salam International Centre for Theoretical Physics,
Strada Costiera 11; P.O.Box 586; 34014 Trieste, Italy
and
Instituto de Estudios Avanzados, Apartado 17606,
Parque Central. Caracas 1015-A, Venezuela.
Direct phone: (int.+39 040) 2240 392.
FAX phone: (int + 39.040) 22 41 63.
E-Mail: chelaf@ictp.trieste.it

FAJARDO, Luis
Electron Microscopy Center,
University of the Andes, P. O. Box 163, Mérida, Venezuela.
Fax: +58-74-403 156
prupal@ula.ve (Fajardo Luis)

MARCANO, Vicente
Electron Microscopy Center,University of the Andes,
P. O. Box 163, Mérida, Venezuela.
Fax: +58-74-403 156
E-mail: vinmar@eudmail.com